D0154302

FLUVIAL HYDROLOGY

FLUVIAL HYDROLOGY

S. Lawrence Dingman

University of New Hampshire

W. H. FREEMAN AND COMPANY
New York

Library of Congress Cataloging in Publication Data

Dingman, S. L.
Fluvial hydrology.

Bibliography: p.
Includes index.
1. Hydraulics. 2. Hydrology. I. Title.
TC160.D45 1984 627′.125 83-20763
ISBN 0-7167-1452-3
ISBN 0-7167-1453-1 (pbk.)

Printed in the United States of America

1 2 3 4 5 6 7 8 9 0 MP 2 1 0 8 9 8 7 6 5 4

To Jane

Contents

Preface

Fluvial Hydrology is intended for the large number of students and others who seek a sound understanding of the physical laws that govern the movement of water in the hydrologic cycle. Until the 1960s hydrology was generally considered to be a minor specialty within the field of civil engineering—a prerequisite for the design of structures that transmit and control water. Since then the central integrating role of the hydrologic cycle in the earth's physical, chemical, biological, and cultural environment has become widely recognized. Correspondingly, the academic status of hydrology has broadened from that of, typically, a course (or less) in the civil engineering curriculum to that of several courses in a department of earth sciences or natural resources.

I have come to believe that a one-semester course in the physics of natural flows is an essential foundation for a sound program in hydrology and fluvial geomorphology. In my earlier teaching of the principles of hydrology, I attempted to devote some lecture and laboratory time to aspects of open-channel flow. This approach proved unsatisfactory because there was not enough time to develop the subject properly. As I considered initiating a separate course, I found no suitable textbook. This book evolved to meet the need for such a text. The term *fluvial hydrology* was coined in order to emphasize the book's concern with the physics of flow in the land phase of the hydrologic cycle as well as its close relation to fluvial geomorphology.

Although some might argue that a sound understanding of the mechanics of natural flows can be achieved only in the standard undergraduate engineering sequence of mechanics and fluid mechanics, my experience has been otherwise. If students have had one-year courses in calculus and physics, their physical intuition can be developed into a solid grasp of the aspects of fluid mechanics that are relevant to natural flows, including the derivations of the partial differential equations that describe one-dimensional unsteady flows in open channels and three-dimensional unsteady flows in saturated and unsaturated porous media. Nevertheless, I have found it desirable to begin this development at a rather elementary level, with a discussion of dimensions, units, significant figures, and types of equations (Chapters 2 and 3), for although students will have been exposed to this material in basic physics courses, they do not usually have much opportunity to apply it systematically in their earth-science or natural-resource courses, and hence are seldom fully conversant with it. In many cases it may be appropriate to assign these early chapters as review reading, and to begin the classwork with Chapter 4 or Chapter 5.

A course in fluvial hydrology is greatly enhanced by (1) working out solutions to problems; (2) discussion of papers from the literature, showing how principles are applied in attacking real problems; and (3) laboratory and/or field experience. In the present text, exercises have been provided at the end of each chapter (except the first); answers to most of these exercises are given at the back of the book. The instructor can develop his or her own approach to exploring the literature; I have tried to assist by providing fairly extensive (but not, of course, exhaustive) citations in the text. In addition, there are annotations of several books and papers that seem to me especially useful reviews or sources of further insight into particular subjects, or that are just particularly interesting.

Laboratory work can best be done by using a small recirculating flume in which students can measure velocity profiles; observe laminar, turbulent, subcritical, and supercritical flows; carry out energy computations; and study water-surface profiles. Many of these things can also be accomplished in the field by using a current meter and surveying equipment. I have found it particularly valuable to require students to make velocity–area stream-gaging measurements and to survey water-surface slopes in small streams. Besides providing useful experience with such common hydrologic measurements, the data can be used to compute Froude and Reynolds numbers, resistance factors, and related flow characteristics. A Darcy tube can be inexpensively constructed to demonstrate principles of flow in porous media.

It will be noted that the SI, cgs, and English systems of units are used more or less randomly throughout the text and exercises. No apology is made for this, since hydrologists—at least in North America—will encounter all three systems in the foreseeable future and must become comfortable with them.

Several people have made direct, identifiable contributions to this work, both wittingly and otherwise. I could not ask for a wiser, more sympathetic, or more helpful colleague over the last decade than Francis R. Hall, who reviewed the last few chapters. Gordon L. Byers has also been a most helpful companion and adviser in the teaching of hydrology. Yen-Hsi Chu helped with his review of Chapter 11, and I have benefited considerably from the comments of several anonymous reviewers. The guidance provided by my students has, of course, been of special value.

A minor but significant portion of this text was completed while I was on sabbatical leave at the Institute of Hydrology in Wallingford, England. I thank the trustees of the Farrington Forestry Fund of the University of New Hampshire for making possible my stay there, and Dr. J. S. G. McCulloch, Director, for making available the facilities of the Institute.

I thank Cindy Grimard for her very competent typing of the original manuscript. The careful editing of Helene De Lorenzo has done much to improve the clarity and consistency of the text. Georgia Lee Hadler's thoughtful guidance of the manuscript preparation process has been invaluable in making this book a reality.

I thank also J. F. Kennedy (University of Iowa), A. V. Jopling (University of Toronto), and C. T. Yang (U.S. Bureau of Reclamation) for providing illustrations from their work and Eileen Smith and Carol Edwards for their efforts in making available photographs from U.S. Geological Survey files.

I owe my interest in hydrology, and much of my understanding of its principles, to brief associations with two teachers. The first was John P. Miller of Harvard University, who enthusiastically introduced me to the field during the year before his tragic and untimely death. The second was Donald R. F. Harleman of the Massachusetts Institute of Technology, whose exquisitely lucid presentations made a single course a major contribution to my education. I can wish no more than to pass on to my students something of the knowledge and inspiration I have gained from these two teachers.

I am more than grateful to my parents, who provided the familial and financial support that made my education possible and who thus have a considerable investment in this book and its author.

This book is dedicated to Jane, my wife. She has read through the text with me several times, and each time has helped to improve it significantly. This book is only one of many things that her love and support have made possible.

Any shortcomings in this text are, of course, my responsibility.

S. Lawrence Dingman
Barrington, New Hampshire
March 1984

I am very fond of brooks, as indeed of all water, from the ocean to the smallest reedy pool. If in the mountains in the summertime my ear but catch the sound of plashing and prattling from afar, I always go to seek out the source of the liquid sounds, a long way if I must; to make the acquaintance and to look in the face of that conversable child of the hills, where he hides. Beautiful are the torrents that come tumbling with mild thunderings down between evergreens and over stony terraces; that form rocky bathing-pools and then dissolve in white foam to fall perpendicularly to the next level. But I have pleasure in the brooks of the flatland, too, whether they be so shallow as hardly to cover the slippery, silver-gleaming pebbles in their bed, or as deep as small rivers between overhanging, guardian willow trees, their current flowing swift and strong in the centre, still and gently at the edge. Who would not choose to follow the sound of running waters? Its attraction for the normal man is of a natural, sympathetic sort. For man is water's child, nine-tenths of our body consists of it, and at a certain stage the foetus possesses gills. For my part I freely admit that the sight of water in whatever form or shape is my most lively and immediate kind of natural enjoyment; yes, I would even say that only in contemplation of it do I achieve true self-forgetfulness and feel my own limited individuality merge into the universal. The sea, still-brooding or coming on in crashing billows, can put me in a stage of such profound organic dreaminess, such remoteness from myself, that I am lost to time. Boredom is unknown, hours pass like minutes, in the unity of that companionship. But then, I can lean on the rail of a little bridge over a brook and contemplate its currents, its whirlpools, and its steady flow for as long as you like; with no sense or fear of that other flowing within and about me, that swift gliding away of time.

THOMAS MANN, *A Man and His Dog,* translated by H. T. Lowe-Porter; from *Death in Venice and Seven Other Stories by Thomas Mann,* A Vintage Book © Alfred A. Knopf, Inc., New York, N.Y. Reprinted by permission.

FLUVIAL HYDROLOGY

CHAPTER 1

Introduction

SCOPE

Hydrology is the science concerned with understanding, describing, and predicting the movement of water on and under the earth's land surface, and the physical, chemical, and biological interactions of water with the earth's terrestrial environment. Thus hydrology is based largely on the application of the physical laws that determine the movement of water through the hydrologic cycle, and is a branch of the broad field of geophysics.

Figure 1.1 identifies the components and processes of the land phase of the hydrologic cycle. The term *fluvial hydrology* refers to those processes that involve the flow of liquid water in the natural environment: infiltration, percolation, overland flow, groundwater flow, and channel flow. Our goal is an understanding of the physical principles that govern those processes.

APPROACH

This book draws on your knowledge of basic mechanics (through first-year university-level physics) and mathematics (through integral calculus) to develop a sound physical intuition—a sense of the relative magnitudes of forces and of the properties and relationships that are significant in various hydrologic flows. Physical intuition consists not only of a large store of factual knowledge, but also of an inventory of patterns that serve as guides to the parts of that knowledge that are relevant to the analysis of a given situation (Larkin et al., 1980). Thus a special attempt is made in this book to emphasize patterns, connections, and analogies.

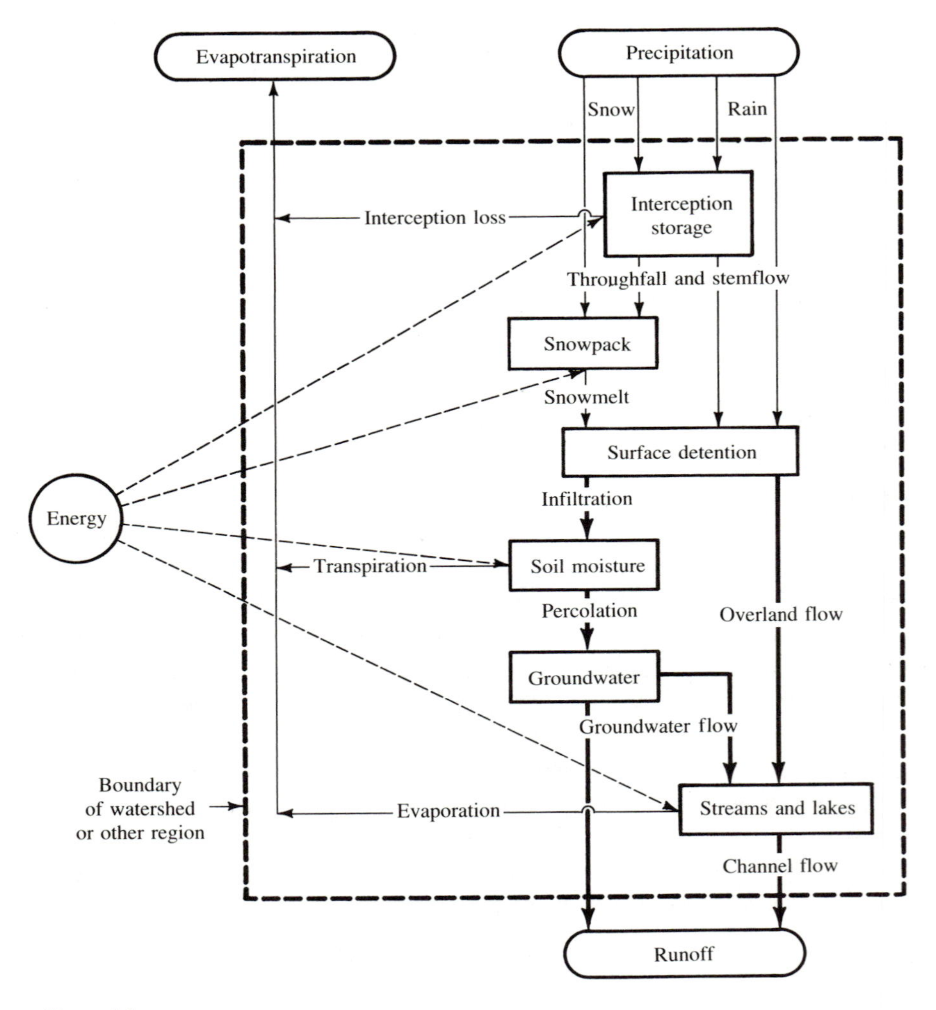

Figure 1.1
Principal pathways (arrows) and storages (boxes) of water in the hydrologic cycle. This text is concerned with the physics of motion in the portions of the cycle indicated by heavy arrows.

Chapters 2 and 3 are specifically devoted to the development of patterns through discussion of the fundamental concept of dimension, the application of this concept in the analysis of equations, and an examination of the principal types of equations that hydrologists and fluvial geomorphologists encounter and use. However, the recognition and understanding of the relevant equations are only part of the requirements for solving scientific and practical problems. You must also be able to obtain the correct numerical solution, within the limits of accuracy and precision dictated by the nature of the problem and the available data. This may seem to be an unnecessary statement of the obvious, but it is worth emphasizing, since you are not likely to get much credit for your grasp of hydrologic concepts if your answer is incorrect because of a mistake in converting from one system of measurement units to another. Thus much of Chapters 2 and 3 is concerned with important aspects of numerical manipulation: significant figures and conversion of units and equations.

Chapter 4 deals with water as a substance, beginning with its molecular structure. This structure determines the physical properties of water, and these properties in turn determine its state, configuration, and rate of movement through the hydrologic cycle. Thought experiments are used to couple your physical intuition to mathematical expressions of fundamental physical relationships that are incorporated in subsequent analyses of hydrologic flows.

Chapters 5–11 are concerned with open-channel flows, which include the wide "sheet flows" that characterize overland flow as well as flows in well-defined channels. Before we consider such flows directly, Chapter 5 introduces some elementary characteristics of water at rest. This review provides an understanding of hydrostatic pressure and of potential energy, concepts that are essential for analyzing flowing water. Next, we consider the basic concept of boundary-layer flows and develop expressions for computing the relative magnitudes of forces in such flows. Our main concern in this chapter is an understanding of the bases for classification of flows, which is an essential component of the pattern-recognition skills mentioned earlier.

One basis for classification depends on the rates of change of velocity and depth with time and with distance. Steady flows are those in which changes with time are negligible; these flows are treated in Chapters 6–10. Chapter 6 discusses steady uniform flows, in which changes with distance are also negligible. Although natural flows are never strictly steady uniform, analysis of such simple cases permits the development of equations of motion. These are expressions of the relations among velocity, depth, channel gradient, and channel resistance that provide the foundation for describing more complex situations.

Chapter 7 develops the equation of energy conservation (the first law of thermodynamics) for steady open-channel flows. This fundamental relation is applied repeatedly in subsequent chapters. The concept of stream power, which expresses the rate of energy expenditure of a stream, is also introduced here.

Chapter 8 examines channel resistance, which determines the rate of energy expenditure by a stream. Resistance is related to the form of the channel at scales ranging from the sediment particles of the bed to the patterns of channel form as observed on a map. Examination of the complex interrelations of channel form and resistance leads naturally to consideration of erosion and sediment transport, where the concept of stream power provides especially useful insight. These topics are of particular interest in fluvial geomorphology and in many engineering problems.

Flows in which depth and velocity change with distance along the channel are called nonuniform. Chapter 9 treats steady nonuniform flows in which the downstream changes are gradual. Analysis of such flows is the basis for treating a large class of engineering problems that require estimating the profile of the water surface over extended river reaches.

Chapter 10 examines three types of steady nonuniform flows in which downstream changes in velocity and depth are sudden rather than gradual. These types are (1) flows that involve abrupt changes in channel width or depth; (2) the standing-wave phenomenon known as the hydraulic jump, which accompanies certain types of changes in width, depth, or slope; and (3) flows over weirs. Weirs are of particular interest to hydrologists because of their use in measuring streamflow rates.

The prediction of streamflow response to rainstorm or snowmelt events is one of the central problems of surface-water hydrology. This prediction involves consideration of changes in both time and space, and Chapter 11 shows how the principles developed in preceding chapters are applied in analyzing unsteady nonuniform flows. All but the most simplified cases of such flows are so complex, however, that these principles must be applied in constructing computer models. These models consist essentially of the basic equations of motion and conservation of mass, written separately for various locations and times. A computer is required to solve these equations simultaneously for a large number of space and time steps that approximate continuous spatial and temporal changes. Thus the emphasis throughout Chapter 11 is on acquiring an awareness of the underlying principles and assumptions so that you can understand the capabilities and limitations of existing models and develop a competence in constructing new ones.

The text concludes with a condensed discussion of flows in saturated and unsaturated porous media. Most hydrologic problems involving such flows require consideration of changes in time and at least two spatial directions. To analyze these flows, the basic principles of flow must again be incorporated into computer models. As in Chapter 11, the main objective of the treatment in Chapter 12 is to build physical intuition. Thus the chapter is largely devoted to applying to porous media flows the concepts developed earlier in our analyses of open-channel flows.

CONCLUDING COMMENT

In addition to helping to build understanding and competence in problem solving, I hope the patterns and connections identified in this book will evoke some appreciation for Albert Einstein's profound observation that "the eternal mystery of the universe is its comprehensibility." One manifestation of this comprehensibility is the power of mathematics in exploring the physical world. It seems little short of miraculous, for example, that we can conduct experiments to determine that the velocity gradient in flowing water is related in a particular way to the applied stress that causes the flow (Chapter 4), use this relationship as a starting point for purely mathematical deductions about the shapes of velocity gradients in flows of various geometries (Chapter 6), and find by observation of such flows that our deductions are correct! The pervasive use of mathematics throughout this text and others like it should thus be viewed not as a burden or distraction, but rather as an attempt to speak the same language that nature speaks, a language that it behooves the student of nature to learn fluently.

The comprehensibility of nature—in particular of the movement of water in the hydrologic cycle—at which Einstein marveled has evolved from practical necessity and intellectual curiosity over the 6000 years of recorded history (Rouse and Ince, 1963; Biswas, 1970). This book attempts to convey this understanding to the student of earth sciences as a basis for analysis of practical problems and as a stimulus to intellectual curiosity. Our achievement of this understanding should, however, be tempered with the realization that there are still fundamental gaps in our knowledge (Dooge, 1983).

CHAPTER 2

Dimensions, Units, and Significant Figures

The study of dimensions, units, and significant figures brings us face to face with some fundamental philosophical questions concerning the nature of physical quantities and their measurement (see, e.g., Ipsen, 1960; Quade, 1967; Pankhurst, 1964; and Taylor, 1974). Furthermore, the techniques of dimensional analysis, which are developed from such theoretical considerations, can be very useful in solving problems in many branches of physics and engineering (Ipsen, 1960; Pankhurst, 1964; Taylor, 1974), and have proved to be an especially powerful tool in fluid mechanics (Rouse, 1961). They have even been successfully applied in economics (de Jong, 1967).

In this book, however, we will not be exploring all these aspects of the topic, interesting as they are. We will instead take a somewhat limited and practically oriented view, and formulate a series of rules for analyzing and manipulating the physical quantities commonly encountered in hydrology and related sciences. These rules form a theoretically sound and logically consistent basis that provides powerful insights into the quantitative relationships that are the essence of the theory and practice of hydrology and fluvial geomorphology. Thus you should learn them as fundamental laws to be applied continually. In fact, unless you do this, you will not truly know what you are reading, writing, or talking about.

DIMENSIONS

Each time we measure a physical quantity, we must express that measurement in two parts: (1) a numerical magnitude; and (2) a unit of measurement. Neither part can stand alone, and we can state the following rule:

Rule 1

The numerical magnitude of a physical quantity has no meaning without an accompanying statement of the unit of measurement.

Obviously, we cannot report a length as 25 without saying whether we're measuring in terms of angstroms, meters, furlongs, or light-years. There is nothing fundamental about units; each is a purely arbitrary choice based on convenience, and a given length can be expressed perfectly satisfactorily as 1 mile, 5280 feet, 1609 meters, 1.609×10^{13} angstroms, or 1.69×10^{-13} light-years.

However, the concept of *length* is fundamental, independent of the units chosen to measure it. It is one of the fundamental *dimensions* of physical science. All measurements of distances, whether they be depths, widths, lengths, heights, or diameters, have the dimension of length. Three other fundamental dimensions are encountered in physical hydrology: mass (or force)[1], time, and temperature. In addition, some numbers are encountered that have no dimensions. We emphasize these concepts in two more rules:

Rule 2

Each physical quantity has a fundamental dimensional character.

Rule 3

Either the dimensional character of a physical quantity is expressed as some combination of length, mass (or force), time, and temperature, or the quantity is dimensionless.

Rule 3 applies to the branches of physics called mechanics and thermodynamics and thus to physical hydrology. We will use the symbols L, M, F, T, and Θ for the dimensions length, mass, force, time, and temperature, respectively, and will give the dimensions of a quantity in square brackets, for example, $[L]$ for length. We will use both the $MLT\Theta$ and $FLT\Theta$ systems in this text; the choice is a matter of convenience.

[1]Either mass or force can be considered fundamental, and the choice is arbitrary. The relation between mass and force is expressed in Newton's second law of motion, as discussed later.

Rule 4a

The dimensional character of each physical quantity is expressed as $[M^aL^bT^c\Theta^d]$ (or $[F^eL^fT^g\Theta^d]$), where $a, b, \ldots, g$ are rational numbers.[2]

Rule 4b

For dimensionless (pure) numbers $a = b = c = \cdots = g = 0$, and the dimensional character is expressed as [1].

The values of $a, b, \ldots, g$ are almost always between -4 and $+4$. In a later table you will find the dimensions of quantities commonly encountered in physical hydrology.

Rules 1–4 are preliminary to the most important rule:

Rule 5

The dimensions of quantities must be subjected to the same mathematical operations as the quantities themselves.

You must train yourself to put this rule into practice any time you read or manipulate an equation. Doing so will be invaluable in helping you understand what you are doing and what others are talking about, as will become more apparent in the discussion of equations in Chapter 3.

For the moment, consider the following equation:

$$y = \tfrac{1}{2}gt^2, \tag{2-1}$$

where y is the distance traveled by a freely falling body in time t and g is the acceleration due to gravity. Rewriting, with the dimensions, we have

$$(y[L]) = (\tfrac{1}{2}[1])(g[LT^{-2}])(t[T])^2.$$

Concentrating on just the dimensions, we have

$$[L] = [1][LT^{-2}][T]^2$$

and thus

$$[L] = [L].$$

In carrying out mathematical operations, you must keep the following in mind:

[2]Rational numbers are integers or ratios of integers.

Rule 6a

The mathematical operations involved in manipulating dimensions are multiplication and exponentiation, and the dimensional symbols can be treated exactly like algebraic symbols.

Rule 6b

Dimensions can be eliminated by means of these mathematical operations (as $[T]$ was above).

Rule 6c

Addition and subtraction can take place only with terms having the same dimensions.

Dimensionless, or pure, numbers are often encountered in physics; examples are $\pi = 3.14159\ldots$ and $e = 2.7182\ldots$. Pure numbers often appear in basic physical laws, such as the multiple $\frac{1}{2}$ and the exponent 2 in Equation 2-1. Similarly, if two distances h and x are related such that h is α times larger than x,

$$h = \alpha x, \tag{2-2}$$

then α is a pure number. This is obvious if the equation is rewritten and α is seen as a ratio of two lengths:

$$\alpha = \frac{h}{x}.$$

Dimensionally we have

$$\frac{[L]}{[L]} = [1].$$

As will be shown shortly, not all multiples in equations are pure numbers, but

Rule 7

All exponents, logarithms, and trigonometric functions are pure (dimensionless) numbers.

As noted earlier, either the $MLT\Theta$ or $FLT\Theta$ system can be used; the choice is a matter of convenience. The equation that allows transposition from one system to the other is Newton's second law, which relates force, F, mass, M, and acceleration, a:

$$F = Ma. \tag{2-3}$$

The dimensional relations are

$$[F] = [M][LT^{-2}] = [MLT^{-2}] \tag{2-4}$$

or

$$[M] = [F][T^2L^{-1}] = [FT^2L^{-1}]. \tag{2-5}$$

For example, from Einstein's work we know that

$$E = Mc^2, \tag{2-6}$$

where E is the energy obtainable from the complete transformation of a mass M, and c is the velocity of light $[LT^{-1}]$. Thus energy has the dimensions $[ML^2T^{-2}]$ in the $MLT\Theta$ system. Equations 2-5 and 2-6 can be used to find that energy has the dimensions $[FL]$ in the $FLT\Theta$ system:

$$[E] = [M][LT^{-1}]^2;$$
$$[E] = [FT^2L^{-1}][LT^{-1}]^2;$$
$$[E] = [FT^2L^{-1}][L^2T^{-2}];$$
$$[E] = [FL]. \tag{2-7}$$

UNITS

All measurements are made in terms of arbitrary *units*. The three systems of units indicated in Table 2.1 have been in widespread use in various branches of science and engineering. The *Système International* (SI) units (Table 2.2) are increasingly

Table 2.1
Units of Measurement of Fundamental Dimensions in Three Systems

Fundamental dimension	*Système International* (SI) unit	Centimeter-gram-second (cgs) unit	English unit
Mass	kilogram (kg)	gram (g)	slug
Force	newton (N)	dyne (dyn)	pound (lb)
Length	meter (m)	centimeter (cm)	foot (ft)
Time	second (s)	second (s)	second (s)
Temperature	kelvin (K)	degree Celsius (°C)	degree Fahrenheit (°F)

Table 2.2
Système International (SI) Units and Prefixes

Unit	Abbreviation	Prefix	Symbol	Multiplication factor
joule (N · m)	J	tera	T	10^{12}
kelvin	K	giga	G	10^{9}
kilogram	kg	mega	M	10^{6}
meter	m	kilo	k	10^{3}
newton	N	milli	m	10^{-3}
pascal (N m^{-2})	Pa	micro	μ	10^{-6}
radian	rad	nano	n	10^{-9}
second	s	piko	p	10^{-12}
watt (J s^{-1})	W	femto	f	10^{-15}
		atto	a	10^{-18}

displacing the other systems in the scientific literature, but the hydrologist frequently encounters the centimeter-gram-second (cgs) system, because it is very convenient for measurements involving water, and the English system, because of long-standing tradition, particularly in engineering-related disciplines.

As indicated in the preceding discussion, it would be sufficient for purposes of physical hydrology to have only four units, one for each of the fundamental dimensions. It has proved convenient, however, to define many other units. Table 2.3 gives the units and dimensions of the quantities commonly encountered in physical hydrology. Those involving length only are classed as geometric (angle is included, although it is a dimensionless ratio of two lengths); those involving length and time or time only are kinematic; those involving mass or force are dynamic; and those involving temperature are thermal (latent heat is included here also).

CONVERSION OF UNITS

Because hydrologists and fluvial geomorphologists often encounter the three systems of units, they must become proficient in converting measurements from one system to another. The method for doing this is straightforward. To convert a quantity of X old units to its equivalent in new units, we proceed as follows:

$$(X \text{ old units})\left(Y\frac{\text{new units}}{\text{old units}}\right) = XY \text{ new units},$$

where Y is a conversion factor that can be found in a table such as Table 2.4. For example,

$$(23.9 \text{ ft})\left(\frac{1 \text{ m}}{3.28 \text{ ft}}\right) = 7.29 \text{ m}.$$

If the quantity is more complex dimensionally, we use this formula:

$$\left(X\frac{\text{old units 1}}{\text{old units 2}}\right)\left(Y\frac{\text{new units 1}}{\text{old units 1}}\right)\left(Z\frac{\text{old units 2}}{\text{new units 2}}\right) = XYZ\frac{\text{new units 1}}{\text{new units 2}},$$

where Y and Z are conversion factors. For example,

$$(6.35 \text{ m s}^{-1})\left(\frac{3.28 \text{ ft}}{1 \text{ m}}\right)\left(\frac{3600 \text{ s}}{1 \text{ h}}\right) = 7500 \text{ ft h}^{-1}.$$

The only trick in such operations is in knowing when to use a conversion factor directly and when to use its inverse. However, if you write out the problem as is done here and algebraically cancel units, the correct choice is clear.

The only case in which conversion is slightly more complicated is the conversion between temperatures in the Fahrenheit (°F) and Celsius (°C) scales, because an additive factor is also involved:

$$°\text{F} = °\text{C}\left(\frac{9°\text{F}}{5°\text{C}}\right) + 32°\text{F};$$

$$°\text{C} = (°\text{F} - 32°\text{F})\frac{5°\text{C}}{9°\text{F}}.$$

Conversion to absolute temperatures merely involves addition:

$$\text{K} = °\text{C} + 273.2; \qquad °\text{R} = °\text{F} + 460.$$

Table 2.4 gives factors for converting cgs and English units to SI units.

NUMERICAL PRECISION AND SIGNIFICANT FIGURES

For practical purposes, we can consider that there are two types of numbers ordinarily encountered in physical problems: *rational numbers,* which can be expressed exactly as the ratio of two integers (and, of course, include the integers

Table 2.3
Dimensions and Units of Physical Quantities Commonly Encountered in Hydrology

Quantity	Common Symbols	Dimensions		Units†	
		$LTM\Theta$	$LTF\Theta$	Common	*SI*
GEOMETRIC					
Angle	α, S, θ	1	1	degree (°)	rad
Area	A	L^2	L^2	ft^2, ac, mi^2, cm^2, ha, km^2	m^2
Length	L, l, x, y, z	L	L	ft, mi, mm, cm, km	m
Volume	V, B	L^3	L^3	ft^3, gal, ac · ft, cm^3, liter	m^3
KINEMATIC					
Acceleration	a, g	LT^{-2}	LT^{-2}	ft s^{-2}, cm s^{-2}	m s^{-2}
Angular acceleration	α	T^{-2}	T^{-2}	degree s^{-2}, rev s^{-2}	rad s^{-2}
Angular velocity	ω	T^{-1}	T^{-1}	degree s^{-1}, rev s^{-1}	rad s^{-1}
Diffusivity	D, K	L^2T^{-1}	L^2T^{-1}	ft^2 s^{-1}, cm^2 s^{-1}	m^2 s^{-1}
Discharge rate	Q	L^3T^{-1}	L^3T^{-1}	ft^3 s^{-1}, gal min^{-1}, liter s^{-1}	m^3 s^{-1}
Kinematic viscosity	ν	L^2T^{-1}	L^2T^{-1}	ft^2 s^{-1}, cm^2 s^{-1}	m^2 s^{-1}
Time	T, t	T	T	min, h, day, yr	s
Velocity	V, u, v, w	LT^{-1}	LT^{-1}	ft s^{-1}, mi h^{-1}, cm s^{-1}	m s^{-1}
DYNAMIC					
Dynamic viscosity	μ	$ML^{-1}T^{-1}$	FTL^{-2}	lb s ft^{-2}, cP, g cm^{-1} s^{-1}, dyn s cm^{-2}	Pa · s
Energy (work)	E, W	ML^2T^{-2}	LF	Btu, ft · lb, cal, kW · H	J

Energy flux density	Q, H	MT^{-3}	$F^{-1}LT^{-1}$	Btu ft^{-2} h^{-1}, cal cm^{-2} s^{-1}	W m^{-2}
Force (weight)	F	MLT^{-2}	F	lb, g, dyn	N
Mass	m, M	M	FT^2L^{-1}	slug, g	kg
Mass density	ρ	ML^{-3}	$FT^2 L^{-4}$	slug ft^{-3}, g cm^{-3}	kg m^{-3}
Momentum	M	MLT^{-1}	FT	slug ft s^{-1}, lb s, dyn s, g cm s^{-1}	kg m s^{-1}, N s
Power	P	ML^2T^{-3}	LFT^{-1}	Btu h^{-1}, cal s^{-1}, ft · lb s^{-1}, hp	W
Pressure	P, e	$ML^{-1}T^{-2}$	FL^{-2}	lb ft^{-2}, lb in.$^{-2}$, dyn cm^{-2}, mb, mm Hg, in. Hg, cm H_2O	Pa
Shear stress	τ	$ML^{-1}T^{-2}$	FL^{-2}	lb ft^{-2}, lb in.$^{-2}$, dyn cm^{-2}	Pa
Surface tension	σ	MT^{-2}	FL^{-1}	lb ft^{-1}, dyn cm^{-1}	N m^{-1}
Weight density	γ	$ML^{-2}T^{-2}$	FL^{-3}	lb ft^{-3}, dyn cm^{-3}	N m^{-3}
THERMAL					
Heat capacity	c	$L^2T^{-2}\Theta^{-1}$	$L^2T^{-2}\Theta^{-1}$	Btu lb^{-1} °F^{-1}, cal g^{-1} °C^{-1}	J kg^{-1} K^{-1}
Heat transfer coefficient	h, q, K	$MT^{-3}\Theta^{-1}$	$FL^{-1}T^{-1}\Theta^{-1}$	Btu ft^{-2} h^{-1} °F^{-1}, cal cm^{-2} s^{-1} °C^{-1}	W m^{-2} K^{-1}
Latent heat	L, H, λ	L^2T^{-2}	L^2T^{-2}	Btu lb^{-1}, cal g^{-1}	J kg^{-1}
Temperature	T, θ	Θ	Θ	°F, °C	K
Thermal conductivity	K	$MLT^{-3}\Theta^{-1}$	$FT^{-1}\Theta^{-1}$	Btu ft^{-1} h^{-1} °F^{-1}, cal cm^{-1} s^{-1} °C^{-1}	W m^{-1} K^{-1}

†See Tables 2.2 and 2.4.

themselves); and *irrational numbers,* which cannot be so expressed. When we come upon rational numbers (such as the exponents in the dimensional relations described earlier, or the fraction $\frac{1}{2}$, which arises in the definitions of quantities such as surface tension and kinetic energy), we should regard them as having infinite precision. Other examples of such infinitely precise numbers are those expressing the number

Table 2.4
Conversion Factors for Non-SI Units

Abbreviation	Unit	Conversion factor	SI unit
ac	acre	$\times$ 4047	$= m^2$
Btu	British thermal unit	$\times$ 1055	= J
cal	calorie	$\times$ 4.187	= J
cm	centimeter	$\times 1 \times 10^{-2}$	= m
cm H_2O	centimeter of water	$\times$ 101.99	= Pa
cP	centipoise	$\times 1 \times 10^{-3}$	= Pa · s
day	day	$\times$ 86,400	= s
°	degree of angle	$\times$ 0.01745	= rad
°C	degree Celsius	+ 273.16	= K
°F	degree Fahrenheit	+ 459.69/1.8	= K
°R	degree Rankine	$\times \frac{5}{9}$	= K
dyn	dyne	$\times 1 \times 10^{-5}$	= N
ft	foot	$\times$ 0.3048	= m
gal	gallon	$\times$ 0.003785	$= m^3$
g	gram	$\times 1 \times 10^{-3}$	= kg
ha	hectare	$\times 1 \times 10^{4}$	$= m^2$
hp	horsepower	$\times$ 746	= W
h	hour	$\times$ 3600	= s
in.	inch	$\times$ 0.00254	= m
in. Hg	inch of mercury	$\times$ 3386	= Pa
liter	liter	$\times 1 \times 10^{-3}$	$= m^3$
mi	mile	$\times$ 1609	= m
mb	millibar	$\times$ 100	= Pa
min	minute	$\times$ 60	= s
lb	pound	$\times$ 4.448	= N
rev	revolution	$\times$ 6.282	= rad
yr	year	$\times$ 31,557,600	= s

of feet in a yard or centimeters in a meter or seconds in a day. Most measured quantities, however, are irrational numbers and have finite precision.

The finite precision of measured quantities can be discussed in both absolute and relative terms. Absolute precision is expressed in terms like "to the nearest x," where x is some measurement unit. For example, if we were to measure a distance to the nearest centimeter, we would have to report it as, say, 21 cm. If we were to report the measurement as 21.0 cm, or 21.00 cm, we would be implying that it had been made to the nearest 0.1 cm or 0.01 cm, respectively. If a measurement is given as, say, 200 m, the precision is not clear, because we don't know if the measurement was to the nearest meter, 10 meters, or 100 meters. One way of avoiding this ambiguity is to use scientific notation and express the quantity as 2×10^2 m, 2.0×10^2 m, or 2.00×10^2 m, as appropriate. Otherwise, additional information, usually in the form of other analogous measurements, is required to clarify the situation.

In relative terms, we express precision as the *number of significant figures* in a quantity; this number is equal to the number of digits beginning with the leftmost nonzero digit and extending to the right to include all digits warranted by the measurement precision. Thus in our first example, 21, 21.0, and 21.00 cm represent two, three, and four significant figures, respectively. In the second example, 2×10^2, 2.0×10^2, and 2.00×10^2 m represent one, two, and three significant figures, respectively.

In adding or subtracting measured values, we must be concerned with absolute precision, and observe the following rule.

Rule 8

The absolute precision of a sum or difference equals the absolute precision of the least absolutely precise number involved in the addition or subtraction.

For example, it is the policy of the U.S. Geological Survey to report measurements of mean daily streamflow with the following precision:

less than 1 $ft^3\ s^{-1}$ to the nearest 0.01 $ft^3\ s^{-1}$

1–9.9 $ft^3\ s^{-1}$ to the nearest 0.1 $ft^3\ s^{-1}$

10–999 $ft^3\ s^{-1}$ to the nearest 1 $ft^3\ s^{-1}$

over 1000 $ft^3\ s^{-1}$ to three significant figures

Suppose the mean flow on two days is given as 102 $ft^3\ s^{-1}$ and 3.2 $ft^3\ s^{-1}$. Then the total for the two days must be reported as 105 $ft^3\ s^{-1}$, not 105.2 $ft^3\ s^{-1}$. If the two flows were 1020 $ft^3\ s^{-1}$ and 3.2 $ft^3\ s^{-1}$, the total would be 1020 $ft^3\ s^{-1}$. As a final example, if we assume U.S. Geological Survey precision and add the following flows (all in cubic feet per second): 27, 104, 12, 2310, 6.4, 0.11, and 256, the computed total of 2715.51 must be rounded to 2720 $ft^3 s^{-1}$.

In multiplication and division, we must be concerned with relative precision, and observe the following rule.

Rule 9

The number of significant figures of a product or quotient equals the number of significant figures of the least relatively precise number involved in the multiplication or division.

Rule 9 is the rule that must be followed in all unit conversions. An example will illustrate how this practice preserves the appropriate expression of precision. Suppose someone indicates that the distance between two points is 9.6 mi. We understand such a statement to mean that the distance is known to within 0.1 mi. If we want to convert this measurement from miles to meters, we would typically find a conversion factor stated as "1 mi = 1609 m." Multiplying 9.6 by 1609 gives 15,446.4 m. If we now state that the distance between the two points is 15,446.4 m, we are implying that the distance is known to within 0.1 m. This is obviously incorrect, since we originally knew the distance only to within 0.1 mi, which is about 161 m. If we follow Rule 9, we would state the distance as 15,000 m. This converted distance is less absolutely precise than the original measurement, because it implies that the distance is known only to within 1000 m, whereas it is actually known to within about 161 m. Generally, we put up with this loss of precision, although an alternative approach would be to approximate the precision of the original measurement by expressing the converted measurement as 15,400 ± 200 m. This is seldom done, however, and Rule 9 should generally be followed.

In hydrology, as in most earth sciences, we are seldom justified in assuming that measured values have relative precisions greater than three significant figures. Thus,

Rule 10

Unless it is clear that greater precision is warranted, assume no more than three-significant-figure precision in hydrologic measurements and computations.

As noted in the description of U.S. Geological Survey practice above, there will be many times when only two-significant-figure precision is warranted.

Special situations arise when we deal with exponentials, logarithms, and trigonometric and other functions, or when we carry out any long involved computation, such as statistical calculations. Typically, we start with actual measurements for which the precision is known. These may then be converted to, say, logarithms and manipulated in some way, and the results then converted back to physical units. In these cases, follow this rule:

Rule 11

In lengthy statistical computations or computations involving conversions to logarithms, trigonometric functions, and the like, do not round off to the appropriate number of significant figures until you get to the final answer.

This is especially important in statistical computations, where it is often necessary to calculate a relatively small difference between two relatively large numbers.

Finally, one additional rule should be kept in mind as you work numerical problems:

Rule 12

Computers and calculators don't know anything about significant figures.

Exercises

The following problems will give you practice in converting measurements from one system of units to another. Use the conversion factors given in Table 2.4, and be sure to observe the rules regarding significant figures.

2-1. Convert 653 $ft^3\ s^{-1}$ to gal min^{-1}, $m^3\ s^{-1}$, and ac · ft yr^{-1}.

2-2. Convert 2.7 mi h^{-1} to cm s^{-1}, ft s^{-1}, and m s^{-1}.

2-3. Convert 10,513 cal cm^{-2} total for 15 days to cal $cm^{-2}\ s^{-1}$, W m^{-2}, and J $cm^{-2}\ day^{-1}$.

2-4. Convert 1.73 mi^2 to ha, ac, and m^2.

2-5. Convert 1.7 in. of rain in 24 h on 9.3 mi^2 to $ft^3\ s^{-1}$, $m^3\ s^{-1}$, and liters s^{-1}.

2-6. Convert 116 ac · ft to m^3, ft^3, and gal.

2-7. Convert 12.1 $m^3\ s^{-1}\ km^{-2}$ to cm s^{-1}, $ft^3\ s^{-1}\ mi^{-2}$, and in. h^{-1}.

2-8. Convert 163 lb ft^{-2} to Pa, lb in.$^{-2}$, dyn cm^{-2}, in. Hg, and mb.

2-9. Convert 1 cal $g^{-1}\ °C^{-1}$ to Btu $lb^{-1}\ °F^{-1}$.

2-10. Convert 0.18 Btu $ft^{-2}\ h^{-1}\ °F^{-1}$ to cal $cm^{-2}\ s^{-1}\ °C^{-1}$ and W $m^{-2}\ K^{-1}$.

Add the following, using the rules for U.S. Geological Survey streamflow data.

2-11. 27 + 1060 + 2.3 + 0.98 + 8700

2-12. 15 + 742 + 10,500 + 6.4 + 83 + 212 + 0.47

2-13. 23,200 + 1040 + 16 + 529,000 + 74 + 1.6

CHAPTER 3

Fundamentals of Physical Equations

In general, each hydrologic problem you encounter will be new in some respects. As your experience with real problems grows, you will learn that solving problems is usually fairly easy; the difficult part is stating the problem precisely.

Setting up problems generally involves the development of a model. Although models are often thought to be relatively complicated, every equation is a model—a concise statement of how specified quantities are thought to be related. Thus, to understand the literature of hydrology and fluvial geomorphology and to formulate and solve problems, you have to be able to recognize and understand the essential characteristics of equations that describe the various parts of the hydrologic cycle.

It is in the analysis of equations that the understanding of dimensions emphasized in Chapter 2 becomes most useful. Thus we begin this chapter with an examination of dimensional analysis of equations, and then go on to emphasize the importance of defining the space and time characteristics of the quantities involved in equations and to discuss the six major types of equations encountered in physical hydrology.

DIMENSIONAL ANALYSIS OF EQUATIONS

Rule 13

An equation that completely and correctly describes a physical relation has the same dimensions on both sides of the equal sign.

Equations that conform to Rule 13 are called *dimensionally homogeneous*. There are, however, two important qualifications of this rule: (1) Not all dimensionally homogeneous equations completely and correctly describe a relationship; and (2) some equations that are not homogeneous are very useful approximations of relationships.

An example of a homogeneous equation that is incorrect is

$$A = 2r^2, \tag{3-1}$$

where r is the radius of a circle ($[L]$) and A is the area of the circle ($[L^2]$). An example of a homogeneous equation that is incomplete is

$$H = Z + Y, \tag{3-2}$$

where H is the energy per weight of flowing water ($[L]$), Z is the height above a horizontal datum ($[L]$), and Y is the depth of flow ($[L]$). The complete equation is

$$H = Z + Y + \frac{V^2}{2g}, \tag{3-3}$$

where V is the mean flow velocity ($[LT^{-1}]$) and g is the acceleration of gravity ($[LT^{-2}]$). However, even though Equation 3-2 is incomplete, it may not be "wrong": In some situations, such as groundwater flow, the flow velocity may be very small, so that $V^2/2g \ll (Z + Y)$. In such a situation, Equation 3-2 gives essentially the same answer as Equation 3-3 and is therefore "correct" for practical purposes.

The following equation is dimensionally inhomogeneous:

$$V = \frac{R^{2/3} S_0^{1/2}}{n}, \tag{3-4}$$

where V is mean stream velocity ($[LT^{-1}]$), R is the hydraulic radius of the stream (defined later; $[L]$), S_0 is the tangent of the channel slope ($[1]$), and n is a roughness factor ($[1]$). Nevertheless, as we will see, Equation 3-4 is the most widely used formula for relating the velocity of an open-channel flow to flow geometry and channel characteristics. It is an incomplete equation; other factors besides R, S_0, and

n influence mean stream velocity. However, the complete relationship is very complicated, and Equation 3-4 is sufficiently accurate for most practical calculations. Such an inhomogeneous equation is also called an *empirical* equation: It is based on observation and experience rather than on a complete formulation of physical principles.

Every time you read or write an equation, you should check for dimensional homogeneity. If an equation is homogeneous, it can be used without change with any consistent system of units (assuming, of course, that it is correct). Thus the following is a homogeneous equation:

$$P = \gamma y, \tag{3-5}$$

where P is the hydrostatic pressure at a point ($[FL^{-2}]$), γ is the weight density of a liquid ($[FL^{-3}]$), and y is the depth of the point ($[L]$). It gives hydrostatic pressure in newtons per square meter when γ is in newtons per cubic meter and y is in meters:

$$(\mathrm{N\ m^{-2}}) = (\mathrm{N\ m^{-3}})(\mathrm{m}).$$

Similarly, Equation 3-5 gives the pressure in pounds per square foot (lb ft^{-2}) when γ is in pounds per cubic foot (lb ft^{-3}) and y is in feet, in dynes per square centimeter (dyn cm^{-2}) when γ is in dynes per cubic centimeter (dyn cm^{-3}) and y is in centimeters, and so on.

If an equation is inhomogeneous, it is essential to follow these rules:

Rule 14a

In an inhomogeneous equation, the units for each variable in the equation must be specified.

Rule 14b

In an inhomogeneous equation, all constants except exponents, logarithms, and trigonometric functions must be changed if the equation is to be used in another system of units.

It is surprising that these rules, particularly Rule 14a, are often not followed in textbooks and professional papers. To avoid confusion you must train yourself to be on guard for such lapses.

CONVERSION OF EQUATIONS

As we said in Chapter 2, hydrologists and fluvial geomorphologists encounter many systems of units in textbooks, manuals, and reports. It is not at all uncommon to read a discussion of floods that speaks of rainfall in inches, stream lengths in miles, water-surface elevations in feet, floodplain areas in acres, and volumes of water in acre-feet. Even stranger, but not rare, are equations with English and metric units mixed, like

$$M = \frac{H}{203B},$$

where H is energy input to a snowpack in calories per square centimeter per day (cal cm^{-2} day^{-1}), B is the "thermal quality" of the snowpack (dimensionless), and M is the resulting snowmelt in inches per day (in. day^{-1}). Although the use of SI units is beginning to be more common, and is insisted upon by some journals, the centimeter-gram-second (cgs) system of units is in many ways more convenient when dealing with water, and is often encountered. Furthermore, English units have been and continue to be in widespread use in this country. This diversity means, unfortunately, that you will often have to convert not only single measurements from one system of units to another, but also equations for use with other units or for comparison with other equations.

The procedure for conversion is straightforward, but must be followed carefully. First, check the equation for dimensional homogeneity. As pointed out earlier, if an equation is homogeneous, it can be used without change in any consistent system of units. A system that mixes inches, feet, and miles as units of length, and acres as units of area, is not consistent, even though all the units are English. The same is true if millimeters, centimeters, meters, and kilometers are used for length, and hectares for area. The following equation is dimensionally homogeneous:

$$\tau_0 = \gamma Y S_0, \tag{3-6}$$

where τ_0 is the shear stress ($[FL^{-2}]$), γ is the weight density of the water ($[FL^{-3}]$), Y is the depth of the water ($[L]$), and S_0 is the slope of the stream bed ($[1]$). Consider a situation where the flow depth is 100 cm and the slope is 0.001. If we take γ in compatible units as 980 dyn cm^{-3}, Equation 3-6 gives $\tau_0 = 98.0$ dyn cm^{-2}. The equation should give the same answer with measurements in another system. In the English system, the depth would be measured as 3.28 ft, and the weight density of water as 62.4 lb ft^{-3}. The slope is dimensionless, so it remains 0.001, and we calculate that $\tau_0 = 0.205$ lb ft^{-2}. Is this the same shear stress as before? Check by converting units by the rule discussed earlier:

$$98.0 \text{ dyn cm}^{-2} \times \left(\frac{1 \text{ lb}}{444{,}800 \text{ dyn}}\right) \times \left(\frac{30.48 \text{ cm}}{1 \text{ ft}}\right)^2 = 0.205 \text{ lb ft}^{-2}.$$

If the equation to be converted is inhomogeneous, first remember that exponents, trigonometric functions, and logarithms will not change (Rule 14b); only the multiplicative and additive constants will change. Then follow this procedure: Multiply each variable of the equation by a factor that would convert the new units to the old units. This may seem backward, but it isn't. For example, suppose we have the empirical equation

$$P = 22.4 + 0.0105Z_s, \tag{3-7}$$

where P is average annual precipitation in inches per year and Z_s is elevation in feet. We want to convert this equation for use with Z_s in meters and the results, P, in centimeters per year. Following the procedure above yields

$$P \text{ cm yr}^{-1} \times \left(\frac{1 \text{ in.}}{2.54 \text{ cm}}\right) = 22.4 + 0.0105Z_s \text{ m} \times \left(\frac{3.28 \text{ ft}}{1 \text{ m}}\right).$$

Eliminating the unit symbols, we have

$$P\left(\frac{1}{2.54}\right) = 22.4 + 0.0105Z_s\left(\frac{3.28}{1}\right),$$

and multiplying through yields

$$P = 22.4(2.54) + 0.0105Z_s(3.28)(2.54);$$

$$P = 56.9 + 0.0875Z_s. \tag{3-8}$$

To check whether this result is correct, see if Equations 3-7 and 3-8 give the same answer. Say $Z_s = 1000$ ft; then Equation 3-7 gives $P = 32.9$ in. yr^{-1}. In the new system of units, the same elevation would be expressed as 305 m. Using this value in Equation 3-8, we find that $P = 83.6$ cm yr^{-1}. Is this the same answer?

$$83.6 \text{ cm yr}^{-1} \times \frac{1 \text{ in.}}{2.54 \text{ cm}} = 32.9 \text{ in. yr}^{-1}.$$

SPACE AND TIME CONSIDERATIONS

Generally, the properties of water that will be discussed in detail in Chapter 4 can be considered to be constants; they do not change from time to time or place to place in a given problem. However, the other quantities of importance to hydrology, such as streamflow rate, rainfall rate, evaporation rate, and depth of water, are highly variable in space and time. Thus, in order to be meaningful, any statement about these kinds of quantities must be accompanied by information about where the quantity was measured and what time period the measurement represents. For example, the statement "I am developing a model to predict streamflow for the Styx River from rainfall" contains very little information. In order to decide whether this model is useful for a particular purpose, you would have to know (1) at exactly what point or points streamflow is being predicted; and (2) whether the model predicts average annual flow, average monthly flow, average daily flow, instantaneous flow, or some other flow. The same kind of information would be needed about the rainfall data to be used as input. Thus the hydrologist must always be fully aware of the space and time characteristics of the quantities involved in any equation or model.

TYPES OF EQUATIONS

Equations of Definition It is often useful to give a single name to the relation between two or more physical quantities. An equation stating such a relation is called an *equation of definition.* For example, the kinematic viscosity ν is defined in Chapter 4 as the dynamic viscosity μ divided by the mass density ρ. This relation is written as

$$\nu \equiv \frac{\mu}{\rho}, \tag{3-9}$$

where the identity sign $\equiv$, indicates that this equation is a definition of ν. Similarly, in discussing the flow of water, we'll see that the cross-sectional area A of the flow divided by the "wetted perimeter" (the distance from one edge of the water surface to the other measured along the channel sides and bottom at right angles to the flow) P is an important geometric parameter. This quantity is called the *hydraulic radius,* and is defined symbolically as

$$R \equiv \frac{A}{P}. \tag{3-10}$$

Many writers do not consistently use the identity sign, so you will often have to study the text in order to identify an equation of definition.

Force-Balance Equations Newton's second law of motion states that the acceleration (rate of change of velocity or direction of motion) of a body is proportional to the resultant, or net, force on the body and is in the same direction as that force. Mathematically, the law is expressed as

$$F = Ma, \tag{3-11}$$

where F is the resultant force, M the mass of the body, and a the acceleration. Many of the derivations of equations in this book are based on an application of this law to situations where acceleration is zero; the body (generally a ''parcel'' of water in our case) is either at rest or moving with a constant velocity and direction.

Equation 3-11 states that if the acceleration of the body is zero, the resultant force must be zero, since mass is constant. We make use of this law when examining hydrostatic (the velocity is zero) and steady-flow (the velocity is constant) conditions by stating that $F = 0$, where F is the net force (i.e., vector sum) of all forces on the body. For example, in the derivation of the equation for the height of the rise of water in a capillary tube in Chapter 4, two forces are acting on the water: (1) the force due to surface tension, which pulls upward (F_{up}); and (2) the force due to gravity, which pulls downward (F_{down}). Thus, using a positive sign for upward-acting forces, we have

$$F = F_{up} - F_{down}. \tag{3-12}$$

When the column of water stops rising,

$$F = 0,$$

so

$$F_{up} - F_{down} = 0$$

and

$$F_{up} = F_{down}. \tag{3-13}$$

To derive the capillary-rise equation, we substitute equivalent expressions for F_{up} and F_{down} in Equation 3-13 and solve it for the height of the rise.

Most flows in nature involve acceleration. It turns out, however, that if the acceleration is not too large, we can ignore it, assume that $F = 0$, and develop equations that are very useful in the description of the real flows. This will be done several times in later sections of this book, and you will encounter many cases in the literature where the force-balance approach is used to develop fundamental equations.

Conservation Equations Conservation equations are fundamental statements of the fact that mass (matter), energy, and momentum (the product of mass times velocity) cannot be created or destroyed in any process (outside the realm of relativistic physics). The general conservation equation can be written as follows: For a given time period and a given region of space,

$$I - Q = S_2 - S_1, \tag{3-14}$$

where I is the quantity of mass, energy, or momentum entering the space (volume); Q is the quantity of mass, energy, or momentum leaving the space; and S_1 and S_2 are the quantities of mass, energy, or momentum stored in the space at the beginning and end of the time period, respectively. Equation 3-14 is simply a formal way of stating the apocryphal Baldy's law: ''Some of it plus the rest of it equals all of it.''

Alternative names for equations of conservation of matter are *continuity equations* or *mass-balance equations;* those applied to energy are commonly called *energy-balance equations,* and those applied to momentum simply *momentum equations.*

Conservation equations can be written for any region of space and any time period that you want to define; but you must remember that the equations are true only for mass ($[M]$), momentum ($[MLT^{-1}]$), and energy ($[ML^2T^{-2}]$), and that the region of space and period of time must be strictly defined for a given problem. If you fail to observe these conditions, the equation you write will be incorrect. Keeping track of dimensions for each term as you study or write a conservation equation will help you avoid many common errors.

Conservation equations are commonly applied to time periods that are long enough to warrant our assuming that $S_2 - S_1 = 0$. For example, if we consider a lake to be the region of space, we can reasonably assume that over several decades the amount of water in the lake has not changed significantly. This assumption is usually warranted for natural reservoirs when a conservation equation is written for long-term average conditions. In such cases, $I = Q$, and the equation can be helpfully simplified.

An alternative formulation of the conservation equation can be written if we take the derivative with respect to time of each term in Equation 3-14: For a given region of space,

$$\frac{dI}{dt} - \frac{dQ}{dt} = \frac{dS}{dt}, \tag{3-15}$$

which states that the rate at which storage is changing is equal to the difference between the rate of inflow and the rate of outflow.

As Equation 3-15 is written, it applies to instantaneous rates of change, so no time period need be specified. If we consider a time period of finite length Δt, then for a given region of space,

$$\frac{\Delta I}{\Delta t} - \frac{\Delta Q}{\Delta t} = \frac{\Delta S}{\Delta t}. \tag{3-16}$$

This equation states that the average rate of inflow during the period Δt minus the average rate of outflow for that period equals the average rate of change in storage for that period. Again, there are many situations where it is reasonable to assume that the quantity of mass, energy, or momentum has not changed significantly over a time period. Thus the average rate of change of storage for that period can be considered equal to zero, and $\Delta I/\Delta t = \Delta Q/\Delta t$.

Conservation equations are very commonly the starting point for developing equations, or models, of hydrologic situations. Such fundamental equations are often written for an infinitesimally small volume of space, usually denoted by $dx\,dy\,dz$, to symbolize infinitesimal lengths in the three directions of Cartesian space, and an infinitesimally small period of time dt.

Once the conservation equation is properly written for a particular problem, we generally proceed by substituting appropriate alternative expressions for the various terms and applying the laws of algebra and (if derivatives are involved) calculus. The expressions to be substituted depend on the details of the problem, and often include equations of motion or diffusion equations, which are discussed next. (Taft and Reisman, 1965, give a valuable systematic review of conservation equations, indicating how they are applied to various problems of physics and engineering.)

Equations of Motion The equations of motion of concern to the hydrologist describe the movement of matter and of heat energy. Matter moves from one point to another if there is a difference in mechanical potential energy between the two points, and thermal energy moves if there is a difference in temperature (thermal potential energy). In all cases, motion is from the point or region of higher potential energy to the point or region of lower potential energy, and the rate of motion is proportional to the spatial rate of change (the *gradient*) of potential energy between the two points.

This tendency can be stated mathematically as

$$V_{12} = K_c\left(\frac{p_1 - p_2}{X_{12}}\right), \tag{3-17}$$

where V_{12} is the rate of movement of matter or energy from point 1 to point 2, p_1 and p_2 are the appropriate potential energies at the two points ($p_1 > p_2$), X_{12} is the distance between the points, and K_c is a proportionality constant. In derivative form, this relation is expressed as

$$V = -K_c \frac{dp}{dx}, \tag{3-18}$$

where the minus sign reflects the fact that the velocity is in the positive x direction if the potential energy decreases in that direction (i.e., if the gradient is negative). The dimensions of the terms in Equation 3-18 depend on whether we are talking about matter or energy. The heat-flow equations covered in the discussion of thermal conductivity in Chapter 4 are examples of equations of motion.

The proportionality constant K_c in Equations 3-17 and 3-18 expresses the *conductivity* of the system for the flow of the particular form of matter or energy under consideration. It depends on both the medium through which flow occurs and the substance that is flowing. As is discussed in detail later, the free-surface flow of water under natural conditions is due to a gradient of gravitational potential energy, that is, the gradient of elevation of the water surface. The exact form of the equation of motion depends on the situation, as will be shown in later chapters. Darcy's law of groundwater flow, discussed fully in Chapter 12, is a good example of this type of equation:

$$q_x = -K_D \frac{dH}{dx}; \tag{3-19}$$

$$K_D = K_I\left(\frac{\gamma}{\mu}\right), \tag{3-20}$$

where q_x is the velocity in the x direction ($[LT^{-1}]$); dH/dx is the gradient of the water surface ($[LL^{-1}]$); and K_D is the hydraulic conductivity ($[LT^{-1}]$), which is a system parameter that depends on the intrinsic permeability K_I of the medium ($[L^2]$), and on the weight density γ ($[FL^{-3}]$), and dynamic viscosity μ ($[FTL^{-2}]$), of the water.

Later, equations of motion of the form of Equation 3-18 will be developed for three types of flow in open channels. We will develop these equations by using the force-balance principle discussed earlier.

It may be of interest to note that Ohm's law, which governs the flow of energy in the form of electricity, is of exactly the same form as other equations of motion:

$$I = \left(\frac{1}{R}\right)E,$$

where I is the current (energy flow) in amperes, E is the voltage drop (potential energy gradient) in volts, and R is the resistance of the circuit (the inverse of the conductivity of the system) in ohms.

Diffusion Equations Diffusion equations describe the movement of matter, momentum, or energy through a medium in response to a gradient of matter, momentum, or energy, respectively. Their general form is much like that of an equation of motion:

$$Q_s = -D_s \frac{ds}{dx}, \tag{3-21}$$

where Q_s is the rate of movement of matter, momentum, or energy through a unit area at right angles to the direction of a gradient of mass, momentum, or energy;

ds/dx represents the gradient of concentration of mass, momentum, or energy in the x direction; and D_s is a diffusion coefficient, or diffusivity, for mass, energy, or momentum in the medium. The Q_s term is generally called a *flux density* (the flow per unit area per unit time); note that according to Equation 3-21 flow is always from a region of high concentration to a region of low concentration.

The dimensions of the diffusivity for all cases are $[L^2T^{-1}]$, whereas those of the other terms depend on whether we're considering matter, energy, or momentum. Specifically, for matter,

$$Q_A = -D_A \frac{d\rho_A}{dx}, \tag{3-22}$$

where Q_A is the flux density of substance A in the x direction ($[ML^{-2}T^{-1}]$), D_A is the diffusivity of substance A in the x direction in the medium of concern ($[L^2T^{-1}]$), and ρ_A is the concentration of substance A ($[ML^{-3}]$). Thus $d\rho_A/dx$ represents the gradient of concentration of subtance A in the x direction. Equations for the diffusion of matter are sometimes called *mass-transfer equations*.

For momentum,

$$\tau = -\nu \frac{d(\rho v)}{dx}, \tag{3-23}$$

where τ is the flux density of momentum in the x direction ($[ML^{-1}T^{-2}]$), ν is the diffusivity of momentum (or kinematic viscosity) in the x direction in the medium ($[L^2T^{-1}]$), ρ is the mass density of the medium, and v is the velocity of the medium in the direction that is orthogonal (i.e., at right angles) to the x direction. Thus (ρv) is momentum per unit volume and $d(\rho v)/dx$ is the gradient of momentum concentration in the x direction. If ρ does not vary with distance in a given problem, it can be considered a constant, and Equation 3-23 can be simplified to

$$\tau = -\nu\rho \frac{dv}{dx}, \tag{3-24}$$

or, from the definition of kinematic viscosity (Equation 3-9), to

$$\tau = -\mu \frac{dv}{dx}, \tag{3-25}$$

which is the form used in discussing viscosity in Chapter 4 and in deriving equations of fluid flow in Chapters 7 and 12. In a sense, equation 3-25 is an equation of motion for momentum; it states that the rate of momentum transfer (shear stress, τ) is proportional to the viscosity and to the velocity gradient, and that transfer is from high-velocity areas to low-velocity areas.

For heat energy,

$$Q_H = -D_H \frac{d(\rho C_p T)}{dx}, \tag{3-26}$$

where Q_H is the flux density of heat energy in the x direction ($[MT^{-3}]$), D_H is the diffusivity of heat energy in the medium ($[L^2T^{-1}]$), ρ is the mass density of the medium ($[ML^{-3}]$), C_p is the heat capacity of the medium ($[L^2T^{-2}\Theta^{-1}]$), and T is the temperature of the medium ($[\Theta]$). Thus $(\rho C_p T)$ is the heat-energy content per unit volume and $d(\rho C_p T)/dx$ is the gradient of heat-energy concentration in the x direction.

As in Equation 3-23, it is often possible to consider the fluid properties ρ and C_p constant, so that Equation 3-26 becomes

$$Q_H = -D_H \rho C_p \frac{dT}{dx}. \tag{3-27}$$

In Chapter 4, an equation of motion for heat flow (the heat-conduction equation) is developed that looks very similar to Equation 3-27. In fact, we can show that the dimensions of $D_H \rho C_p$ are $[MLT^{-3}\Theta^{-1}]$, the same as thermal conductivity (see Table 2.3). Thus here again, a diffusion equation turns out to be identical to an equation of motion when fluid properties are constant.

Regression Equations Regression equations are relationships between physical quantities that are established by statistical analysis of data. They are not statements of fundamental physical laws and hence are qualitatively different from force-balance, conservation, motion, and diffusion equations. This difference is reflected in the fact that regression equations are generally dimensionally inhomogeneous. We will not discuss the detailed statistical methods by which such equations are established, but will examine their basic characteristics.

The general form of a regression equation is

$$y = a + b_1x_1 + b_2x_2 + \cdots + b_nx_n, \tag{3-28}$$

where y (the dependent variable) is any physical quantity; $x_1, x_2, \ldots, x_n$ are n other physical quantities (the independent variables); a is the regression constant; and b_1, $b_2, \ldots, b_n$ are regression coefficients. Examples of regression equations are

$$P = 85.4 + 0.0746Z_s, \tag{3-29}$$

where P is average annual precipitation (cm yr^{-1}) and Z_s is elevation (m);

$$\rho = 1 - (3 \times 10^{-5})(T^2/4 - 2T + 4), \tag{3-30}$$

where ρ is the mass density of water (g cm^{-3}) and T is the water temperature (°C);

$$w = 14Q^{0.26}, \tag{3-31}$$

where w is stream width (m) and Q is streamflow (m^3 s^{-1}); and

$$V = \frac{1.49R^{2/3}S_0^{1/2}}{n}, \tag{3-32}$$

where V is average stream velocity (ft s^{-1}), R is hydraulic radius (ft), S_0 is stream slope (dimensionless), and n is channel roughness (dimensionless). Note that because these equations are dimensionally inhomogeneous, we must state the system of units they are designed for (Rule 14a).

Regression equations are empirical, that is, they are developed from actual measurements of the dependent and independent variables, rather than from fundamental principles of physics. The general procedure for establishing a regression equation is as follows: First, decide on the dependent variable and independent variables. Next, choose the precise form of the relationship; a wide variety of forms is possible if transformations of the variables are used: For example, Equation 3-32 can be written in the form of the general regression equation (Equation 3-28) by a logarithmic transformation:

$$\log V = \log 1.49 + 0.667 \log R + 0.500 \log S_0 - \log n. \tag{3-33}$$

(Transformations involving trigonometric and other functions are also possible.) As the final step in the process, make a series of concurrent measurements of the dependent and independent variables. These observations constitute a statistical sample, and statistical analysis of this sample provides values of the regression constant and coefficients that best fit that sample.

Once a relationship is established, a number of statistical tests can be applied to determine how "good" the regression equation is and how likely it is that the relationship you obtained would occur purely by chance in the absence of a true relationship between the variables. These tests are described in most statistics books (e.g., Crow et al., 1960; Draper and Smith, 1981; Winkler and Hays, 1975; Yevjevich, 1972) and must be understood and reported whenever a regression equation is presented.

The most important things to understand about regression equations are that (1) the variables and the form of the equation are determined by the researcher rather than by nature; (2) the numerical values of the regression constant and coefficients are determined by the sample (i.e., the set of measurements) used in the analysis, and different values will result if different samples are analyzed; and (3) the development of a statistically strong relationship between a dependent variable and one or more independent variables does not imply a direct cause-and-effect relationship. If these considerations are kept in mind, regression equations can be very effective tools in hydrologic analysis. However, the classic paper by Wallis (1965) forcefully illustrates the pitfalls of inferring physical relations from the results of regression analyses.

Exercises

The following problems provide practice in the somewhat tedious but very useful skill of converting equations for use with different systems of units.

3-1. Convert the following equation for use with SI units:

$$U_c = n(24a + b)$$

where U_c is consumptive use by vegetation (in inches per day), n is soil porosity (dimensionless), a is the measured rate of rise in the water table over a specified time period (in inches), and b is the net change in water table elevation during the day (also in inches).

Convert the following equations for use with the units shown.

3-2. $E = 15(1 + 0.1W)(e_w - e_a)$

Symbol	Original unit	New unit
E (evaporation rate)	in. month^{-1}	cm day^{-1}
W (wind velocity)	mi h^{-1}	m s^{-1}
e_w (vapor pressure of water surface)	in. Hg*	mb
e_a (vapor pressure of air)	in. Hg*	mb

*in. Hg × 33.86 = mb.

3-3. $R = (0.085 + 0.734q_i)P$

Symbol	Original unit	New unit
R (storm runoff)	in.	mm
q_i (antecedent discharge)	ft^3 s^{-1}	liter s^{-1}
P (storm rainfall)	in.	mm

3-4. $Q = \left(\dfrac{44{,}000}{A + 170} + 20\right)A$

Symbol	Original unit	New unit
Q (peak flood discharge)	ft^3 s^{-1}	m^3 s^{-1}
A (drainage area)	mi^2	km^2

3-5. $R_B = 0.61\left(\frac{T_0 - T_a}{e_0 - e_a}\right)\frac{P}{1000}$

Symbol	Original unit	New unit
R_B (Bowen ratio)	dimensionless	dimensionless
T_0 (water-surface temperature)	°C	K
T_a (air temperature)	°C	K
e_0 (surface water-vapor pressure)	mb	Pa
e_a (air water-vapor pressure)	mb	Pa
P (atmospheric pressure)	mb	Pa

3-6. $\tau_0 = \gamma R S_0$

Symbol	Original unit	New unit
τ_0 (boundary shear stress)	dyn cm^{-2}	lb ft^{-2}
γ (weight density of water)	dyn cm^{-3}	lb ft^{-3}
R (hydraulic radius)	cm	ft
S_0 (channel slope)	dimensionless	dimensionless

3-7. $M = K(0.0084u_0)(0.22T_a + 0.78T_d) + 0.029T_a$

Symbol	Original unit	New unit
M (snowmelt)	in. day^{-1}	mm day^{-1}
K (areal coefficient)	dimensionless	dimensionless
u_b (wind speed)	mi h^{-1}	m s^{-1}
T_a (average daily air temperature)	°F	°C
T_d (average daily dewpoint)	°F	°C

3-8. $V = 2.5(gRS_0)^{1/2} \ln\left(\frac{y}{y_0}\right)$

Symbol	Original unit	New unit
V (velocity)	cm s^{-1}	ft s^{-1}
g (acceleration due to gravity)	cm s^{-2}	ft s^{-2}
R (hydraulic radius)	cm	ft
S_0 (slope)	dimensionless	dimensionless
y (height above bottom)	cm	ft
y_0 (constant of integration)	cm	ft

3-9. $C = 95Q^{0.7}$

Symbol	Original unit	New unit
C (concentration of suspended sediment)	mg liter^{-1}	mg liter^{-1}
Q (discharge)	ft^3 s^{-1}	m^3 s^{-1}

CHAPTER 4

Structure and Properties of Water

Forces acting on water cause it to move through the hydrologic cycle, and the physical properties of water determine the quantitative relations between those forces and the resulting motion. Forces as well as accelerations and velocities are *vector quantities;* that is, they have both magnitude and direction. All fluid properties and pressure, temperature, length, area, and volume are *scalar quantities* and have magnitude only. You should keep in mind the fundamental distinction between vector and scalar quantities as you study this book. However, as Rouse (1961, p. 35) has pointed out, "knowledge of vector analysis is not essential to the study of fluid motion, for the variation of a vector may be fully described by the changes in magnitude of its three [directional] components." Thus it will not be necessary to make explicit use of vector notation and vector algebra in most of our analyses; we will generally deal only with the magnitudes of the vector quantities while making clear their directions by means of careful definitions and the use of diagrams.

The physical properties of water are quantified by measuring its response to known forces applied in specific experimental situations. In this chapter we examine a series of such situations—thought experiments—and develop from them simple quantitative relations that will be invoked repeatedly in subsequent chapters, when conditions of more direct hydrologic relevance are described.

Applying the principles developed in Chapter 3, we know that these quantitative expressions must be dimensionally homogeneous if they completely and correctly describe the physical relationships. Thus we can determine the dimensional character of a given property by straightforward dimensional analysis of the expressions developed from our thought experiments. The properties that reflect the response of water to forces acting on it are classified as *dynamic* if their dimensions include force (or mass), and as *kinematic* if they do not (see Table 2.3).

Another class of properties, characterizing the movement and storage of heat energy in water, is also important in hydrology, and thought experiments developing expressions for these *thermal* properties are also described in this chapter. With the exception of latent heat, the dimensions of thermal properties include temperature.

The physical and thermal properties of water are determined by its atomic and molecular structures, and so we begin with an examination of these structures. We will see that water is a very unusual substance, and that consequently the numerical magnitudes of many of its properties are anomalous. Interestingly, the strangeness of water is the reason it is so common at the earth's surface; the profound importance of its unusual properties in shaping the earth-surface environment is briefly noted at the end of the chapter.

ATOMIC AND MOLECULAR STRUCTURES

In the periodic chart, the elements are arranged in order of atomic weight. Atomic weight is determined by the number of protons (particles with an electrical charge of +1 and a mass of 1 atomic weight unit) and neutrons (particles with an electrical charge of 0 and a mass of 1 atomic weight unit) in the nucleus. Each electron that surrounds the nucleus has a charge of −1 and essentially 0 mass, and the number of electrons is equal to the number of protons, with the result that the atom has a net charge of 0. The electrons can be thought of as occupying a series of shells, and the number of electrons in the outermost shell, which can vary from one to eight, largely determines the chemical characteristics of each element. The chart of the elements is periodic because the elements in each column, called a *group,* have the same number of electrons in their outer shells and hence have similar chemical properties.

The chemical composition of water is H_2O; the water molecule is formed by the combination of two hydrogen atoms (Group Ia, with one electron in the outer shell) and one oxygen atom (Group VIa, with six electrons in the outermost shell). Figure 4.1 shows these two atomic species. The outer shell of oxygen can hold eight electrons, and hence has two vacancies. These vacant positions can readily accept the single electrons that orbit the hydrogen nuclei, resulting in the *covalent bonds* shown in Figure 4.2. (Figures 4.1 and 4.2 are highly schematic two-dimensional drawings, but they illustrate the basic principles involved.)

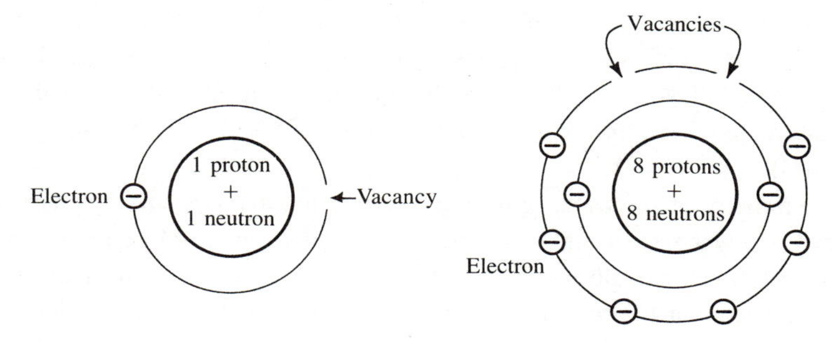

Figure 4.1
Schematic diagram of a hydrogen atom (left) and an oxygen atom (right).

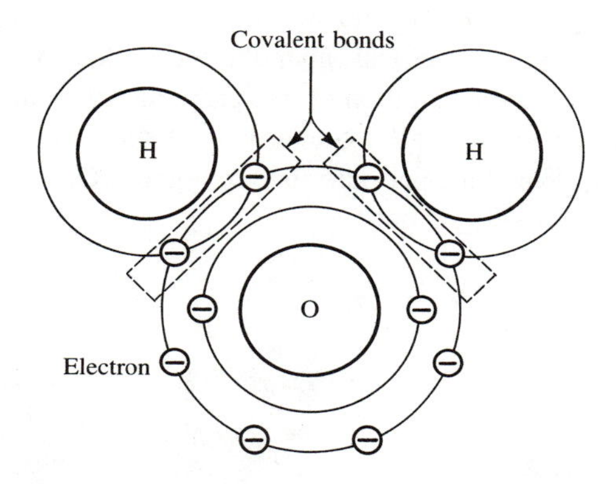

Figure 4.2
Schematic diagram of a water molecule.

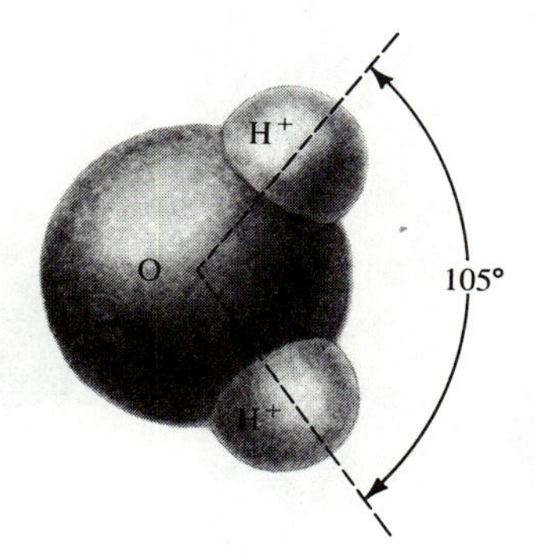

Figure 4.3
Diagram of a water molecule showing the angle between the hydrogen atoms. (From K. S. Davis and J. A. Day, *Water, the Mirror of Science,* © 1961, p. 95. Reprinted by permission of Doubleday & Co., New York, and Heinemann Educational Books, Ltd., London.)

The most important features of the water molecule are that (1) the covalent bonds between the hydrogens and the oxygens are very strong (i.e., much energy is needed to break these bonds); and (2) the hydrogen atoms are always attached on one "side" of the oxygen atom, with an angle of about 105° between the two hydrogens (Figure 4.3).

The asymmetry of the water molecule causes it to have a positively charged end and a negatively charged end, like the poles of a magnet. Most of the unusual properties of water are ultimately the result of its being made up of these *polar* molecules. The polarity produces an attractive force between the positively charged end of one molecule and the negatively charged end of another, as shown in Figure 4.4. This force, called a *hydrogen bond,* is absent from most other liquids.

Although the hydrogen bond has only about one-twentieth the strength of the covalent bond (Stillinger, 1980), it is far stronger than the intermolecular bonds that are present in liquids with symmetrical, nonpolar molecules. We get an idea of this strength when we examine the variation of two properties, the melting and/or freezing temperature and the boiling and/or condensation temperature, of the hydrides of the elements of Group VIa of the periodic chart: oxygen ($^{16}_{8}O$), sulfur ($^{32}_{16}S$), selenium

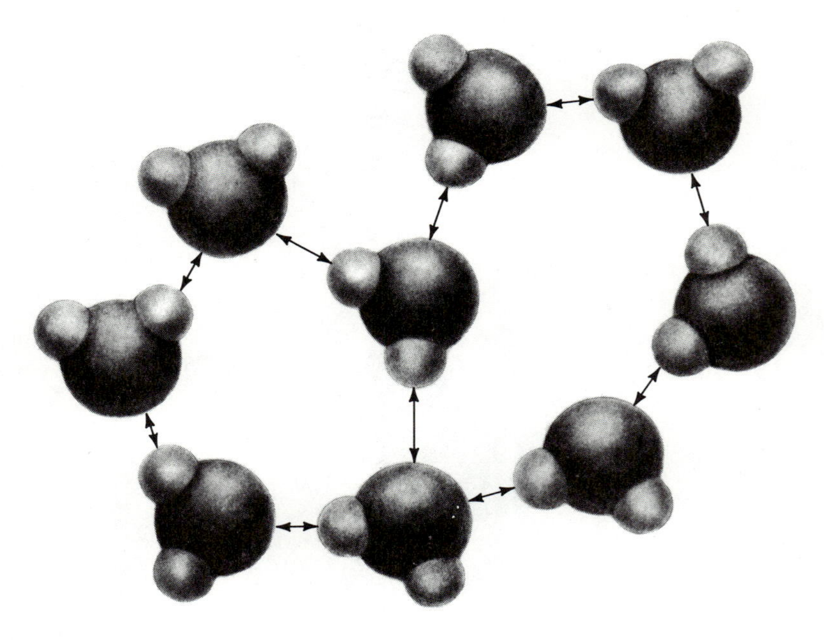

Figure 4.4
Diagram of water molecules in the liquid state. Arrows indicate hydrogen bonds between the oppositely charged ends of adjacent molecules.

($^{79}_{34}Se$), and tellurium ($^{128}_{52}Te$) (Table 4.1). (According to convention, the atomic weight is written as a superscript and the atomic number as a subscript to the left of the symbol for each element.) These elements are all characterized by an outer electron shell that can hold eight electrons but has two vacancies. If no other forces were acting, the melting and/or freezing and boiling and/or condensation temperatures would be expected to rise as the atomic weight increased. These expectations are fulfilled, except—strikingly—in the case of H_2O. The main factor in this departure from expectations is the H-bonds, which attract one molecule to another and can be overcome only by a great deal of vibratory energy (a high temperature). Because of its abnormally high melting and boiling temperatures, water is one of the few substances that exists in all three physical states—solid, liquid, and gas—at earth-surface temperatures.

Table 4.1
Melting and Boiling Temperatures of Hydrides of Group VIa Elements
Data from Davis and Day (1961).

Compound	Molecular weight	Melting temperature (°C)	Boiling temperature (°C)
H_2O	18	0	100
H_2S	34	−82	−61
H_2Se	81	−64	−42
H_2Te	130	−51	−4

At temperatures below 0°C the vibratory energy of water molecules is sufficiently low that the H-bonds lock the molecules into a regular three-dimensional lattice (Figure 4.5). In this lattice, each molecule is hydrogen-bonded to four adjacent molecules, with a distance of about 2.74 angstroms (1 angstrom = 10^{-8} cm) between molecules (Stillinger, 1980). The angle between the hydrogen atoms in each molecule remains virtually unchanged at 105°, but each is oriented such that a puckered honeycomb of perfect hexagons is visible when the lattice is viewed from one direction. Thus normal ice is a hexagonal crystal, and snowflakes show infinite variation on a theme of sixfold symmetry.

When ice is warmed to 0°C, further additions of heat energy cause melting, in which about 15% of the H-bonds break. The energy required to bring about complete melting is called the *latent heat of fusion,* and is discussed later in this chapter. Because of the rupturing of some of the H-bonds, the rigid ice lattice partially collapses, and a given number of molecules take up less space in the liquid phase than in the solid. As a result, the density of ice is 0.917 times that of liquid water at 0°C. Very few substances have a lower density in the solid state than in the liquid, and this property is of immense importance: If rivers and lakes froze from the bottom up rather than from the top down, biological and hydrologic conditions in higher latitudes would be markedly altered.

Although melting always occurs when ice is warmed to 0°C (at earth-surface

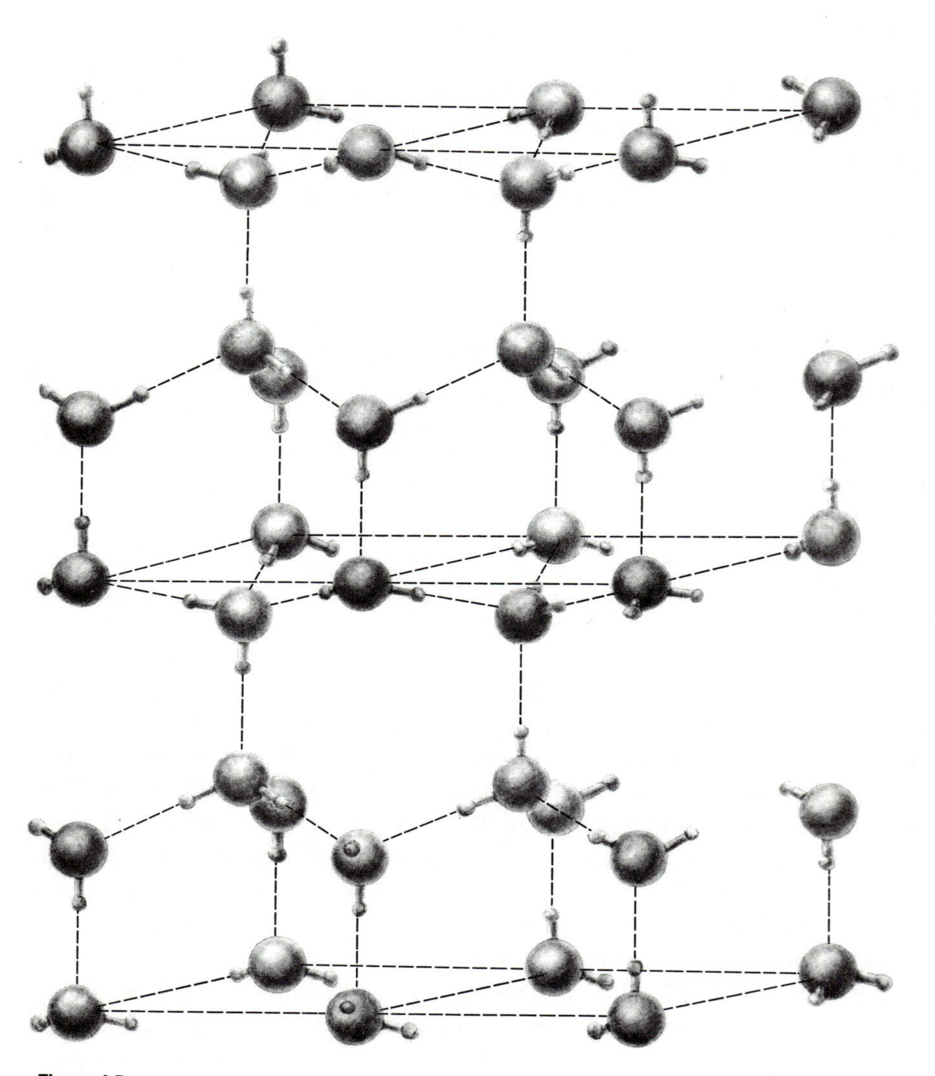

Figure 4.5
The crystal lattice of ice, showing its hexagonal structure. (From L. C. Pauling, *The Nature of the Chemical Bond,* © 1939 and 1940 by Cornell University. Third edition © 1960 by Cornell University. Used by permission of Cornell University Press, Ithaca, New York.)

pressures), freezing may not always take place when liquid water is cooled to 0°C. If the liquid contains no impurities and is not in contact with already formed ice, it is possible to supercool it to temperatures as low as −41°C. This resistance to freezing appears to be due to the formation of various types of nonhexagonal hydrogen-bonded polyhedra, which prevent the formation of the ice lattice (Stillinger, 1980). If ice particles (or common impurities like clay minerals, which have a crystal structure like that of ice) are present, they provide templates that act as growth nuclei to trigger the formation of ice at 0°C. Significant supercooling is quite common in clouds, where effective growth nuclei may be lacking, and is an important factor in the development of cloud droplets into snowflakes and raindrops (see Day and Sternes, 1970). In rivers, supercooling of 0.01–0.1°C often occurs during the formation of ice in fast-flowing stretches (see Michel, 1971).

Liquid water is largely a random three-dimensional network of hydrogen bonds, with many of the bonds deformed from their ice-lattice configurations (Stillinger, 1980). As warming proceeds above 0°C, the deformation increases as the molecules continue to pack closer together, causing density to continue to increase. Above 3.98°C, however, greater and greater numbers of the H-bonds rupture. Nonbonded neighbor molecules tend to be located farther apart than when they are bonded, so that above 3.98°C density decreases with temperature, as in most other substances. The anomalous density maximum at 3.98°C also has important biological and hydrologic consequences, since it is responsible for the overturn of lakes in temperate and cold climates.

At temperatures between 0°C and 100°C, some molecules with greater than average energy sever all hydrogen bonds with their neighbors and fly off from the bulk liquid. Thus evaporation can take place at all temperatures below the boiling point, and the loss of these high-energy molecules results in a lowering of the average energy, and hence of the temperature, of the remaining liquid. Once the entire liquid has been heated to 100°C, further additions of energy cause the breaking of virtually all the remaining H-bonds, and the liquid is entirely transformed into a gas consisting of relatively widely spaced, mostly nonbonded individual H_2O molecules. The energy used, or lost, in vaporization at and below the boiling point is called the *latent heat of vaporization,* and is discussed later in this chapter.

PROPERTIES OF WATER

In this section we define the kinematic, dynamic, and thermal properties of water. In most cases, we use thought experiments to couple your physical understanding of these properties to their quantitative definitions. These definitions will be invoked repeatedly throughout the rest of this book.

For most hydrologic purposes we need be concerned only with how the numerical magnitudes of these properties change with temperature. Other factors, such as pressure, usually have a negligible effect on these magnitudes; however, high concentrations of dissolved or suspended solids can have important effects that must be accounted for under certain conditions. Magnitudes of each property at 273.2 K (0°C, 32°F) in the SI, cgs, and English systems of units are given in the following subsections. To find their magnitudes at other temperatures, use the multipliers given in Table 4.2; this table presents a convenient overview of the relative changes of each property with temperature. Alternatively, Heggen (1983) gives useful empirical equations for estimating the values of most of these properties as functions of temperature.

Table 4.2
Relative Values of Water Properties as Functions of Temperature
Data from Weast (1971); ρ represents mass density; γ, weight density; μ, dynamic viscosity; ν, kinematic viscosity; σ, surface tension; C_p, heat capacity; λ_v, latent heat of vaporization; k_θ, molecular thermal conductivity.

Temperature (°C)	ρ, γ	μ	ν	σ	C_p	λ_v	k_θ
0	1.0000	1.000	1.000	1.000	1.000	1.000	1.000
5	1.0001	0.8500	0.8500	0.9907	0.9963	0.9953	1.016
10	0.99986	0.7314	0.7315	0.9815	0.9940	0.9904	1.031
15	0.99926	0.6374	0.6379	0.9722	0.9924	0.9857	1.046
20	0.99836	0.5607	0.5616	0.9630	0.9915	0.9810	1.062
25	0.99720	0.4983	0.4997	0.9524	0.9910	0.9763	1.077
30	0.99580	0.4463	0.4482	0.9418	0.9907	0.9715	1.093

Some of the following analyses (and some of those in Chapter 12) examine water movement on small—even microscopic—scales. These scales are, however, many orders of magnitude larger than that at which individual molecules can be discerned. Thus, although we have seen that molecular structure determines the physical properties described in the following subsections, from here on we can treat water as a *continuum* of matter.

Mass Density *Mass density* (ρ) is defined as the mass per unit volume of a substance; its dimensions are $[ML^{-3}]$. Because gravitational force and momentum are both proportional to mass, mass density appears in most equations describing fluid motion.

In the cgs system of units, the gram is defined as the mass of 1 cm^3 of pure water at its temperature of maximum density, 3.98°C. At 0°C,

$$\rho = 999.87 \text{ kg m}^{-3} = 0.99987 \text{ g cm}^{-3} = 1.9397 \text{ slugs ft}^{-3}.$$

As shown in Table 4.2, the variation with temperature is relatively slight, and for most hydrologic purposes we can use the value $\rho = 1000$ kg m^{-3} (= 1.00 g cm^{-3} = 1.94 slugs ft^{-3}). The principal exception to this statement occurs when we are concerned with modeling the seasonal overturn of lakes, which is an important limnological process that depends on the minute variations of density with temperature (see Hutchinson, 1957). Water is for all practical purposes incompressible, so its mass density is unaffected by pressure except at extreme ocean depths. The addition of dissolved or suspended solids to water increases the mass density of the solution or suspension in proportion to the mass density of the solids and to their concentration.

Weight Density Weight is the force exerted on a mass by gravitational attraction, and the *weight density* (γ) is the weight per unit volume of a substance. Its dimensions are $[FL^{-3}]$. The properties γ and ρ are related via Newton's second law (i.e., force equals mass times acceleration) as follows:

$$\gamma = \rho g, \tag{4-1}$$

where g is the acceleration due to gravity ($[LT^{-2}]$).

For hydrologic purposes, $g = 9.800$ m s^{-2}, so at 0°C

$$\gamma = 9799 \text{ N m}^{-3} = 979.9 \text{ dyn cm}^{-3} = 62.46 \text{ lb ft}^{-3}.$$

Because of Equation 4-1, the variability of γ with temperature and other factors is identical to that of ρ.

The gram is commonly used as a unit of force as well as of mass. When used in this way, the "gram" represents the weight of 1 g of mass at the earth's surface, so that 1 "gram" of force is equivalent to 979.9 dyn.

The *specific gravity* of a substance is the ratio of its weight density to the weight density of water at 3.98°C.

Dynamic Viscosity Conceptually, *dynamic viscosity* is the internal friction of a fluid that resists forces tending to cause flow. The physical meaning of the property can be understood by considering the experiment illustrated in Figure 4.6. Imagine that the overall dimensions of the experiment are small (on the order of a few centimeters) and that we have some miniature current meters for measuring the fluid velocity between the two boundaries. The lower boundary is fixed and of indefinite extent, whereas the upper one is movable and has an area A. The entire system is initially at rest, and we begin the experiment by pulling on the upper boundary with a steady force F, causing it to move with a constant velocity. One of the important properties of water is called the *no-slip condition,* which means that the layer of

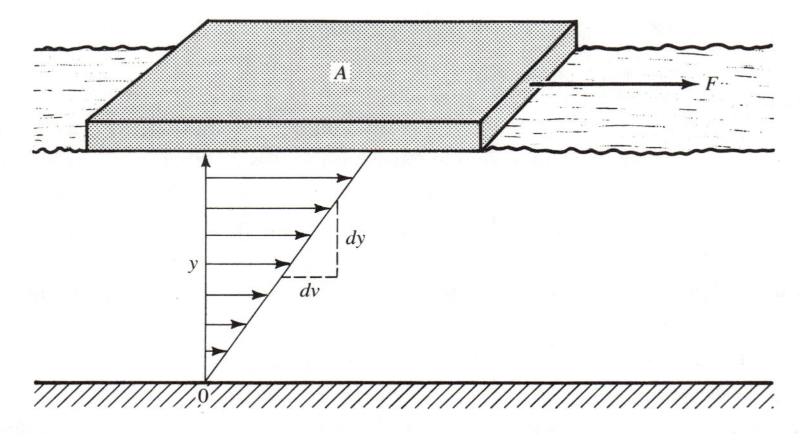

Figure 4.6
Thought experiment for viscosity, showing the velocity gradient dv/dy developed when a force F is applied to the movable upper boundary of the area A.

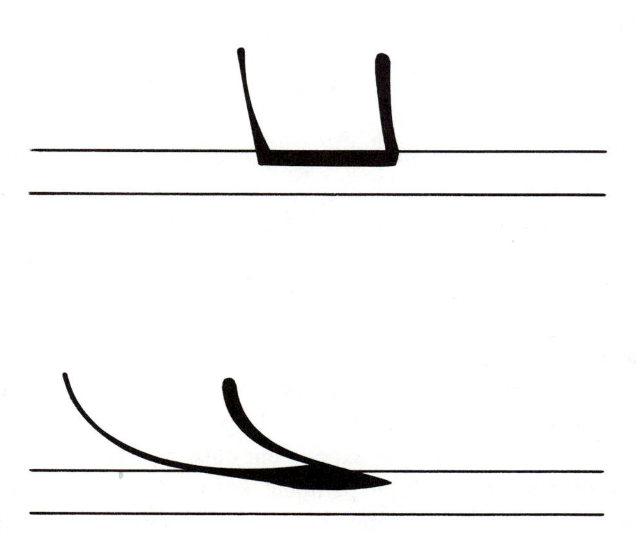

Figure 4.7
Demonstration of the no-slip condition. (a) A U-shaped line of colored glycerine inserted in a stationary body of water. (b) Deformation of glycerine a short time after flow has begun. Note that the horizontal portion of the U adjacent to the bottom has not moved. (Redrawn from a photograph in A. H. Shapiro, *Shape and Flow,* © 1961, p. 112. Used by permission of Doubleday & Co., New York.)

water in immediate contact with a boundary has the same velocity as the boundary (Figure 4.7). Thus, in the thought experiment, the velocity of the water next to the lower boundary is zero, whereas that of the uppermost layer is the same as the velocity of the upper boundary. The current meters would show a linear increase of velocity, as shown in Figure 4.6.

The entire experiment consists of repeating the foregoing many times, each time varying the force F and recording the velocity gradient dv/dy induced in the water (y is the distance from the lower boundary). A plot of the relation between the applied force per unit area F/A and the velocity gradient would look like Figure 4.8. There is a direct proportionality between F/A and dv/dy. The proportionality constant is defined as the dynamic viscosity μ:

$$\mu \equiv \frac{F/A}{dv/dy} \tag{4-2}$$

and

$$\frac{F}{A} = \mu \frac{dv}{dy}. \tag{4-3}$$

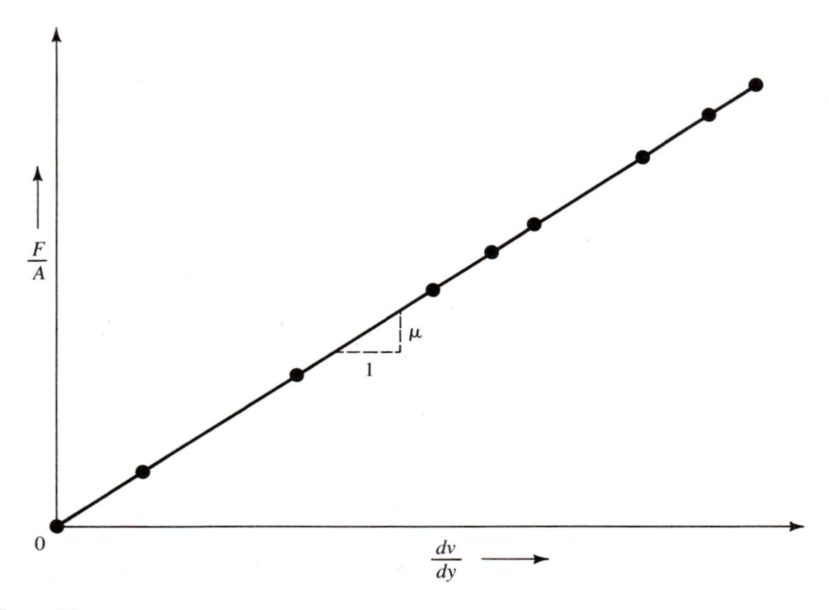

Figure 4.8
Relation between applied shear stress F/A and resulting velocity gradient dv/dy for several trials of viscosity thought experiment.

Since the force was applied in the same direction as the flow, it is called a *shear force*. Shear force divided by area is called a *shear stress*, usually denoted by τ. Thus, finally,

$$\tau = \mu \frac{dv}{dy} \tag{4-4}$$

and

$$\frac{dv}{dy} = \left(\frac{1}{\mu}\right)\tau. \tag{4-5}$$

Equation 4-5 says that the velocity gradient induced by a shear stress is directly proportional to the shear stress. As μ increases, a larger shear stress must be applied in order to induce the same velocity gradient. Note also that any shear stress greater than zero induces a velocity gradient. If the flow of water between the two boundaries is thought of as the sliding of layers over each other, as in a stack of cards (Figure 4.9), μ can be thought of as the friction between the layers.

Interestingly, not all fluids behave according to Equation 4-5. Those that do, like water, are called *Newtonian fluids*. In some fluids (e.g., many plastics) the velocity gradient is nonlinearly related to the shear stress; and we will see later that even for water Equation 4-5 applies only when the dimensions of the system are small and when the induced velocities remain small.

The dimensions of μ are readily determined by solving Equation 4-3:

$$\frac{[F]}{[L^2]} = [\mu]\frac{[LT^{-1}]}{[L]};$$

$$[\mu] = [FTL^{-2}].$$

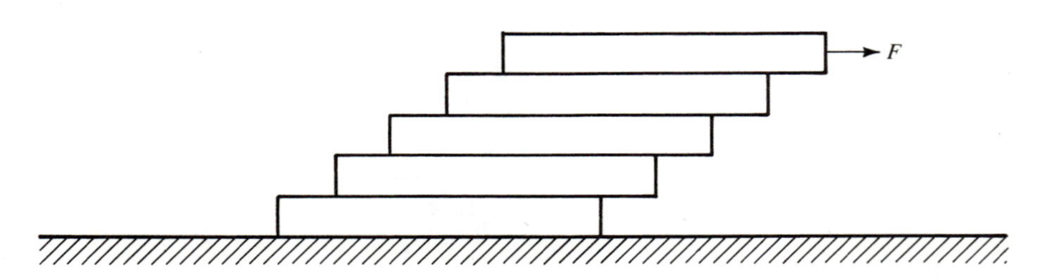

Figure 4.9
Deformation of a stack of cards when a force F is applied to the top card.

Thus the units of dynamic viscosity in the SI system are newton seconds per square meter (N s m^{-2}) or, since the pascal is defined as 1 N m^{-2}, pascal seconds (Pa · s).

At 0°C,

$$\mu = 1.787 \times 10^{-3} \text{ Pa} \cdot \text{s} = 1.787 \times 10^{-2} \text{ dyn s cm}^{-2}$$
$$= 3.735 \times 10^{-5} \text{ lb s ft}^{-2}.$$

The cgs unit is often called the *poise* after the French scientist Poiseuille (1799–1869), who studied the flow of blood.

Dynamic viscosity decreases rapidly as temperature increases, as shown in Table 4.2. The addition of suspended solids to water increases the viscosity of the mixture, but the effect is not significant unless concentrations are very high (as in mud flows). Most dissolved solids also increase viscosity, but some decrease it (see Weast, 1971); these effects can usually be ignored for hydrologic purposes.

Kinematic Viscosity The ratio of dynamic viscosity to mass density occurs in equations describing many flow situations; this common ratio is called the *kinematic viscosity* (ν):

$$\nu \equiv \frac{\mu}{\rho}. \tag{4-6}$$

Recall from Chapter 3 that kinematic viscosity can be thought of as the diffusivity of momentum (Equation 3-23).

The units of kinematic viscosity are readily determined by writing

$$[\nu] = \frac{[\mu]}{[\rho]} = \frac{[FTL^{-2}]}{[ML^{-3}]}$$

and, from Equation 2-4,

$$[\nu] = [L^2T^{-1}].$$

The values of ν at 0°C are

$$\nu = 1.787 \times 10^{-6} \text{ m}^2 \text{ s}^{-1} = 1.787 \times 10^{-2} \text{ cm}^2 \text{ s}^{-1}$$
$$= 1.926 \times 10^{-5} \text{ ft}^2 \text{ s}^{-1}.$$

Surface Tension Like viscosity, surface tension is best understood by visualizing a thought experiment. Consider a device consisting of an inverted U-shaped wire defining three sides of a rectangular area, with the fourth side formed by a straight piece of wire that can slide along the arms of the U (Figure 4.10). The size of the area is a few square millimeters, about that of the mesh of a window screen. When the device is dipped into water and taken out, a film of water is retained inside the opening. If we imagine that the sliding wire moves without friction, we would observe that it is pulled toward the top of the inverted U (Figure 4.10a). Clearly a force is responsible for this movement: the force caused by the mutual attraction of water molecules due to their hydrogen bonds. Molecules in contact with the wire are attracted to it and to other nearby water molecules. There is a net force directed into the body of the fluid, and if the sliding wire is not too heavy it is pulled upward by this force, and the surface area of the suspended water decreases.

Now imagine that, in order to measure the magnitude of this force, we suspend from the slide wire a tiny weight W^* that just balances the upward force caused by the H-bonds (Figure 4.10b). At equilibrium the upward force F^* is equal to W^*, and the surface tension σ is defined as

$$\sigma \equiv \frac{F^*}{2x}, \tag{4-7}$$

where x is the length of the slide wire. Thus the *surface tension* is the force tending to decrease the surface area of the liquid divided by the distance over which this force acts. In this case the distance is $2x$ because there are two surfaces, an upper and a lower. Note that the dimensions of surface tension are $[FL^{-1}]$.

Another way of looking at surface tension is as the work (energy) required to increase the area of a surface by a unit amount. In our surface-tension experiment, with the weight W^* attached the surface area could be increased by adding to W^* a very small increment of weight, dW. The surface would continue to increase as long as this increment remained in place, but would become fixed as soon as it was removed. Suppose the wire were moved a distance y downward in this way (Figure 4.10c). The areas of the two surfaces of the fluid would increase by an amount xy, and so the total increase in area would be $2xy$. If dW is infinitesimally small, the work done in increasing the area by $2xy$ is $W^*y = F^*y$ (work equals force times distance). Thus the ratio of work done to increase in area is

$$\frac{F^*y}{2xy} = \frac{F^*}{2x} = \sigma. \tag{4-8}$$

Increasing the surface area of water involves moving molecules from the body of the liquid to the surface. This movement requires energy because the H-bonds among water molecules always cause a net inward force at the surface; a molecule in the body of the liquid is attracted equally to the molecules that surround it in all directions, whereas one at the surface is attracted only to those next to it in the

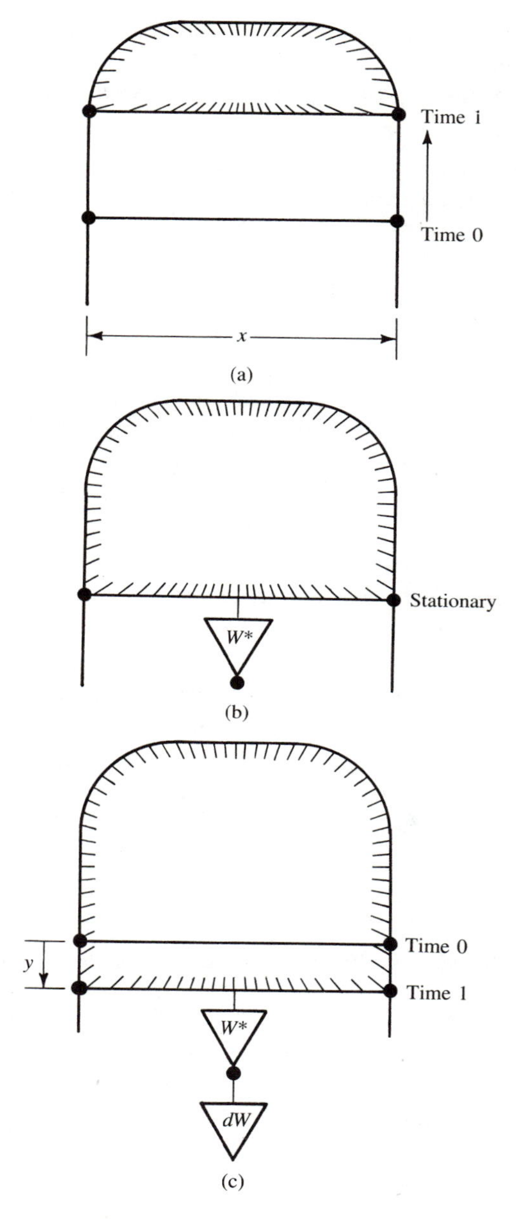

Figure 4.10
Thought experiment for surface tension, showing (a) the motion of a slide wire between time 0 and time 1 due to surface-tension forces. (b) In the second stage of the surface-tension thought experiment a weight W^* has been attached to the slide wire to balance the upward surface-tension forces. (c) In the third stage of the experiment a small increment dW of weight added to the slide wire has pulled it down a distance y.

surface and those in the body of the fluid (Figure 4.11). Thus in any water surface there is a net inward force that tends to make the surface area as small as possible. Since a sphere has the smallest possible area per unit volume, water drops are spherical unless distorted by other forces.

We see from Equation 4-8 that surface tension has the dimensions $[FL^{-1}]$; its units in the SI system are thus newtons per meter (N m^{-1}). At 0°C,

$$\sigma = 7.56 \times 10^{-2} \text{ N m}^{-1} = 75.6 \text{ dyn cm}^{-1} = 5.18 \times 10^{-3} \text{ lb ft}^{-1}.$$

Surface tension decreases rapidly as temperature increases (Table 4.2). Dissolved solids may raise or lower surface tension (see Weast, 1971), and these effects can be important, especially when we consider the behavior of water in clay soils.

Capillarity is a phenomenon closely related to surface tension. If water is placed in a narrow cylindrical container made of a material that has little attraction for water molecules (i.e., the water molecules do not form H-bonds to the material), the water surface, or *meniscus,* will tend to a minimum and will have a shape like that shown in Figure 4.12a. If the container is of a material to which water molecules are strongly attracted, however, molecules at the surface will be pulled up and will pull their neighbors up as well because of intermolecular attractions. The water surface will then look like that shown in Figure 4.12b.

The strength of the attraction between the water molecules and the molecules of the container wall is reflected in the *contact angle* θ_c between the surface and the wall. When this attraction is high, $\theta_c = 0°$ and the meniscus is tangent to the wall and is concave toward the air. As the water–wall adhesion decreases, θ_c increases, and when there is neither attraction nor repulsion, $\theta_c = 90°$ and the meniscus is at right angles to the wall. When there is a repulsive force between the water and the wall, $\theta_c > 90°$ and the meniscus is convex toward the air. When θ_c is near zero for a given substance, water is said to *wet* the surface; when θ_c is near 180° for a substance, that substance is termed *nonwetting,* or *hydrophobic*. Data on contact

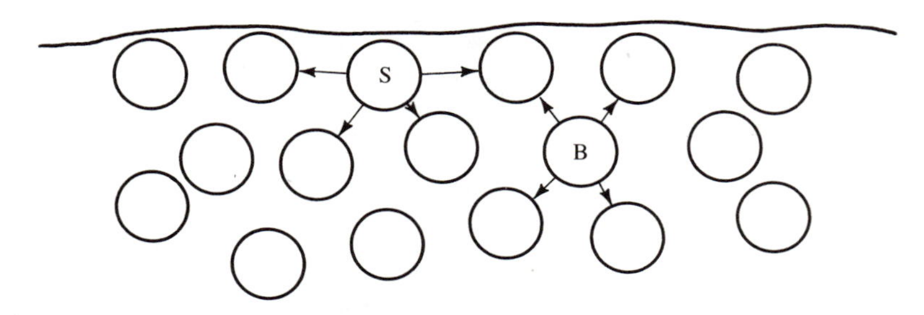

Figure 4.11
Intermolecular forces acting on typical surface (S) and nonsurface (B) molecules.

angles (also called *wetting angles*) are difficult to find, but values for water in contact with a few solids (with air containing water vapor adjacent to the liquid and solid surfaces) are given in Table 4.3.

In the situation shown in Figure 4.12b the upward force caused by the attraction of the molecules of the container wall to the molecules in the water surface continues to act even after the surface area is increased. If this upward force is not balanced

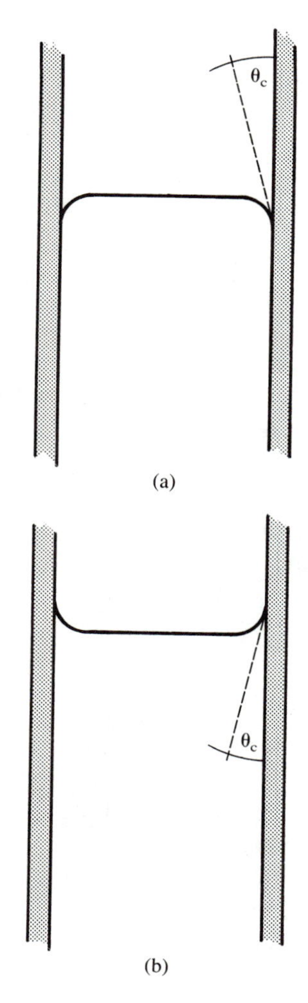

Figure 4.12
Shape of the water surface (meniscus) in a container (a) to which water molecules are not attracted and (b) to which water molecules are attracted by hydrogen bonds; θ_c is the contact angle between the meniscus and the wall.

Table 4.3
Contact Angles (θ_c) for Water against Various Solids

Data are from Dorsey (1940), except those for ice, which are from Jellinek (1972). Values of θ_c are in degrees.

Solid	θ_c	$\cos\theta_c$
Glass	0	1.0000
Most silicate minerals	0	1.0000
Ice	20	0.9397
Platinum	63	0.4540
Gold	68	0.3746
Talc	86	0.0698
Paraffin	105–110	−0.2588 to −0.3420
Shellac	107	−0.2924
Carnauba wax	107	−0.2924

by an equal force directed downward, the water surface will continue to move upward and will cause the entire body of liquid to move upward (again as a result of attractions among the water molecules). This phenomenon is called *capillary rise.* The final height of capillary rise is related to the radius of the container as follows: The upward force F_{up} is equal in magnitude to the force due to surface tension, which resists a change in surface area while the column of water is rising. Since

$$\sigma\cos\theta_c = \frac{F_{up}}{\text{distance over which force acts}} = \frac{F_{up}}{\text{circumference of container}} = \frac{F_{up}}{2\pi R},$$

where R is the radius of the container, we have

$$F_{up} = 2\pi R\sigma\cos\theta_c. \tag{4-9}$$

The downward force that balances this upward force when upward motion ceases is the weight of the column of water:

$$F_{down} = (\text{volume of cylinder of water})\left(\frac{\text{weight}}{\text{unit volume of water}}\right);$$

$$F_{down} = \pi R^2 h\gamma, \tag{4-10}$$

where h is the height of the column of water and γ is the weight density of water. Then, at equilibrium,

$$F_{up} = F_{down};$$

$$2\pi R\sigma \cos \theta_c = \pi R^2 h\gamma;$$

$$h = \frac{2\sigma \cos \theta_c}{\gamma R}. \tag{4-11}$$

Thus the height of capillary rise of a column of water is inversely proportional to the radius of the column. As an exercise, check to see if Equation 4-11 is dimensionally homogeneous.

In later discussions of the flow of water in porous media (Chapter 12), it will be useful to relate the pressure difference across a meniscus to the radius of the capillary opening. In a capillary tube such as that shown in Figure 4.13[1], the water is

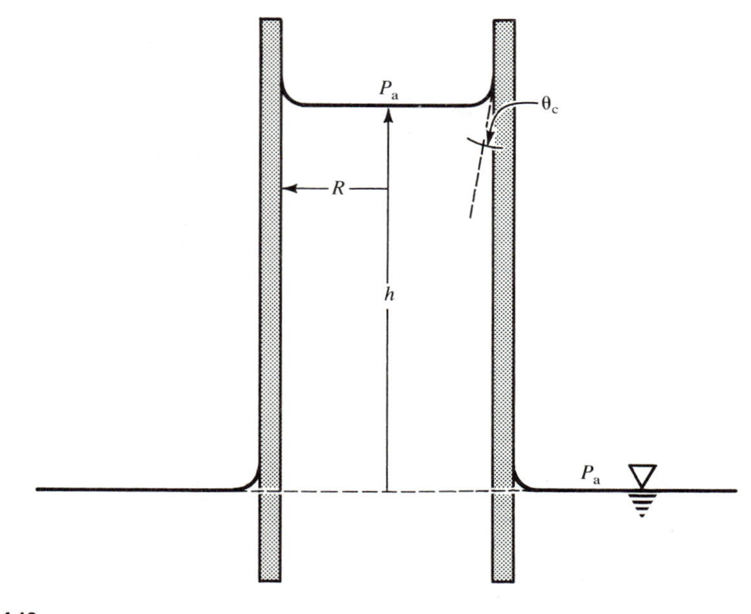

Figure 4.13
Definition sketch for the computation of the height h of capillary rise in a circular tube of radius R; θ_c is the contact angle between the meniscus and the wall, and P_a is atmospheric pressure.

[1]The inverted triangle in Figure 4.13 is often conventionally used to indicate the location of a free water surface at atmospheric pressure. We will use this symbol in figures throughout the rest of this book. No one seems to have given this symbol a name; if it needs one, a suggestion might be to call it a *hydrat* to convey the idea of an interface between water and atmosphere.

suspended from the meniscus, which in turn is attached to the walls by hydrogen bonds. Thus the water is under tension, which is defined as negative pressure. The weight of the water suspended beneath the plane that is tangent to the lowest point of the meniscus is given by Equation 4-10. Since the area of that plane is πR^2, the pressure P_s on the plane due to the suspended water is

$$P_s = \frac{F_{\text{down}}}{\text{area}} = -\frac{\gamma \pi R^2 h}{\pi R^2} = -\gamma h. \tag{4-12}$$

The total pressure P on the plane is found as follows:

$$P = P_a + P_s, \tag{4-13}$$

where P_a is atmospheric pressure. Combining Equations 4-12 and 4-13, we obtain

$$P = P_a - \gamma h. \tag{4-14}$$

Now Equation 4-11 can be substituted into Equation 4-14:

$$P = P_a - \frac{2\gamma\sigma \cos \theta_c}{\gamma R} = P_a - \frac{2\sigma \cos \theta_c}{R}. \tag{4-15}$$

The pressure difference ΔP across the meniscus is

$$\Delta P = P_a - P,$$

and Equation 4-15 can be rearranged to provide the relation we are seeking:

$$\Delta P = \frac{2\sigma \cos \theta_c}{R}. \tag{4-16}$$

Thus the pressure difference across a meniscus, like the height of capillary rise, is inversely proportional to the radius of a capillary opening. (See Remson and Randolph, 1962, for a more elaborate derivation of Equation 4-16.)

Heat Capacity *Heat capacity* (C_p) is the property that relates a temperature change of a substance to a change in its energy content and is defined as the amount of energy required to raise the temperature of a unit mass of substance by a unit amount, that is, the change in thermal energy (ΔE) per mass (M) per change in temperature (ΔT). Thus its dimensions are

$$\frac{\Delta E}{M\,\Delta T} \frac{[ML^2T^{-2}]}{[M][\Theta]} = [L^2T^{-2}\Theta^{-1}].$$

In the SI system, heat capacity is measured in joules per kilogram per kelvin ($J\ kg^{-1}\ K^{-1}$). In the cgs system the unit of energy, the calorie, is defined as the amount of energy required to raise the temperature of 1 g of pure water from 14.5°C to 15.5°C. Since the heat capacity varies only slightly with temperature (Table 4.2), the heat capacity of water for hydrologic purposes has the convenient value of 1 cal g^{-1} in the cgs system. Similarly, the British thermal unit (Btu) is the amount of energy required to raise the temperature of 1 lb (0.03084 slug) of water from 63°F to 64°F. At 0°C,

$$C_p = 4216\ J\ kg^{-1}\ K^{-1} = 1.007\ cal\ g^{-1}\ °C^{-1}$$
$$= 32.43\ Btu\ slug^{-1}\ °F^{-1} \quad (1.007\ Btu\ lb^{-1}\ °F^{-1}).$$

Because of the hydrogen bonding among water molecules, the heat capacity of water is high relative to that of most other liquids (Table 4.4). The heat capacity of ice is almost exactly half the value for liquid water (Dorsey, 1940; Eisenberg and Kauzmann, 1969).

The term *specific heat* means the ratio of the heat capacity of a substance to that of water. It is sometimes used colloquially as a synonym for heat capacity.

Latent Heats Consider 1 g of ice at a temperature below 0°C; the ice is in a covered container and is heated at a constant rate (i.e., subjected to a constant rate of energy input). The temperature of the substance in the container can be measured and recorded continuously.

The results of this thought experiment are shown in Figure 4.14. As the ice absorbs heat energy, its temperature rises steadily at a nearly constant rate equal to

Table 4.4
Comparison of Properties of Common Liquids
Data from Davis and Day (1961).

Liquid	Melting temperature (°C)	Boiling temperature (°C)	Latent heat of freezing (cal g^{-1})	Latent heat of vaporization (cal g^{-1})	Heat capacity (cal g^{-1} °C^{-1})
Acetone	−95.0	56.5	23.4	124.5	0.506
Ethyl alcohol	−117.3	78.5	24.9	204	0.535
Benzene	5.5	80.1	30.3	94.3	0.389
Carbon tetrachloride	−22.8	76.8	4.2	46.4	0.198
Mercury	−38.9	356.6	2.8	70.6	0.033
Sulfuric acid	−10.5	330.0	24.0	122.1	0.270
Turpentine	—	159.0	—	68.6	0.411
Water	0	100	79.7	539.6	1.007

the inverse of the heat capacity of ice: 2°C cal^{-1} (0.478 K J^{-1}). The energy absorbed by the ice causes its molecules to vibrate faster, and this vibration is reflected in the temperature increase. When the temperature reaches 0°C, the temperature ceases to rise. As discussed earlier, further addition of energy at this temperature is used to break some of the H-bonds of the rigid ice structure, rather than to increase molecular vibration.

Melting will continue until 79.7 cal (334 J) have been absorbed, after which only liquid water is present in the container. At this point, as energy is added at the same rate, we observe that the temperature rise resumes at a less steep but nearly constant rate of about 1°C cal^{-1} (0.239 K J^{-1}), reflecting the higher heat capacity of liquid water. This rise continues until a temperature of 100°C is reached, at which point all the intermolecular H-bonds begin to break, and molecules leave the liquid as H_2O vapor. Again, no temperature change would be observed while this change of state continued. Very nearly 540 cal (2260 J) would be required to complete the vaporization of the original gram of water, after which the temperature would once again resume its increase.

If the experiment just described were reversed—if we were to *remove* heat at a constant rate—the graph in Figure 4.14 would not change at all. As heat was extracted, 539.8 cal (2260 J) would be withdrawn at 100°C without a change in temperature before the vapor was all changed to liquid, and 79.7 cal (334 J) would be withdrawn at 0°C before all the water froze. The quantities of energy input or output that accompany the melting or freezing and vaporization or condensation of a unit mass are known as *latent heats*.

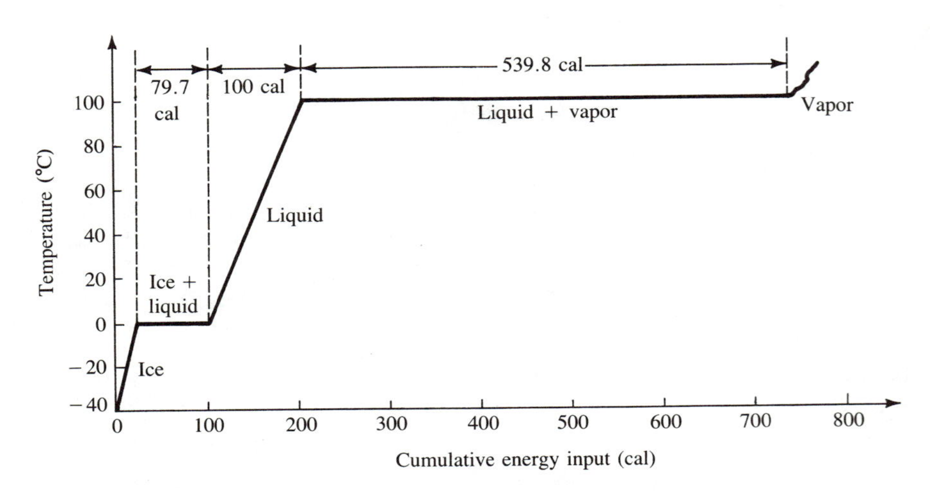

Figure 4.14
Temperature as a function of cumulative energy input for a thought experiment in which 1 g of water is heated from −40°C until it is completely vaporized.

Melting or freezing always occurs at or very close to 0°C (273.2 K, 32°F) in surface-water bodies, so we need not be concerned with the temperature variation of the latent heat of fusion (λ_f). In the three unit systems:

$$\lambda_f = 3.34 \times 10^5 \text{ J kg}^{-1} = 79.7 \text{ cal g}^{-1}$$
$$= 4620 \text{ Btu slug}^{-1} \quad (144 \text{ Btu lb}^{-1}).$$

In an earlier discussion, we saw that water molecules are continually exchanged between water and air at all temperatures below 100°C. As water temperature decreases, more energy is required to sever the H-bonds, so the latent heat of vaporization (λ_v) increases from 539.8 cal g^{-1} (2260 J kg^{-1}) at 100°C to its value at 0°C:

$$\lambda_v = 2.495 \times 10^6 \text{ J kg}^{-1} = 595.9 \text{ cal g}^{-1}$$
$$= 34{,}570 \text{ Btu slug}^{-1} \quad (1074 \text{ Btu lb}^{-1}).$$

Table 4.2 shows the variation of λ_v over the range of temperature of hydrologic interest. For practical calculations, a value of 590 cal g^{-1} (2.47×10^6 J kg^{-1}) is often assumed.

The term *sublimation* is used for direct changes between the solid and vapor states in either direction. The latent heat of sublimation is $79.7 + 595.9 = 675.6$ cal g^{-1} (2.829×10^6 J kg^{-1}) when the ice surface is near 0°C.

It is important to remember that latent heat exchanges accompany all changes of state of water in the environment; they are vitally important in many hydrometeorologic processes.

Thermal Conductivity Figure 4.15 depicts a quantity of water retained between two plates a small distance Δx apart. The area of water in contact with each plate is A. Suppose that we have a means of measuring temperature at several points between the plates and of measuring the heat-flow rate through the water. Originally, the temperature on both plates and throughout the water is at a value T_1. Then we raise the temperature of one plate to T_2 and periodically measure the temperature within the water. If the two plate temperatures are maintained at T_1 and T_2 for a sufficiently long time, we will find that the temperature at each point in the water will increase to a particular value and then remain at that value. If these equilibrium temperatures are plotted as a function of the distance x measured from plate 2 to plate 1, we find a linear temperature decrease with distance:

$$\frac{dT}{dx} = -\frac{T_2 - T_1}{\Delta x}. \tag{4-17}$$

If this experiment were repeated several times with varying values of T_2 and T_1, we would find that Equation 4-17 always holds and that the heat-flow rate per unit area is directly proportional to the temperature gradient:

$$\frac{Q}{A} \propto -\frac{dT}{dx}. \tag{4-18}$$

The constant of proportionality k_θ is the *thermal conductivity,*

$$k_\theta = \frac{Q/A}{-dT/dx};$$

$$\frac{Q}{A} = -k_\theta \frac{dT}{dx}. \tag{4-19}$$

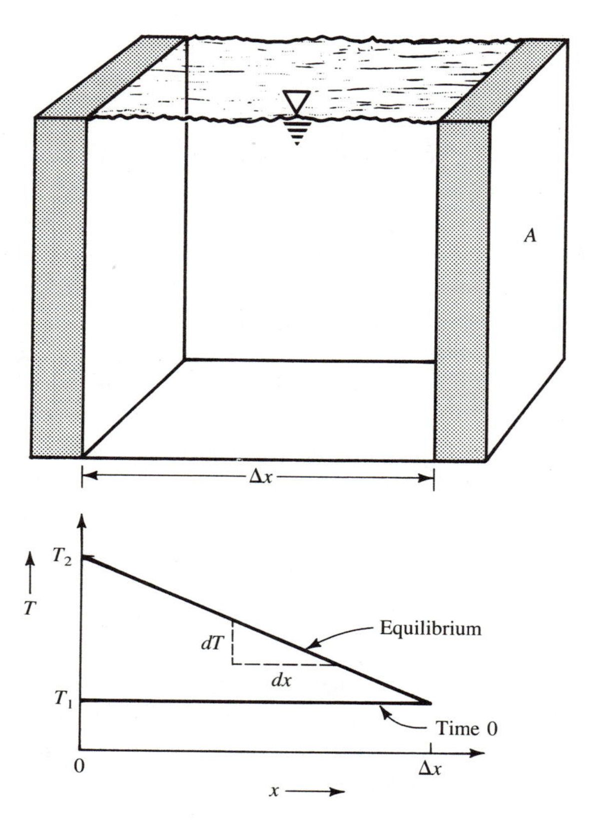

Figure 4.15
Definition sketch for a thermal conductivity thought experiment. The area of both plates is A; the distance between them is Δx. At time 0 both plates are at temperature T_1. The temperature of one plate is then raised to T_2, and after sufficient time a linear equilibrium temperature gradient dT/dx is established.

You can verify that the dimensions of k_θ are $[MLT^{-3}\Theta^{-1}]$ or $[FT^{-1}\Theta^{-1}]$; its values at 0°C are

$$k_\theta = 3.37 \times 10^{-3}\ \mathrm{J\ s^{-1}\ m^{-1}\ K^{-1}} = 8.06 \times 10^{-2}\ \mathrm{cal\ s^{-1}\ cm^{-1}\ {}^\circ C^{-1}}$$
$$= 5.42 \times 10^{-3}\ \mathrm{Btu\ s^{-1}\ ft^{-1}\ {}^\circ F^{-1}}.$$

Thermal conductivity increases gradually with temperature, as shown in Table 4.2.

It must be emphasized that k_θ in the thermal conductivity experiment and in Equation 4-19 is the *molecular* thermal conductivity. That is, the heat energy in the thought experiment moved through the water from molecule to molecule, while the molecules themselves had no net movement. In many situations, heat transfer in fluids may also take place by the net movement of groups of hotter molecules displacing colder ones. As discussed later, this net movement happens when turbulence is present in the flow. For the time being, it is sufficient to note that heat transfer via turbulence is much more rapid than by molecular conduction, but that an equation of exactly the same form as Equation 4-19 describes the turbulent case.

Thermal Diffusivity In many heat-flow problems, the following ratio occurs:

$$\frac{k_\theta}{\rho C_p} \equiv D_H, \quad (4\text{-}20)$$

where k_θ is the molecular thermal conductivity, ρ is mass density, and C_p is heat capacity. This ratio is defined as the *thermal diffusivity* D_H. As noted in Chapter 3, its dimensions are $[L^2T^{-1}]$.

IMPORTANCE OF WATER'S ANOMALOUS PROPERTIES

In discussing the importance of hydrogen bonding in the structure of water we noted that water's unusually high melting and boiling temperatures are due to its very high intermolecular attractions (Table 4.1). Table 4.4 shows that the surface tension, latent heats, and heat capacity of water are also much higher than those of other liquids. Now that the definitions of those properties are clear, it should be apparent that these anomalously high values are also due to hydrogen bonding.

The significance of these properties as regards the biophysical environment of the earth is profound. First, water is one of only a few substances that can exist in the liquid phase under earth-surface conditions, and its anomalously high melting and boiling points make this possible. Liquid water plays a major role in moderating the earth's climate because of the extraordinary thermal inertia that water manifests; that is, very large amounts of energy must be transferred in order to cause phase changes and temperature changes in water. The strength of intermolecular bonding also permits columns of water to be drawn up to great heights, and so makes possible the existence of trees. Water is, of course, the medium in which the chemical reactions of life processes occur, and the influence of polar molecules on the properties of water as a solute is also profound. These and other aspects of the impact of water's unusual properties on physical, chemical, and biological processes are discussed further by Hutchinson (1957), Davis and Day (1961), van Hylckama (1979), and Trincher (1981).

Exercises

4-1. Using the data in Table 4.1, plot on a single graph the melting and boiling temperatures of the hydrides of the Group VIa elements versus their molecular weights. Connect the melting temperatures and boiling temperatures by smooth curves and extrapolate to show what those temperatures should be for water in the absence of hydrogen bonding.

4-2. Plot the variations of the properties in Table 4.2 as functions of temperature.

4-3. Suppose that the experiment depicted in Figure 4.6 were done with a distance of 1.00 cm between the fixed and movable plates, and a shear stress of 0.0500 dyn cm^{-2} applied to the movable plate. Make a graph showing the velocity versus the distance from the bottom for temperatures of 0°C and 30°C. Plot the velocities every 0.2 cm.

4-4. Show by means of a diagram why the surface tension must be multiplied by the cosine of the contact angle θ_c in order to compute the upward force in Equation 4-9.

4-5. The range of pore sizes in silts and sands is from 10^{-4} cm to 10^{-1} cm. Make a graph of the height of capillary rise that corresponds to the diameters of tubes covering that range. Plot the relation for water at temperatures of 0°C and 30°C on the same graph, and assume that $\theta_c = 0°$. Use three- by five-cycle log–log paper.

4-6. What is the height of capillary rise of water in a paraffin tube with a radius of 0.1 cm? What is the pressure difference across the meniscus in this situation?

4-7. Derive a formula giving the height of capillary rise between two vertical parallel plates of width W and spaced a distance d apart. Neglect any effects at the edges of the plates.

4-8. Derive a formula for the pressure difference across the meniscus for the parallel plates in Exercise 4-7.

4-9. Water in a tall tree may be drawn up to a height of 30 m. What radius of capillary tube would be required to effect a capillary rise of that height? (Assume that $\theta_c = 0°$.) Compute the tension in the water at the top of such a tube, and express this in atmospheres (1 atm = 1.01×10^6 dyn cm^{-2}). Incidentally, the tensile strength of water has been measured at 160 atm (Dorsey, 1940).

4-10. If all the latent heat from the melting of a 30-g ice cube were contributed by the 200 g of water in a glass, how much would the temperature of the water be lowered?

CHAPTER 5

Open-Channel Flows: Forces and Classification

In this chapter we discuss the principal classifications of open-channel flows. These classifications are the key to building the physical intuition that allows you to identify the most important features of flows, and thus provide powerful insight for analysis and modeling.

As discussed in Chapter 3, motion occurs when a fluid is subjected to a gradient of mechanical potential energy. Mechanical potential energy is due to the forces of gravity and hydrostatic pressure (itself ultimately due to gravity), and these two forces are always present in stationary and moving fluids. We begin this chapter by examining the magnitude and direction of hydrostatic pressure in fluids at rest. We then derive expressions for the magnitude of potential energy at any point in a fluid; these expressions allow us to compute the gradients of potential energy that induce flow in open channels.

Once motion begins, other forces come into play that tend to resist the motion or change its direction. The relative magnitudes of all these flow-inducing and flow-resisting forces determine the behavior of a flow. The most important resisting forces in open-channel flows are due to the relative proximity of a solid boundary, so we explore in some detail how these forces arise. With this knowledge we can derive expressions for calculating the relative magnitudes of all mechanical forces acting on a given open-channel flow.

The final section of this chapter shows how specific ratios of force magnitudes are used as a basis for classifying flows. Another classification, based on the rates of spatial and temporal changes in velocity and depth, is also discussed here; these are our criteria for distinguishing the types of flows considered in Chapters 6–11.

HYDROSTATIC PRESSURE

Magnitude Consider a portion of a horizontal plane of area A immersed in a standing body of water (Figure 5.1). The hydrostatic pressure on this plane is due to the weight of the water and atmosphere directly above it. The weight of the water column is γyA, where γ is the weight density of the water and y the height of the column. The resulting water pressure P_w at the plane is simply force divided by area, or

$$P_w = \frac{\gamma yA}{A} = \gamma y. \tag{5-1}$$

The total pressure P at the plane is then

$$P = P_w + P_a = \gamma y + P_a, \tag{5-2}$$

where P_a is the atmospheric pressure.

As will be shown in the next section, pressure is critically important in considering the flow of water because it is one component of potential energy. Recall from Chapter 3 that motion from one point to another occurs when there is a difference in potential energy between the points (Equation 3-17). Thus we are almost always concerned with pressure differences rather than with actual pressures. Since atmospheric pressure can usually be considered a constant for any given problem, it plays no role in these differences and is left out of the formulation. Thus we are usually concerned with the *gage pressure*, $P_w = \gamma y$, rather than with the true total pressure as given in Equation 5-2. A free water surface is always at atmospheric pressure or zero gage pressure, $P_w = 0$. As noted in Chapter 4, such a surface is usually indicated by the triangular symbol in Figure 5.1.

Direction Consider a wedge-shaped portion of a motionless body of water that has a free surface (Figure 5.2). The left face of the wedge is oriented vertically (in the y direction), the bottom face is perpendicular to the left edge (in the x direction), and the slanting face makes an angle θ with the bottom face. The areas of these faces are A_x, A_y, and A_s, respectively. Let F_x, F_y, and F_s be the pressure forces exerted at right angles to the three faces. Since there is no motion or acceleration within the fluid, we know from Newton's second law (Equation 3-11) that the resultant force on any element of the fluid must be zero. By simple geometry, the vertical component of the force F_s is $F_s \cos \theta$, and the horizontal component is $F_s \sin \theta$. Thus at equilibrium

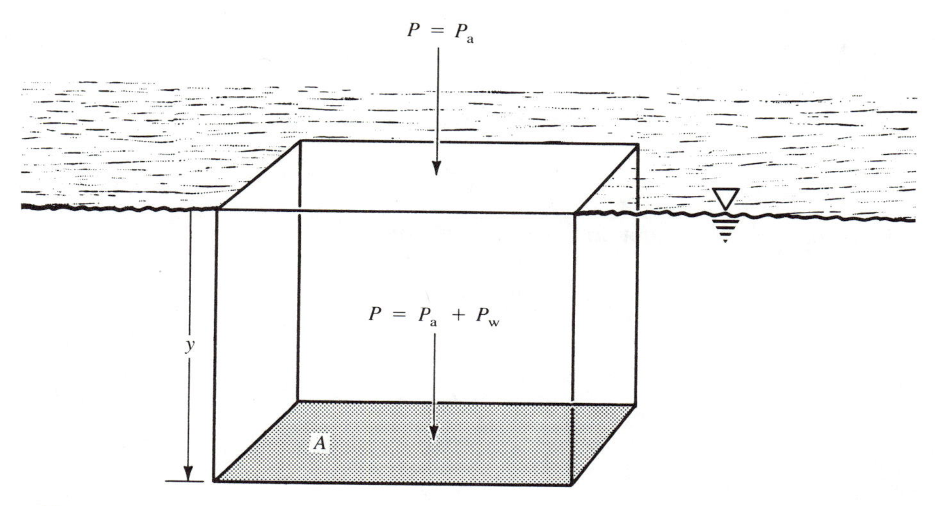

Figure 5.1
Definitions of terms for computing pressure under hydrostatic conditions (Equations 5-1 and 5-2): A is the area of a plane at a distance y below the surface; P_a is atmospheric pressure; P_w is water pressure; and P is total pressure.

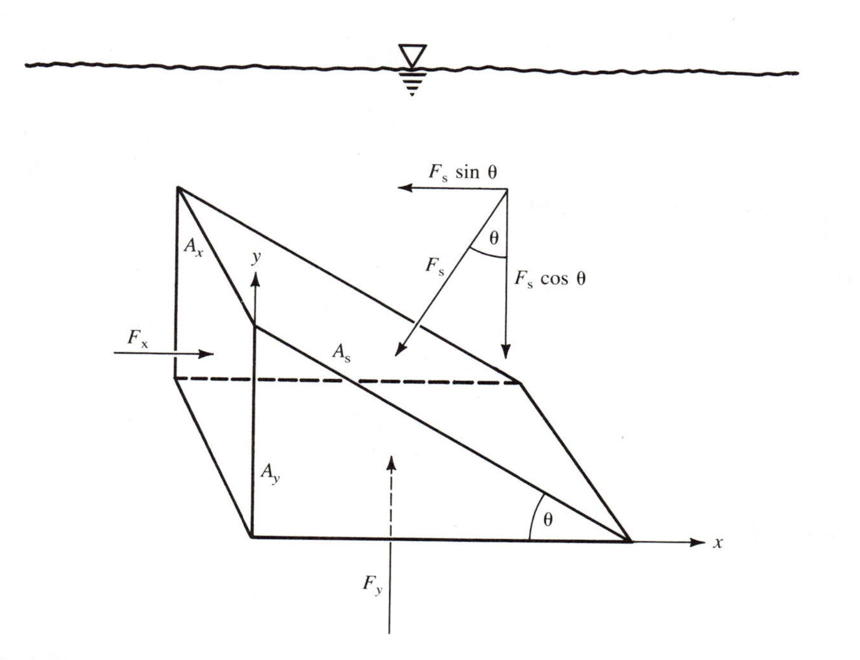

Figure 5.2
Forces on a stationary wedge-shaped parcel of water (Equations 5-3–5-6).

$$F_x = F_s \sin \theta \tag{5-3a}$$

and

$$F_y = F_s \cos \theta . \tag{5-3b}$$

Pressure is force per unit area, so we can write

$$\frac{F_x}{A_x} = P_x , \tag{5-4a}$$

$$\frac{F_y}{A_y} = P_y , \tag{5-4b}$$

and

$$\frac{F_s}{A_s} = P_s , \tag{5-4c}$$

where P_x, P_y, and P_s are the pressures on the respective faces. Again from geometry,

$$A_s \sin \theta = A_x \tag{5-5a}$$

and

$$A_s \cos \theta = A_y . \tag{5-5b}$$

Substituting Equations 5-3a and 5-5a in 5-4a, and 5-3b and 5-5b in 5-4b, gives

$$P_x = \frac{F_x}{A_x} = \frac{F_s \sin \theta}{A_s \sin \theta} = \frac{F_s}{A_s} = P_s$$

and

$$P_y = \frac{F_y}{A_y} = \frac{F_s \cos \theta}{A_s \cos \theta} = \frac{F_s}{A_s} = P_s ,$$

and therefore

$$P_x = P_s = P_y . \tag{5-6}$$

Since the wedge-shaped element in Figure 5.2 can be made as small as we wish, and the angle θ can have any value, Equation 5-6 leads to an important conclusion: At any point in a fluid, the pressure is equal in all directions.

MECHANICAL POTENTIAL ENERGY

You may recall from the study of physics that the elevation of a body above a horizontal datum (e.g., sea level) represents the *gravitational potential energy* of that body. Again, since we are generally interested in *differences* in potential energy, the location of the datum is immaterial as long as it remains constant for a particular problem. From our discussion of the equations of motion in Chapter 3, we know that a body always tends to move to a lower elevation in a gravitational field. This is as true of a volume element, or *parcel,* of water on an inclined plane as it is of a solid block.

More precisely, the gravitational potential energy E_{pg} of a body or parcel is equal to the weight of the parcel times its elevation z above the datum:

$$E_{pg} = \gamma B z, \tag{5-7}$$

where γ is the weight density of the material and B is its volume. (Check to see if Equation 5-7 is dimensionally homogeneous.)

In a fluid, there is a second component of potential energy due to hydrostatic pressure. This can be demonstrated by reference to Figure 5.3, where according to Equation 5-1 the water at point C is under a pressure of γy_C and that at point D is under a pressure of γy_D. The elevation of the two points is identical. Since $y_C > y_D$, the pressure at C is greater than that at D, and obviously water will move from C toward D in response to this pressure difference. This thought experiment thus demonstrates the existence of *pressure potential energy.*

To formulate the expression for pressure potential energy, consider the body of water in Figure 5.4. The water is at rest, and two parcels of water, each with a volume B, are shown at different elevations. The vertical distance downward from the surface to an arbitrary horizontal datum is denoted by Y^*, and y_C and y_D are the distances from the surface down to the midpoints of the two parcels. Since there is no motion, it must be true that all parcels in this body of water have the same potential energy (see the discussion of the equations of motion in Chapter 3). Thus, for the two parcels identified in Figure 5.4,

$$E_{pC} = E_{pD}, \tag{5-8}$$

where E_{pC} and E_{pD} are the total potential energies of parcels C and D, respectively. Earlier in this section we saw that one component of the total potential energy is the gravitational potential energy, given by Equation 5-7. Thus we can rewrite Equation 5-8 as

$$\gamma B(Y^* - y_C) + E_{xC} = \gamma B(Y^* - y_D) + E_{xD}, \tag{5-9}$$

where E_{xC} and E_{xD} are the nongravitational components of potential energy of parcels C and D, respectively. Expanding and rearranging Equation 5-9 gives

$$\gamma B(y_D - y_C) = E_{xD} - E_{xC},$$

and this expression leads to the conclusion that

$$E_{xC} = \gamma B y_C \tag{5-10a}$$

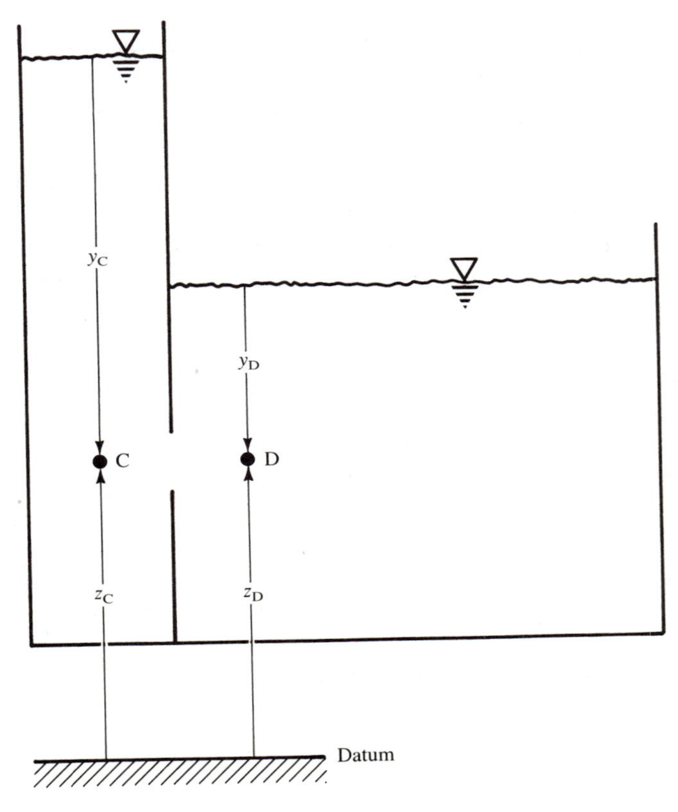

Figure 5.3
Definitions of terms for thought experiment demonstrating the existence of pressure potential energy in a liquid.

and

$$E_{xD} = \gamma B y_D. \tag{5-10b}$$

Thus, referring to Equation 5-1, we can say that the nongravitational component of potential energy at any point depends only on the pressure γy at that point. This component is therefore the pressure potential energy, designated E_{pp}, and we can state generally that

$$E_{pp} = \gamma B y, \tag{5-11}$$

where y is the distance measured vertically downward from the water surface to the point of interest.

Now we can use Equations 5-7 and 5-11 to write the expression for total potential energy E_p at any point:

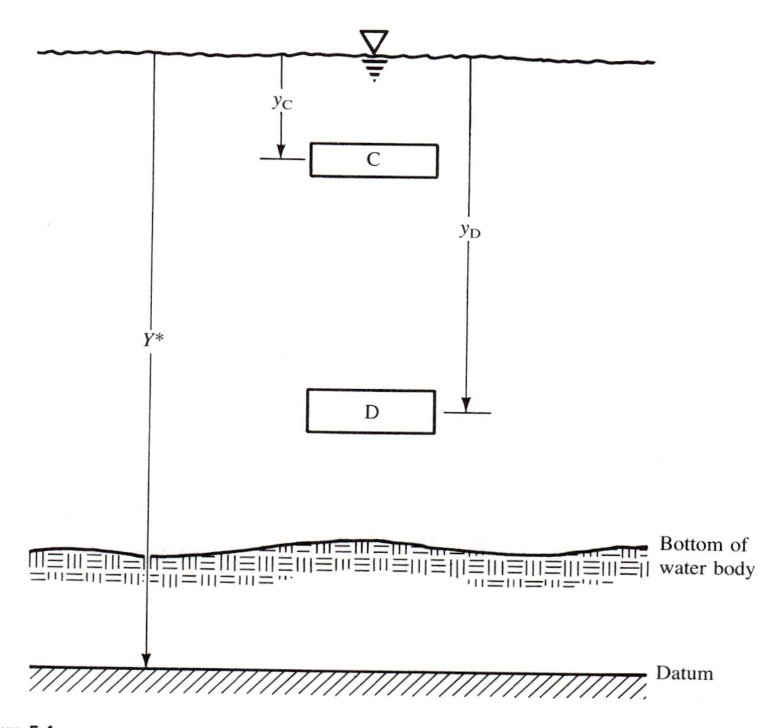

Figure 5.4
Definitions of terms for computing the magnitude of total potential energy in a liquid; C and D are two arbitrary parcels of water at depths y_C and y_D, respectively.

$$E_p = E_{pp} + E_{pg} = \gamma By + \gamma Bz. \quad (5\text{-}12)$$

Equation 5-12 gives the total potential energy of a parcel of liquid with a given volume B. This expression can be made much more useful if it is generalized by dividing each term by the weight γB of the parcel, which yields

$$H_p = z + y, \quad (5\text{-}13)$$

where the potential energy of a fluid parcel divided by its weight is called *potential head*, H_p, and has the dimensions $[L]$. Thus the vertical distance z above a datum of a point in a liquid is the *elevation head* at that point. If the free water surface is horizontal, the distance y of the point below that surface is the *pressure head* at that point.

From Figure 5.5 it is clear that H_p is the same at all points in a water body with a horizontal surface. Thus no gradients of potential energy exist, and hence no motion occurs, in such a body.

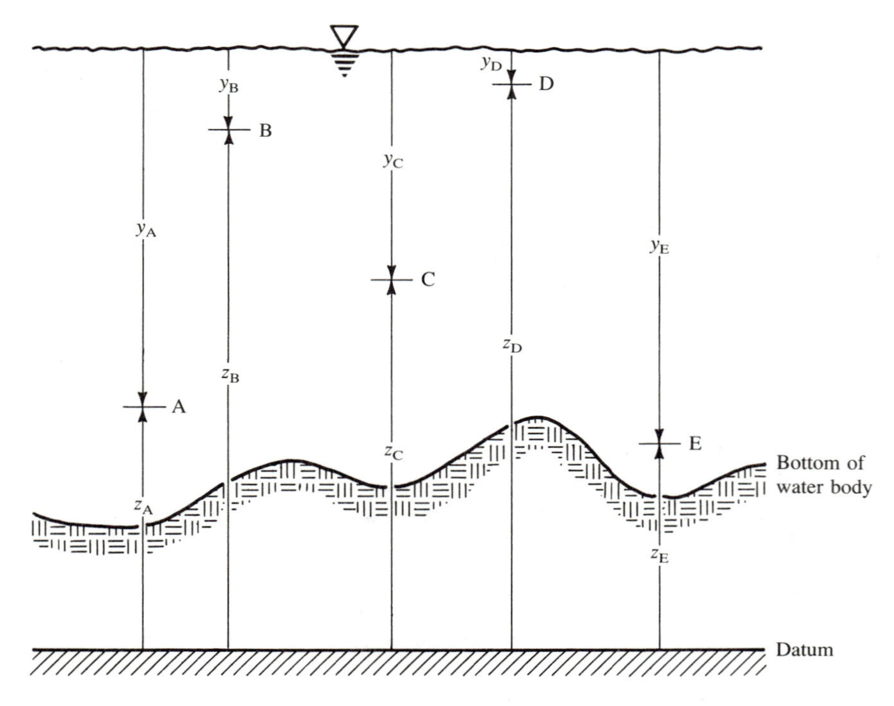

Figure 5.5
Diagram showing the equality of total head at all points in a liquid under hydrostatic conditions. Points A, B, C, D, and E are arbitrary points, and $y_A + z_A = y_B + z_B = y_C + z_C = y_D + z_D = y_E + z_E$.

Another situation of interest is shown in Figure 5.6. The points A, B, and C all have the same elevation head z, and points A and B are the same distance below the horizontal free water surface and hence have the same pressure head y. Above point C is a vertical tube with a valve at its base, and the valve is initially closed (Figure

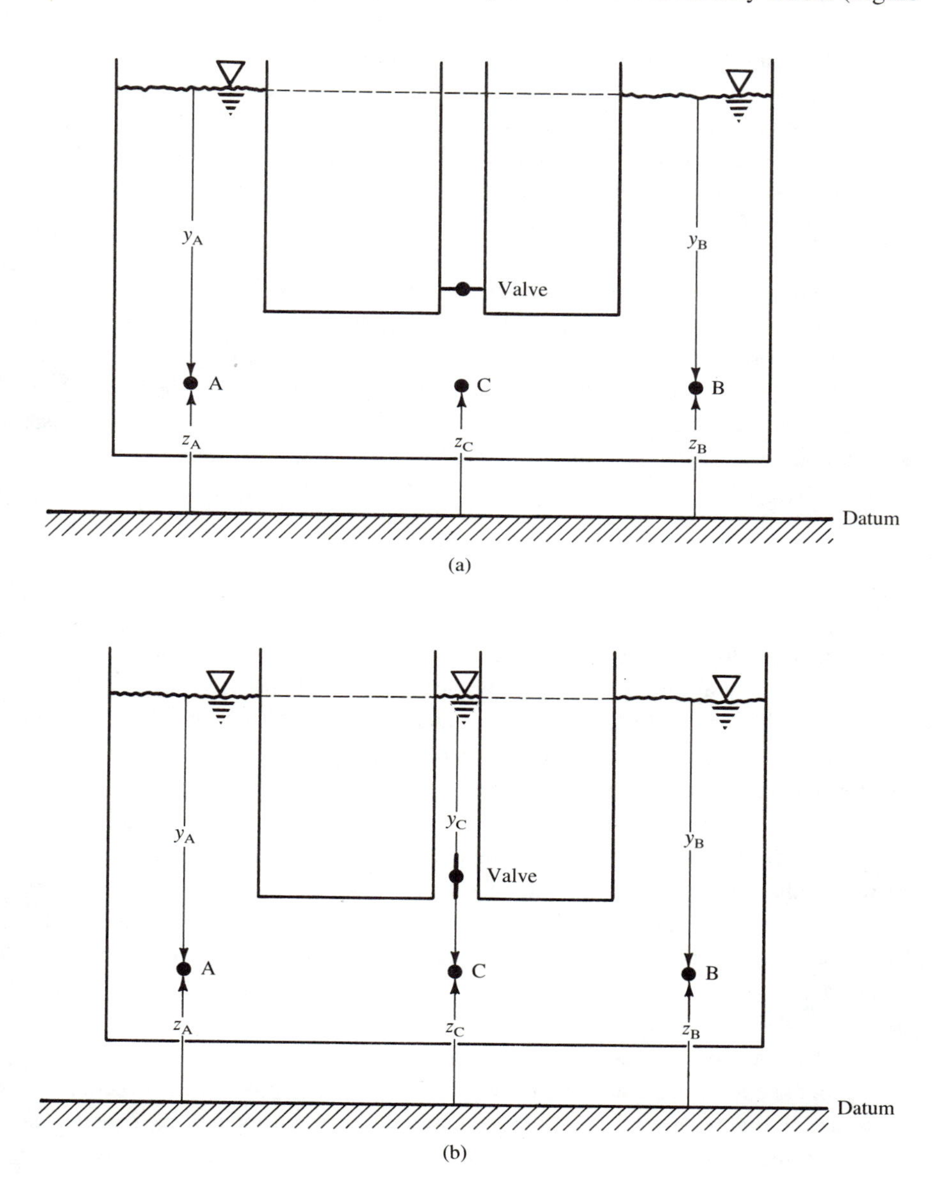

Figure 5.6
Thought experiment illustrating the concept of the piezometric, or total, potential head line (dashed line): (a) valve above point C closed; (b) valve above point C open.

5.6a). Intuitively we know that there will be no motion in this situation, so the potential head at point C must be the same as at A and B (and everywhere else in the liquid). This implies that the pressure head at C is the same as at A and B, even though there is no free surface directly above C. This is verified when the valve is opened, for water will rise in the tube until it reaches the same height as that above A and B (Figure 5.6b). Thus in a case like that in Figure 5.6a the pressure head is determined as the depth below the level of the free water surface that would exist if the geometry of the system permitted such a surface. The general term for the level of a free water surface at atmospheric pressure, whether visible or imaginary, is the *piezometric surface, piezometric head line,* or *potential head line.* Thus y in Equation 5-13 should henceforth be understood to represent the vertical distance below the piezometric surface.

In a stationary system, the potential head line is horizontal, as indicated by the dashed lines in Figures 5.6a and 5.6b. In the discussions of flows in subsequent chapters, the potential head line always slopes downward in the direction of motion, as required by Equation 3-17. We will see, however, that the level of the free surface in a tube inserted into the flow will be slightly below the water level that exists vertically above the point. In most natural flows, though, the distance between the water surface and the piezometric surface is negligibly small.

BOUNDARY LAYERS AND MOMENTUM TRANSFER

Consider the flow of water illustrated in Figure 5.7. At the left side of the diagram the velocity is equal throughout the flow, and is called the *free-stream velocity* V_0. The absence of a vertical velocity gradient means that there is no shear stress within the flow and thus that viscosity has no effect on it (see Equation 4-4). To the right the flow encounters a horizontal boundary. Because of the no-slip property of water, the velocity of the layer immediately next to that boundary is zero. As indicated in Chapter 4, if the flow velocity and depth are relatively small, viscous friction between adjacent layers will act to retard the flow for some distance above the boundary. Thus a vertical velocity gradient is created, extending to some height. The zone in which this gradient exists is called a *boundary layer.* The top of this layer cannot be precisely located, so the *boundary-layer thickness* δ is usually defined as the vertical distance y above the bottom at which the velocity v equals 99% of the free-stream velocity. Thus when $y = \delta$, $v = 0.99V_0$.

The thickness of the boundary layer increases in the downstream direction, as illustrated in Figure 5.7. Initially, the increase in δ is proportional to the square root of the distance from the leading edge of the boundary, and the parcels of water continue to move as parallel laminae, as illustrated in Figure 4.9. This condition is called the state of *laminar flow,* and the zone of the flow in which the velocity gradient exists is a *laminar boundary layer.*

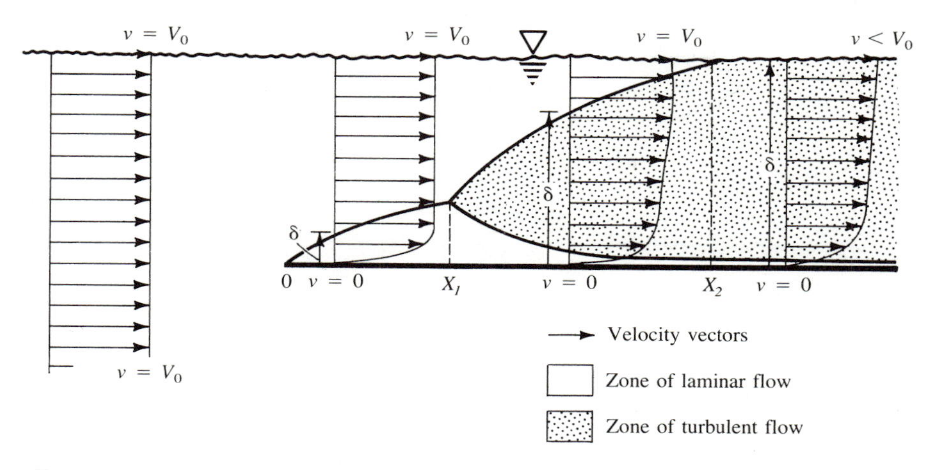

Figure 5.7
Development of a boundary layer with accompanying downward transfer of momentum. At the left the flow is unaffected by any boundary, and the velocity everywhere equals V_0. Friction caused by the horizontal boundary retards the flow, inducing a downward transfer of momentum and creating a boundary layer with a thickness of δ. A laminar boundary layer is developed between 0 and X_1. Turbulence arises at X_1, and a turbulent boundary layer develops between X_1 and X_2. Downstream of X_2 the turbulent boundary layer is fully developed throughout the flow, with a thin zone of laminar flow near the bottom. This condition is typical of most streams. After Chow (1959).

At some point along the flow, the layer will become thick enough for wavelike fluctuations to develop in the formerly parallel laminae. The location of this point moves upstream as V_0 increases and downstream as the viscosity increases. A bit farther downstream the fluctuations become more pronounced and irregular, and ultimately the laminae lose their identity and the flow is characterized by eddies. Parcels of water now move in a highly irregular path, while retaining a net downstream velocity. This condition is the state of *turbulent flow*. Figure 5.8 schematically illustrates the transition from laminar to turbulent flow, and Figure 5.9 shows photographs of the two states of flow.

Once the flow becomes turbulent the boundary layer begins to grow more rapidly, such that δ becomes approximately proportional to the 0.8 power of the distance. As shown in Figure 5.7, the eddies also spread downward, and the zone of laminar flow becomes confined to a very thin layer near the bottom. A *turbulent boundary layer* is now established.

All open-channel flows of hydrologic interest are boundary-layer flows, with the boundary layer extending from the bottom to the surface as shown at the far right in Figure 5.7. Later in this chapter we will see that turbulent flows are far more common than laminar flows.

Details of the velocity distributions of each flow state are examined in Chapter 6.

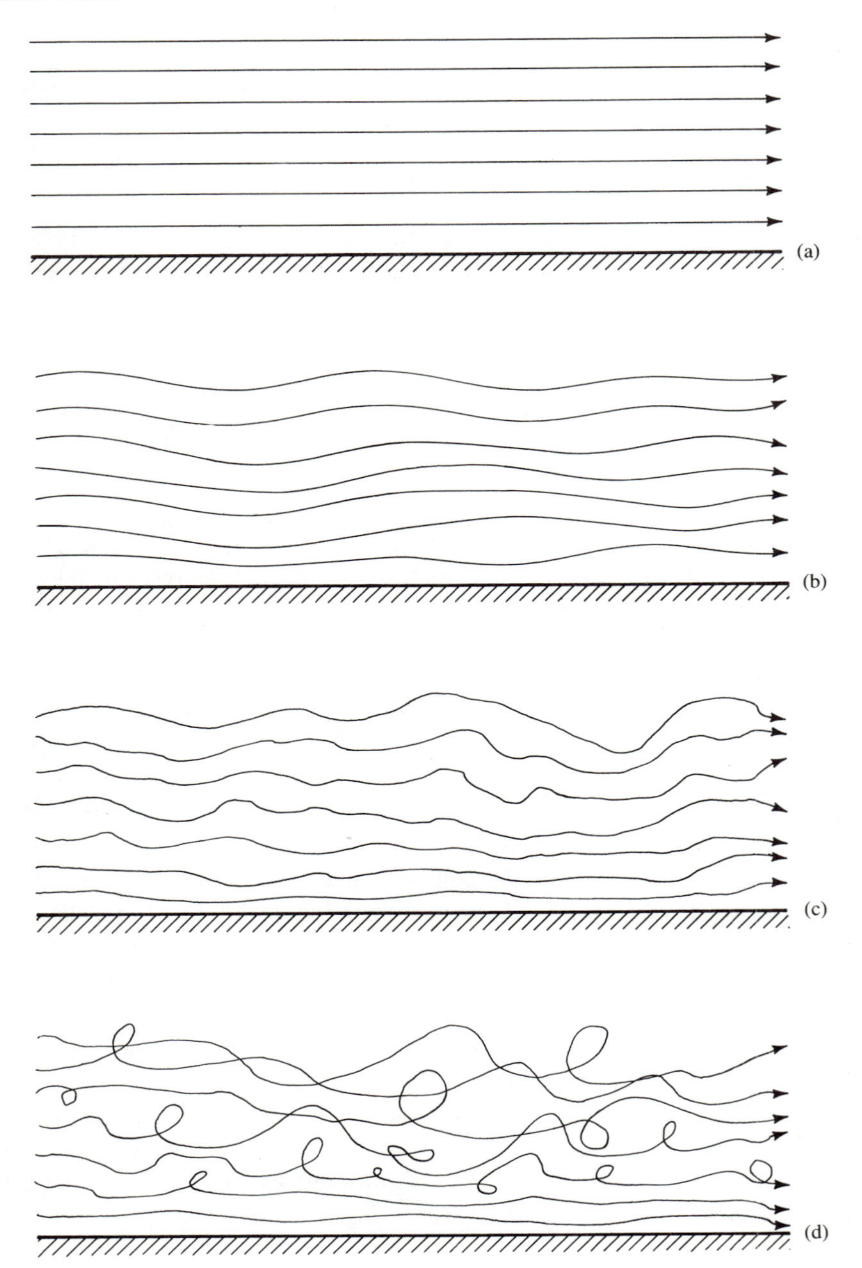

Figure 5.8
Schematic diagram showing the paths of individual water particles as flow changes from the laminar state (a) to the fully turbulent state (d). Flow in (b) and (c) is transitional between the two regimes.

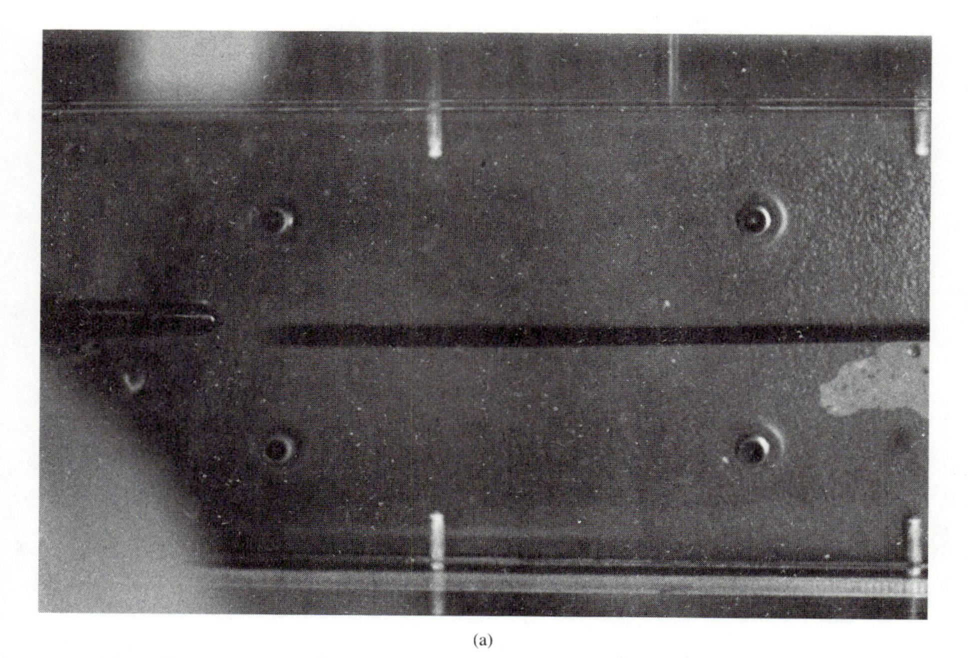

(a)

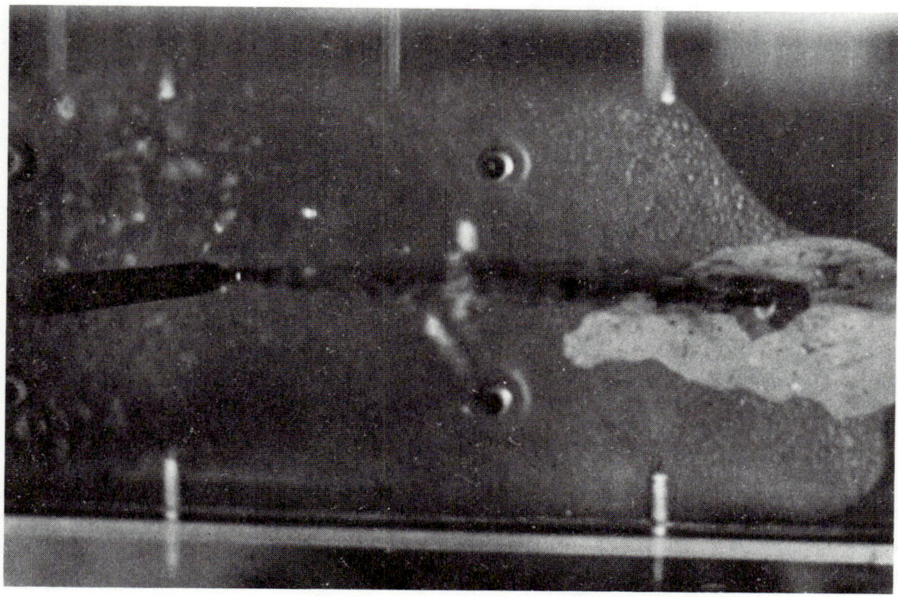

(b)

Figure 5.9
Dye injected into flows from the tube at the left shows the existence of a laminar state in (a) and a turbulent state in (b).

For the present, we look at this velocity gradient as a ''deficiency'' in velocity compared to what would exist without the boundary. We could also say that there has been a loss of momentum (mass times velocity) from the boundary layer, or a transfer of momentum downward from the original flow ''into'' the boundary. Thus in any flow the existence of a velocity gradient reflects the transfer of momentum from zones with higher velocities to those with lower velocities; it also reflects the existence of a shear force within the flow.

The ultimate source of this shear force is the frictional resistance of the boundary. Viscosity transmits this resistance via the frictional drag of adjacent layers when the flow is laminar, but a different mechanism operates in turbulent flows. In these, water particles near the bed are also retarded by viscous friction, but move to higher levels as eddies. The presence of slower-moving particles at a higher level reduces the average velocity at that level; eddies from that level move still higher, and the effect is propagated upward to create an upward-increasing velocity gradient. Of course, faster particles also move downward into slower levels, but because of the frictional drag of the boundary, the net effect is a retardation of the flow and a downward momentum transfer.

A number of important differences between laminar and turbulent flows will become more apparent when expressions for velocity gradients in the two flow states are developed in Chapter 6. For one thing, we will find that the magnitude of the fluid viscosity has a strong influence on laminar flows but only a minor one on turbulent flows, even though viscosity is the ultimate cause of turbulence. We will also see that turbulence is a much more efficient agent of momentum transfer than viscosity alone.

FORCES IN OPEN-CHANNEL FLOWS

The following sections discuss in detail the nature of the forces in open-channel flows and develop expressions for calculating their magnitudes. It will be useful in these analyses to use a form of vector notation to distinguish those forces that (1) act in the direction of the flow and thus induce or maintain motion (shown by an arrow over the letter symbol that points to the right); (2) act in the direction opposite to the flow and thus resist the downstream motion (shown by an arrow over the letter symbol that points to the left); and (3) act at right angles to the flow and hence affect only the direction of the flow (shown by a dot over the letter symbol). Once the formulas giving the relative magnitudes of the forces have been developed, the directional nature of each force will be clear and we need not retain the vector notation.

In order to examine the forces involved in natural flows, we write Newton's second law as

$$\Sigma\vec{F} = M\frac{d\vec{V}}{dt}, \tag{5-14}$$

where $\Sigma\vec{F}$ is the net downstream-directed force acting on a parcel of water of mass M, $\vec{V}$ is the velocity of the parcel, and t is time. A major task of this chapter is to identify the forces contributing to $\Sigma\vec{F}$ and to develop expressions to compute their relative magnitudes. Doing this will help us to identify the important forces to be included in subsequent derivations of equations of motion, and may help simplify those derivations by enabling us to identify forces that can be neglected because of their relative unimportance.

In the thought experiment to illustrate viscosity (Figure 4.6), the water surface was horizontal and the motion-inducing stress was applied by the experimenter by means of a movable upper boundary. Earlier in this chapter we saw that there is no flow in an open channel with a horizontal surface. If the surface is sloping, however, there is a downslope component of force due to gravity acting on each water parcel. As shown in Figure 5.10, this component, $\vec{F_d}$, is given by

$$\vec{F_d} = \gamma B \sin\theta, \tag{5-15}$$

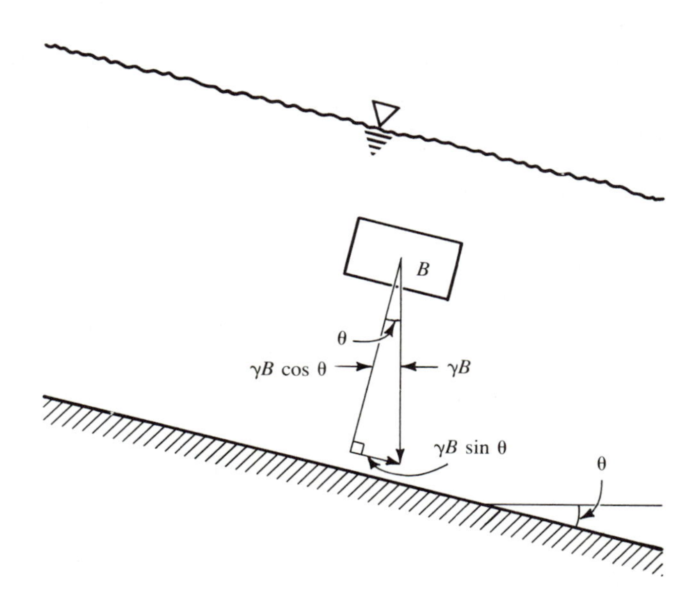

Figure 5.10
Diagram illustrating the computation of the downslope component of gravitational force on a parcel of water of volume B; γ is the weight density of water and θ the slope angle of the bottom.

where B is the volume of the parcel, γ is the weight density of water, and θ is the slope angle. It is this component of gravitational force that induces natural open-channel flows.

Equation 5-14 states that if $\Sigma\overrightarrow{F} > 0$, $d\overrightarrow{V}/dt > 0$ and the parcel will accelerate continuously. In natural flows, accelerations are usually small and short-lived, so forces other than gravity must be present such that, usually,

$$\Sigma\overrightarrow{F} \simeq 0. \tag{5-16}$$

From the discussion in the preceding section it is clear that one of these forces is due to frictional resistance transmitted by viscosity and, often, turbulence. This *frictional force* is a shear force that acts to retard the flow, and is denoted by $\overleftarrow{F_f}$. Because it will be of interest to compare the magnitudes of the frictional forces $\overleftarrow{F_v}$ due to viscosity alone and those ($\overleftarrow{F_t}$) due to turbulence, we will examine these quantities separately. For now, we note that

$$\overleftarrow{F_f} = \overleftarrow{F_v} + \overleftarrow{F_t}. \tag{5-17}$$

If there is a gradient in the elevation of the water surface with respect to the general downslope direction, a water parcel will also be subject to a *pressure-gradient force* $\overleftrightarrow{F_p}$. As indicated by the two arrows, this force may either accelerate the flow or retard it, depending on the direction of the gradient.

Finally, there are three forces (discussed in some detail later) that act at right angles to flows: the surface-tension force $\dot{F_s}$; the centrifugal force $\dot{F_r}$ due to a curvature of the flow path; and the Coriolis force $\dot{F_C}$ due to the rotation of the earth. These are included in our analysis because they can be important in determining the configurations of some flows.

In order to simplify and generalize the following discussion, we will derive the relative magnitudes of forces for the macroscopic parcel of water shown in Figure 5.11 and will divide these magnitudes by the mass of that parcel. Thus we will be calculating and comparing forces per unit mass (accelerations due to the various forces) rather than the forces themselves. Thus Equation 5-14 becomes

$$\frac{\Sigma\overrightarrow{F}}{M} = \frac{d\overrightarrow{V}}{dt}. \tag{5-18}$$

Gravitational Force The gravitational force acting on a parcel of water is expressed in Equation 5-15. Applying the same reasoning to the volume in Figure 5.11, we obtain

$$\overrightarrow{F_g} = \gamma w X Y \sin\theta. \tag{5-19}$$

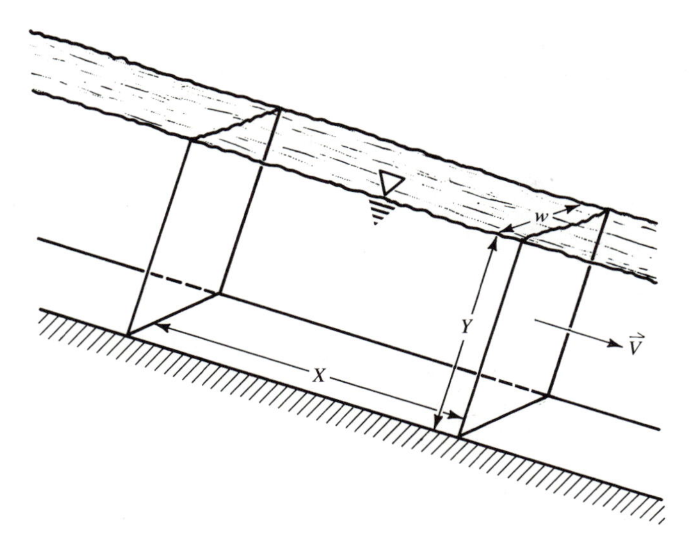

Figure 5.11
Definitions of terms for computing the relative magnitudes of forces in open-channel flows.

The mass of the volume is ρwXY, so the gravitational force per unit mass is

$$\frac{\overrightarrow{F_g}}{M} = \frac{\gamma wXY \sin\theta}{\rho wXY} = \frac{\gamma}{\rho} \sin\theta = \overrightarrow{g \sin\theta}. \tag{5-20a}$$

Although Equation 5-20a is correct for our analysis, it is standard practice in fluid mechanics to represent the gravitational force by the acceleration g due to gravity, omitting the multiplication by $\sin\theta$. We will follow this practice in this chapter in order to develop some standard formulas. Thus, with this convention,

$$\frac{\overrightarrow{F_g}}{M} = \vec{g}. \tag{5-20b}$$

The implications of using Equation 5-20a rather than 5-20b are explored at the end of Chapter 6.

Viscous Force According to Equation 4-4, the shear stress $(\overleftarrow{\tau})$ resisting the motion in a viscous flow is

$$\overleftarrow{\tau} = \mu \frac{d\vec{v}}{dy}, \tag{5-21}$$

where y is the direction at right angles to the flow. In the macroscopic flow element of Figure 5.11 the "representative" velocity gradient can be expressed as $\overrightarrow{V}/Y$, so that Equation 5-21 becomes

$$\overleftarrow{\tau} = \frac{\mu\overrightarrow{V}}{Y}. \tag{5-22}$$

The shear stress acts on a plane parallel to the flow, so in Figure 5.11 the viscous force $\overleftarrow{F_v}$ is

$$\overleftarrow{F_v} = \overleftarrow{\tau}Xw = \frac{\mu\overrightarrow{V}Xw}{Y}. \tag{5-23}$$

Again dividing by the mass, we obtain

$$\frac{\overleftarrow{F_v}}{M} = \frac{\mu\overrightarrow{V}Xw}{Y\rho wXY} = \frac{\mu\overrightarrow{V}}{\rho Y^2} = \frac{\nu\overrightarrow{V}}{Y^2}, \tag{5-24}$$

which gives the viscous force per unit mass. However, because the representative bulk velocity gradient $\overrightarrow{V}/Y$ was used in this development, it should be understood that Equation 5-24 is actually *proportional,* not strictly equal, to the viscous force per unit mass. We develop an expression for the actual viscous forces at the end of Chapter 6.

Turbulent Force As noted earlier in this chapter, turbulence acts analogously to viscosity to retard boundary-layer flows. This effect of turbulence, which can be thought of as an upstream-directed force, is sometimes referred to as the *inertial force*. To formulate an expression for its relative magnitude, we can begin with an expression analogous to Equation 5-21:

$$\overleftarrow{\tau} = \varepsilon \frac{d\overrightarrow{v}}{dy}, \tag{5-25}$$

where ε is called the *eddy viscosity* and the other symbols are as in Equation 5-21.

The eddy viscosity is proportional to the rate of momentum transfer via turbulent eddies. Although a more detailed formulation of this parameter will be made in Chapter 6, we can reason for the present that it should depend on the concentration of momentum in the flow and on the vertical scale of the turbulent eddies. The former is represented by $\rho\overrightarrow{V}$, and it is reasonable to assume that the latter is proportional to the mean depth Y. Thus for present purposes we represent ε as

$$\varepsilon = \rho\overrightarrow{V}Y. \tag{5-26}$$

Note that this expression has the same dimensions as dynamic viscosity, but rather than being a property of the fluid, like viscosity, it depends mostly on the characteristics of the flow.

Substituting Equation 5-26 into Equation 5-25 and retaining $\overrightarrow{V}/Y$ as an expression of the velocity gradient yields

$$\overleftarrow{\tau} = \rho\overrightarrow{V}Y\left(\frac{\overrightarrow{V}}{Y}\right) = \rho\overrightarrow{V}^2. \tag{5-27}$$

Multiplying Equation 5-27 by the area wX over which the shear stress acts gives the turbulent force $\overleftarrow{F_t}$,

$$\overleftarrow{F_t} = \rho\overrightarrow{V}^2 wX,$$

and dividing by the mass M gives finally

$$\frac{\overleftarrow{F_t}}{M} = \frac{\rho\overrightarrow{V}^2 wX}{\rho wXY} = \frac{\overrightarrow{V}^2}{Y} \tag{5-28}$$

as the expression for turbulent force per unit mass in the flow. Again, we should understand Equation 5-28 as being proportional, rather than strictly equal, to the inertial or turbulent force per unit mass. The discussion at the end of Chapter 6 will develop a more exact expression for this force.

Pressure-Gradient Force In an open channel, pressure is proportional to distance below the surface (Equation 5-1). Thus if the distance to a plane parallel to the bottom at the upstream end of a water parcel is different from that at the downstream end, there will be a pressure gradient along the channel and a net force on the volume. If the pressure increases downstream, the force will be directed upstream, and will retard the flow; if the pressure decreases downstream, the force is directed downstream, and will accelerate the flow. In Figure 5.11 there is a plane bottom, and a pressure gradient will exist if the depth differs at the upstream and downstream ends. Denoting this depth difference by ΔY, we have

$$\overleftrightarrow{F_p} = \gamma\, \Delta Y\, wY, \tag{5-29}$$

where $\overleftrightarrow{F_p}$ is the pressure-gradient force on the volume and Y is now the average of the depths at the two ends. Dividing Equation 5-29 by the mass of the volume gives

$$\frac{\overleftrightarrow{F_p}}{M} = \frac{\gamma\, \Delta Y\, wY}{\rho wXY} = \frac{\gamma\, \Delta Y}{\rho X} = \frac{g\, \Delta Y}{X} \tag{5-30}$$

as the expression for the pressure-gradient force per unit mass.

If we were to consider a flow with no free surface, as in a pipe, the pressure difference ΔP would not be related to depth. In this case,

$$\overleftarrow{\vec{F}_p} = \Delta P\, wY \tag{5-31}$$

and

$$\frac{\overleftarrow{\vec{F}_p}}{M} = \frac{\Delta P\, wY}{\rho wXY} = \frac{\Delta P}{\rho X}. \tag{5-32}$$

Surface-Tension Force Surface-tension force acts only along the two upper boundaries of the volume shown in Figure 5.11, a total distance of $2X$, and is directed at right angles to the flow. This force is designated $\dot{F}_s$, and following the discussion of surface tension in Chapter 4,

$$\dot{F}_s = \sigma 2X. \tag{5-33}$$

Dividing by the mass gives the expression for surface-tension force per unit mass:

$$\frac{\dot{F}_s}{M} = \frac{\sigma 2X}{\rho wXY} = \frac{2\sigma}{\rho wY}. \tag{5-34}$$

Centrifugal Force Newton's first law of motion states that a body in motion tends to move in a straight path at a constant velocity unless acted upon by an outside force. Thus a parcel of water moving in a curved path is subject to an apparent force acting to straighten the path, and the direction in which this apparent force acts is at right angles to the mean velocity. It is shown in elementary physics that the magnitude $\dot{F}_r$ of this force is

$$\dot{F}_r = M\frac{\overrightarrow{V^2}}{r}, \tag{5-35}$$

where r is the radius of curvature of the path. Thus the *centrifugal force* per unit mass is simply

$$\frac{\dot{F}_r}{M} = \frac{\overrightarrow{V^2}}{r}. \tag{5-36}$$

Coriolis Force As a result of the earth's rotation, an acceleration is produced on any body that is moving in a plane that is not parallel to the axis of rotation. Known as the *Coriolis acceleration,* it produces a force called the *Coriolis force,* after the French mathematician Gaspard de Coriolis (1792–1843), who first analyzed the phenomenon.

The magnitude of this acceleration is proportional to the velocity of the body, to the angular velocity of the earth's rotation, and to the sine of the latitude at which the motion occurs:

$$\frac{\dot{F}_C}{M} \propto \omega \overrightarrow{V} \sin \phi, \tag{5-37}$$

where ω is the angular velocity of the earth ($15° \text{ h}^{-1}$ or 0.73×10^{-4} rad s^{-1}) and ϕ is the latitude. It can be shown that the proportionality constant in Equation 5-37 is 2, so that

$$\frac{\dot{F}_C}{M} = 2\omega \overrightarrow{V} \sin \phi. \tag{5-38}$$

As noted earlier, this force acts at right angles to the flow direction, to its right in the Northern Hemisphere and to its left in the Southern Hemisphere.

Daily and Harleman (1966) and Eagleson (1970) give detailed mathematical analyses of the Coriolis acceleration, and helpful intuitive discussions can be found in McDonald (1952) and Miller and Thompson (1970).

Summary of Force-per-Unit-Mass Expressions The preceding analyses show that the forces affecting a parcel of water in an open channel can be classified as (1) those that accelerate motion (gravity, pressure gradient); (2) those that resist motion (pressure gradient, viscous, turbulent); and (3) those that act at right angles to the motion (surface tension, Coriolis, centrifugal). The expressions for these forces per unit mass are summarized in Table 5.1. With them, Equation 5-14 can be written as

$$\overrightarrow{F_g} - \overleftarrow{F_v} - \overleftarrow{F_t} \pm \overleftrightarrow{F_p} - \dot{F}_s - \dot{F}_r - \dot{F}_C = M\frac{d\overrightarrow{V}}{dt}, \tag{5-39}$$

where a minus sign is used for any force that does not contribute to the downslope acceleration. In the following examination of the relative magnitudes of these forces, it is shown that $\dot{F}_s$ and $\dot{F}_C$ can usually be neglected in open-channel flows of hydrologic interest. In most cases, $\dot{F}_r$ is also negligible, and there are many flows in which $\overleftrightarrow{F_p}$ plays only a minor role. Thus, in most hydrologic situations,

$$\overrightarrow{F_g} - \overleftarrow{F_v} - \overleftarrow{F_t} = M\frac{d\overrightarrow{V}}{dt}. \tag{5-40}$$

Furthermore, as noted in Equation 5-16, accelerations are often very small, so we often have the situation in which

$$\overrightarrow{F_g} \simeq \overleftarrow{F_v} + \overleftarrow{F_t}; \tag{5-41}$$

for laminar flow

$$\overrightarrow{F_g} \simeq \overleftarrow{F_v}, \tag{5-42}$$

and for turbulent flow

$$\overrightarrow{F_g} \simeq \overleftarrow{F_t}. \tag{5-43}$$

Chapter 6 is devoted to analyses of conditions in which Equations 5-42 and 5-43 apply.

Table 5.1
Expressions for Computing the Relative Magnitudes of Forces in Open-Channel Flows

Action	Force	Expression (per unit mass)
Motion inducing	Gravitational	$g^{\dagger}$
	Pressure	$\frac{\gamma\ \Delta Y}{\rho X} = \frac{g\ \Delta Y}{X}$
Motion resisting	Pressure	$\frac{\gamma\ \Delta Y}{\rho X} = \frac{g\ \Delta Y}{X}$
	Viscous	$\frac{\mu V}{\rho Y^2} = \frac{\nu V}{Y^2}$
	Turbulent	$\frac{V^2}{Y}$
Right angles to motion	Surface tension	$\frac{2\sigma}{\rho W Y}$
	Coriolis	$2\omega V \sin \phi$
	Centrifugal	$\frac{V^2}{r}$

†Downslope component $= g \sin \theta$

Relative Forces in Actual Flows It is now of interest to examine the relative magnitudes of the various forces in a range of open-channel flows. To make the comparisons most general, we will assume that the changes in depth are negligible and that the flows are straight, so that there are no net pressure-gradient or centrifugal forces.

Table 5.2 gives the characteristics of some open-channel flows over a range of widths and depths covering six orders of magnitude, and shows the per-unit-mass values for gravity, viscous, turbulent, surface-tension, and Coriolis forces for those flows. Note that gravitational force dominates and turbulent force is second in importance in all but the two smallest flows. Viscous forces are negligible in flows

Table 5.2

Forces per Unit Mass in Open-Channel Flows of Various Sizes

Fluid properties at 10°C are assumed.

	Typical flow characteristics			Forces per unit mass (m s^{-2})				
	Width (m)	Depth (m)	Velocity (m s^{-1})	Gravitational	Viscous	Turbulent	Surface tension	Coriolis†
Vertical trickle	0.007	0.002	0.7	9.8	0.23	250	1×10^{3}	6.6×10^{-5}
Flow from a parking lot	25	0.002	0.1	9.8	0.033	5	3×10^{-3}	9.4×10^{-6}
Model river	0.5	0.05	0.1	9.8	5.2×10^{-5}	0.2	6×10^{-3}	9.4×10^{-6}
Small stream	5	0.5	0.5	9.8	2.6×10^{-6}	0.5	6×10^{-5}	4.7×10^{-5}
Medium river	25	2	1	9.8	3.3×10^{-7}	0.5	3×10^{-6}	9.4×10^{-5}
Large river	100	5	1	9.8	5.2×10^{-8}	0.2	3×10^{-7}	9.4×10^{-5}
Gulf Stream	4×10^{4}	2×10^{3}	1	9.8	3.3×10^{-13}	5×10^{-4}	2×10^{-12}	9.4×10^{-5}

†At latitude 40°.

larger than that from a typical parking lot; this means that in most surface flows of interest to hydrologists the downslope component of gravity is balanced by the turbulent force; that is, the flows are turbulent.

The boundaries of the trickle (as down a window during a rain) are due to surface tension, and the importance of that force is reflected in the magnitude of F_s/M in that flow. For flows larger than the model river (laboratory flume), however, surface-tension force is negligible relative to the other forces.

The relative magnitude of the Coriolis force remains small in all flows, but increases as the flows get larger. In the Gulf Stream, the downslope component of gravity is very small (about of the same magnitude as the turbulent force), and the Coriolis force becomes comparable to these forces. The relative importance of the Coriolis force causes the Gulf Stream to turn gradually to the right in its course up the east coast of North America and thence across the North Atlantic. Interestingly, it has been suggested that *meanders,* the semiregular sinuous bends of many lowland rivers, are due to the action of the Coriolis effect. The numbers in Table 5.2 show that this explanation cannot be correct, since the Coriolis force is too small to have any effect on river flows. Similarly, the direction of rotation of the vortex you observe when you pull the plug in a bathtub or sink is in general determined by random residual motions of the water; the Coriolis force is of negligible relative magnitude unless special experimental conditions are established (Sibulkin, 1983).

CLASSIFICATION OF FLOWS

There are several ways in which the flows of interest to hydrologists can be usefully classified. One important way is to make the distinction indicated in the title of this chapter: Flows in which one boundary is a free water surface at atmospheric pressure (open-channel flows) can be contrasted to those in which all the boundaries are solid. In natural systems, the latter type exists only in groundwater flows between confining layers. Thus the surface-water hydrologist is concerned only with free-surface flows, which are called open-channel flows even if they occur in sheets that are very wide relative to their depth rather than in well-defined channels.

We noted earlier that the no-slip condition is an important feature of flowing water, and that in virtually all open-channel flows of hydrologic interest the retarding effect of a boundary (the boundary friction) is transmitted throughout the flow. Such flows, then, are characterized by a velocity gradient at right angles to the boundary, with a velocity of zero at the boundary and a maximum value at or near the free surface. Flows of this type are called *boundary-layer flows* and are contrasted to *potential flows,* in which there are no boundary effects. Interestingly, although flow through the pores of a granular medium like soil is a boundary-layer flow at the microscopic level, the entire flow through such a medium can be treated as a potential flow when considered macroscopically.

Thus the next several chapters concern boundary-layer flows in open channels. The discussion of these flows is based on a three-way classification that depends on (1) the relative magnitudes of the gravity, viscous, and turbulent forces; (2) the rate of change of the mean velocity or mean depth with respect to time; and (3) the rate of change of the mean velocity or mean depth in the downstream direction. Since it is virtually impossible for velocity and depth to change independently of each other, they are interchangeable in the classification.

Relative Magnitudes of Forces In light of the preceding discussions, it should not be surprising that the relative magnitudes of the viscous and turbulent forces determine whether the flow state is laminar or turbulent. This ratio can be evaluated from Equations 5-24 and 5-28:

$$\frac{F_t/M}{F_v/M} = \frac{V^2Y^2}{Y\nu V} = \frac{VY}{\nu}. \tag{5-44}$$

(The arrows indicating the direction of action of vector quantities are not shown in this and subsequent developments.)

An English hydraulician, Osborne Reynolds (1842–1912), first recognized the importance of this dimensionless ratio in determining the flow state in 1883; it is called the Reynolds number and is denoted by $\mathbb{R}$. In open-channel flows it is preferable to alter its definition somewhat to account more completely for the channel geometry. This is done by defining the hydraulic radius R as

$$R \equiv \frac{A}{P}, \tag{5-45}$$

where A is the cross-sectional area of the flow (i.e., the area at right angles to the mean velocity) and P is the *wetted perimeter* of the flow. This latter is the distance from one edge of the flow surface to the other, measured along the boundary (Figure 5.12). Thus the definition of the *Reynolds number* for open-channel flow is more properly given by Equation 5-46 than by Equation 5-44:

$$\mathbb{R} \equiv \frac{VR}{\nu}. \tag{5-46}$$

For channels that are wide relative to their depth, however, there is little difference in the values of R and Y (see Exercise 5-6).

Reynolds found by experiment that when $\mathbb{R} < 500$, viscous force dominates, and the flow is laminar. In this range, any disturbances to the flow are damped out by the viscous friction, and the parallel laminae are reestablished. When $\mathbb{R} > 500$ the turbulent force becomes significant, and the flow is not strictly laminar. If such a flow is disturbed, the disturbance may grow into full turbulence or may largely

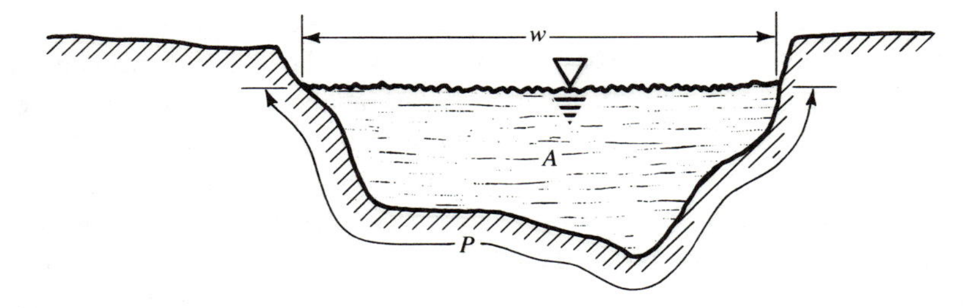

Figure 5.12
Definitions of terms for computing the hydraulic radius and hydraulic (mean) depth; w is the water-surface width, A is the cross-sectional area, and P is the wetted perimeter.

subside, depending on the frequency, amplitude, and persistence of the disturbance. Thus between Reynolds numbers of 500 and 2000, open-channel flow is considered to be in the transitional state. When $\mathbb{R} > 2000$, turbulent force dominates the flow and it is fully turbulent. Most natural flows are well within the turbulent state (Figure 5.13; Exercise 5-5).

When the Reynolds number is applied to other flow situations, the characteristic length and velocity used in the definition may change. For example, for pipe flows, the appropriate length is the diameter of the pipe, whereas in flows through soil pores the appropriate length is the diameter of the pores (which in practice is approximated by the mean diameter of the soil grains). Thus the most general definition of the Reynolds number is

$$\mathbb{R} \equiv \frac{Vl}{\nu}, \tag{5-47}$$

where V is a characteristic velocity and l is a characteristic length of the flow being considered. Other forms of the Reynolds number will be defined in Chapter 8 when sediment erosion and transportation are considered. The values of $\mathbb{R}$ at which the flow state changes depend on the geometry of the system and on the exact definitions of V and l in Equation 5-47.

A second classification of flows is based on the ratio of turbulent to gravitational forces. From Equations 5-16 and 5-29, this ratio is

$$\frac{F_t/M}{F_g/M} = \frac{V^2}{gY}. \tag{5-48}$$

The square root of this ratio, rather than the ratio itself, is the usual way of expressing the relative magnitudes of the turbulent and gravitational forces. This is

(a)

(b)

Figure 5.13
Virtually all natural open-channel flows are turbulent. The presence of turbulent eddies is demonstrated at the boundaries of clear and sediment-laden water in these two Alaskan rivers: (a) the Yukon River near Galena; (b) a tributary of the Susitna River near Montana.

convenient because, as will be discussed more fully in Chapter 11, the speed C_g relative to the water of a small gravity wave in shallow water is

$$C_g = (gY)^{1/2}. \tag{5-49}$$

Thus the dimensionless *Froude number* $\mathbb{F}$ is defined as

$$\mathbb{F} \equiv \frac{V}{(gY)^{1/2}}. \tag{5-50}$$

This is named for William Froude (1810–1879), another English hydraulician.

The Froude number is an important criterion for flow classification because when $\mathbb{F} < 1$ the wave velocity exceeds the flow velocity, and a wave caused by a disturbance or obstacle in the flow can travel upstream. Such flows are said to be in the *subcritical regime*. When $\mathbb{F} > 1$ waves cannot be propagated upstream, and the flow is in the *supercritical regime*. When $\mathbb{F} = 1$ the flow is said to be *critical*. Subcritical flows are also described as *tranquil*, and supercritical flows as *rapid* or *shooting*. The Froude number will also appear later in discussions of energy in flowing water, and its full significance will be more apparent when the profiles of open-channel flow are analyzed in Chapter 9. Note that the depth, rather than the hydraulic radius, is used in computing the Froude number. In channels of irregular cross section, the *mean depth*, or *hydraulic depth*, is used, where

$$Y \equiv \frac{A}{w}, \tag{5-51}$$

and A is the cross-sectional area and w the *top width* of the flow.

Rates of Change in Time and Space The basic characteristics of every natural flow (its velocity, depth, and width) change with time and in the three spatial dimensions. It is futile to attempt a complete physical description of such variability, and the scientist seeks instead to identify the aspects that are essential to a description, or model, of a phenomenon as opposed to those that are inapplicable or of negligible importance. Your success in problem solving will depend largely on your ability to make this kind of discrimination.

The classification in Table 5.3, which is based on the magnitudes of temporal and spatial changes, is fundamental and should be incorporated into your vocabulary. This classification applies to flows whether they are laminar, transitional, or turbulent; subcritical or supercritical. Note that only changes in the mean values of velocity and depth at a given cross section of the flow are considered. The infinite variations of turbulence are ignored, and only spatial changes in the downstream direction are included.

Table 5.3

Classification of Open-Channel Flows Based on Temporal and Spatial Changes of Mean Velocity and Mean Depth

After Chow (1959); V represents mean velocity; Y, mean depth; t, time; and x, distance along the channel.

Steady flow: velocity and depth are constant during the time period considered; that is, $dV/dt = 0$; $dY/dt = 0$.

Uniform flow: Velocity and depth are constant over a portion of the channel considered; that is, $dV/dx = 0$; $dY/dx = 0$.

Varied flow: $|dV/dx| > 0$, $|dY/dx| > 0$.

Gradually varied flow: $|dV/dx|$, $|dY/dx|$ relatively small

Rapidly varied flow: $|dV/dx|$, $|dY/dx|$ relatively large

Unsteady flow: $|dV/dt| > 0$, $|dY/dt| > 0$.

Unsteady uniform flow: $|dV/dx| = 0$, $|dY/dx| = 0$ (very rare).

Unsteady varied flow: $|dV/\mathrm{dx}| > 0$, $|dY/dx| > 0$.

Gradually varied unsteady flow: $|dV/dx|$, $|dY/dx|$ relatively small

Rapidly varied unsteady flow: $|dV/\mathrm{dx}|$, $|dY/dx|$ relatively large

Figure 5.14 shows schematically the various flow types. Steady uniform flow, in which depth and velocity do not change in space or time, is of course the simplest type to analyze. Although no natural flows are strictly steady uniform, the equations of motion for such flows are the basis for analyzing steady gradually varied flows and some cases of unsteady gradually varied flows. These equations of motion, which are developed in the next chapter, relate the flow velocity, depth, and frictional resistance; they differ for laminar, transitional, and turbulent flows. In rapidly varying flows, the changes in velocity and depth usually occur over such short distances that frictional resistance can be ignored, and the flow is treated as a potential flow (Chow, 1959). In Table 5.3 unsteady uniform flow is described as very rare because it is virtually impossible for the depth to change with time while maintaining exactly the same slope.

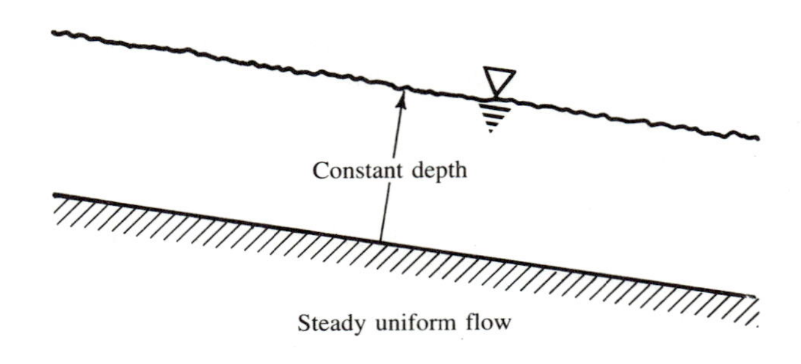

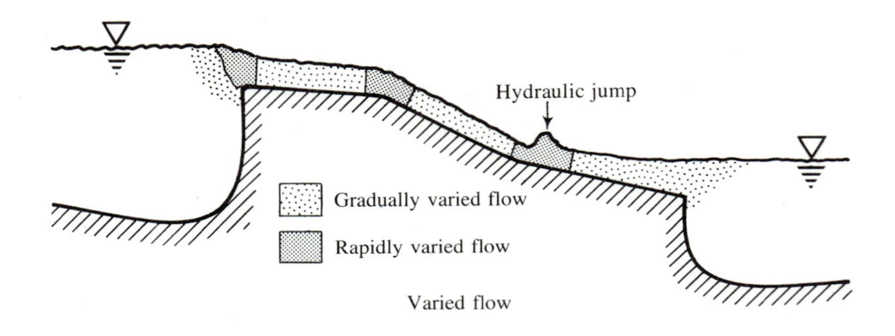

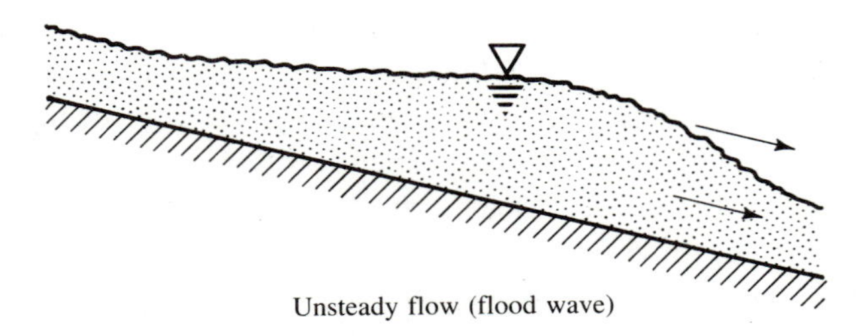

Figure 5.14
Schematic diagrams showing the occurrence of types of open-channel flows of hydrologic importance. Dots indicate gradually varied flow; shading indicates rapidly varied flow. Hydraulic jumps are a standing-wave phenomenon that is discussed in Chapter 10. Adapted from Chow (1959).

Exercises

5-1. Lake Baikal, in the eastern Soviet Union, is 1620 m deep at its deepest point. What is the pressure at this depth, expressed in atmospheres?

5-2. At 325 m, the highest dam in the world is the Rogunsky Dam in the Soviet Union. What hydrostatic pressure must this dam withstand at its base?

5-3. Suppose it is required that the water pressure at fire hydrants in a town be at least 40 lb in.$^{-2}$. The surface of the town's reservoir is 100 ft above the highest hydrant. Will this reservoir provide the necessary pressure?

5-4. Calculate the Reynolds and Froude numbers for the flows in Table 5.2, assuming the channels are rectangular.

5-5. On a piece of three- by three-cycle logarithmic graph paper, label the vertical axis for depths (assumed equal to the hydraulic radius) from 0.1 to 100 cm and the horizontal axis for mean velocities from 1 to 100 cm s^{-1}. Delineate on the graph ''fields'' that show the ranges of velocities and depths for which open-channel flows are laminar, transitional, or turbulent; and subcritical or supercritical.

5-6. What is the formula for hydraulic radius as a function of the channel width w and depth Y for a rectangular channel? For ratios of width to depth of 5, 10, 20, and 40, compute the ratio of hydraulic depth to hydraulic radius.

5-7. Derive a formula for the area, hydraulic depth, and hydraulic radius for a symmetrical triangular channel (vertex downward) as a function of the vertex angle and the maximum depth. (See Chow, 1959, for geometric formulas for channels of various regular shapes.)

5-8. The data in the following table define the cross section of a natural channel. Plot the cross section and compute the area, hydraulic depth, and hydraulic radius when the maximum depth of water in the channel is 0.5, 1.0, 2.0, 3.0, and 4.1 m. On arithmetic graph paper, plot these cross-section parameters as a function of maximum depth. Save your results (Exercise 6-8).

Distance from left bank (m)	Distance below edge of banks (m)	Distance from left bank (m)	Distance below edge of banks (m)
0	0	11.0	3.7
1.0	1.0	12.0	3.6
2.0	1.2	13.0	3.8
3.0	1.3	14.0	4.0
4.0	1.4	15.0	4.1
5.0	1.6	16.0	4.0
6.0	1.7	17.0	3.6
7.0	2.4	18.0	2.9
8.0	3.0	19.0	2.0
9.0	3.4	20.0	0
10.0	3.7		

5-9. Suppose the mean flow velocity V in the channel of Exercise 5-8 is related to the hydraulic radius R as

$$V = 3.31R^{0.67}$$

where V is in meters per second and R is in meters. At what value of maximum depth will the flow become critical?

CHAPTER 6

Uniform Flows in Open Channels

SIGNIFICANCE OF UNIFORM FLOW

To appreciate the significance of uniform flow we must have a precise definition of acceleration. To develop this definition we introduce the concept of a *streamline,* which is an imaginary line constructed within a flow such that it is tangent to the average velocity vector at each point it touches. Any number of streamlines can be constructed within a given flow; each indicates the local flow direction at a particular instant. To understand this concept more fully, imagine that we can identify a small volume of water (called a parcel) and can trace its movement in a steady laminar flow. If we were to draw the velocity vectors of this parcel at successive instants we would find that they were always tangent to the line tracing the parcel's path; therefore that line is a streamline. In turbulent flows the velocity at each point fluctuates from instant to instant as a result of eddies. The flow at any point thus cannot be strictly steady, and the path of a parcel is irregular. To develop a useful definition of a streamline for turbulent flows we take the time average of the velocity at each point. As noted in Chapter 5, the flow is considered steady if this average velocity does not change over time. A streamline in a steady turbulent flow is thus tangent to the time-averaged velocity vector at each point, and traces the average downstream path of a parcel rather than its actual path (Figure 6.1). In unsteady flows, the streamlines shift from instant to instant and thus do not even conceptually represent the paths of parcels (Daily and Harleman, 1966).

Consider any streamline in a flow. The time-averaged velocity at any point on the streamline is designated v, and the acceleration a is defined as the time rate of change of velocity:

$$a \equiv \frac{dv}{dt}, \tag{6-1}$$

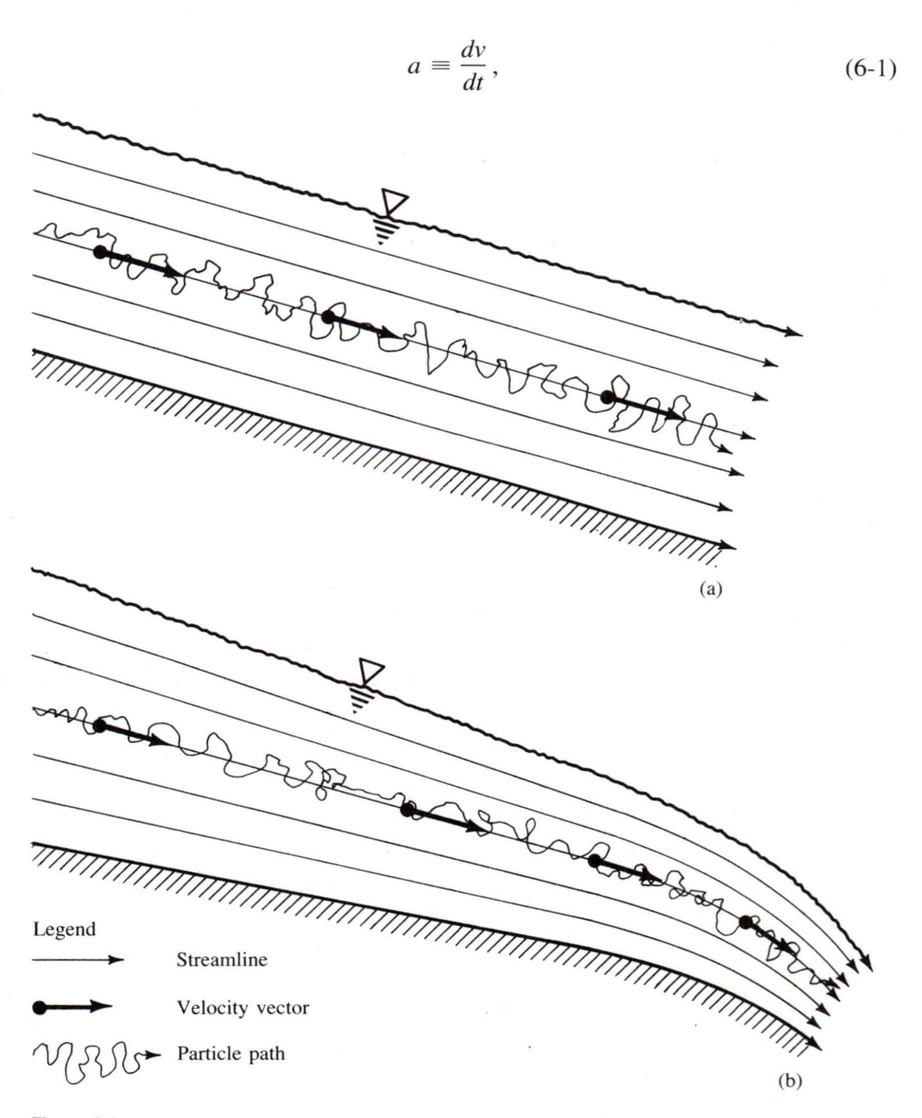

Figure 6.1
Streamlines, velocity vectors, and particle paths in (a) a uniform flow and (b) a steady nonuniform flow. Note that the channel bottom and water surface are both streamlines.

where t is time. The velocity at any point may change with time, and the velocity of any parcel may change as it moves along the streamline. Thus acceleration has two components. To express these components mathematically, we can write the functional dependence of velocity on both time and space as

$$v = f(t, x), \tag{6-2}$$

where x is the distance along the streamline. From the rules of differentials, Equation 6-2 implies that

$$dv = \frac{\partial v}{\partial x} dx + \frac{\partial v}{\partial t} dt, \tag{6-3}$$

where dv, dx, and dt are very small increments (differentials) of the mean velocity, distance, and time, respectively; and $\partial v/\partial x$ and $\partial v/\partial t$ are the rates of change of velocity with distance at a given time and of velocity with time at a given location, respectively.[1] Substituting Equation 6-3 into 6-1 yields

$$\frac{dv}{dt} = \frac{\partial v}{\partial x}\frac{dx}{dt} + \frac{\partial v}{\partial t}\frac{dt}{dt}. \tag{6-4}$$

However, $dt/dt = 1$, and by definition $dx/dt = v$, so Equation 6-4 can be written as

$$a = \frac{dv}{dt} = v\frac{\partial v}{\partial x} + \frac{\partial v}{\partial t}. \tag{6-5}$$

The right-hand side of Equation 6-5 contains the two components of acceleration. The first, $v(\partial v/\partial x)$, is called the *convective acceleration;* it represents the change of velocity in the downstream direction (represented by convergence, divergence, or curvature of streamlines), and its magnitude is greater than zero in varied flows. The second component, $\partial v/\partial t$, is called the *local acceleration;* it represents the change of velocity with time at a given point, and its magnitude is greater than zero in unsteady flows.

By definition, convective acceleration is zero in a uniform flow; and since flows cannot remain spatially uniform while undergoing changes with time, it is also true that local acceleration is zero in uniform flows. Thus uniform flows are characterized

[1] The "curly d" notation is employed when a dependent variable (e.g., v) varies with respect to more than one independent variable (e.g., x and t), and the derivatives so written are called *partial derivatives.* Their mathematical treatment follows the same rules that apply to ordinary ("complete") derivatives.

by zero total acceleration. From Newton's second law, this means that the net forces on a given parcel will be zero in such flows. Thus in the following sections the principle of force balancing is applied in developing the equations of motion for uniform laminar and turbulent flows, as discussed in Chapter 3.

Another important characteristic of uniform flows is that the stream surface in such flows is parallel to the channel bottom (neglecting small-scale irregularities of the bed). If this were not so, depth would be changing in the downstream direction, and by definition such a flow is varied.

UNIFORM LAMINAR FLOW

Velocity Profile To develop the equation for the vertical distribution of velocity in a uniform laminar open-channel flow, we need only two basic principles: (1) Because the flow is steady and uniform, acceleration is zero, and thus the net force on each parcel of flowing water must be zero (balance of forces); and (2) because the flow is laminar, Equation 4-5, which relates the velocity gradient and the applied stress, is applicable.

Figure 6.2 is a water-surface profile of a steady uniform laminar flow. Consider

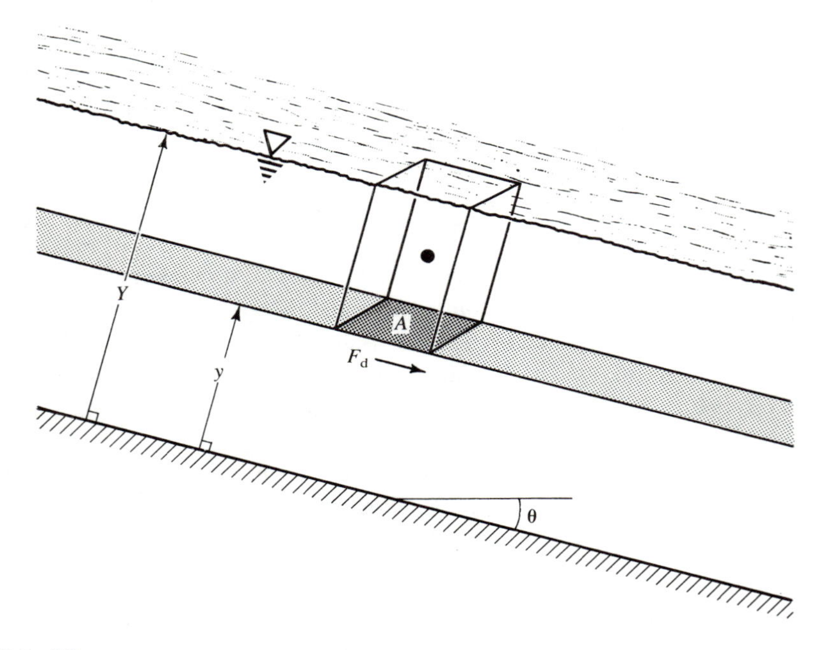

Figure 6.2
Definition sketch for the development of the vertical velocity profile in a uniform laminar flow.

an arbitrary plane in the flow, parallel to the surface and channel bottom, and the forces on a parcel of water of unit area extending from the plane to the surface. The weight of this parcel is

$$(Y - y)\gamma A, \tag{6-6}$$

where Y is the total flow depth, y is the distance from the bottom to the plane, A is the area of the parcel parallel to the plane, and γ is the weight density of water. As in Figure 5.10, the downslope component F_d of this weight is

$$F_d = \gamma(Y - y)A \sin\theta, \tag{6-7}$$

where θ is the slope of the channel bottom (Figure 6.3). Since the flow is steady (no acceleration), this force must be opposed by an equal force in the opposite direction, F_u. In a sliding block, this opposing force is provided by sliding friction; in laminar flow it is provided by the internal friction between sliding layers of fluid, that is, the viscosity. From Equation 4-5 this force is

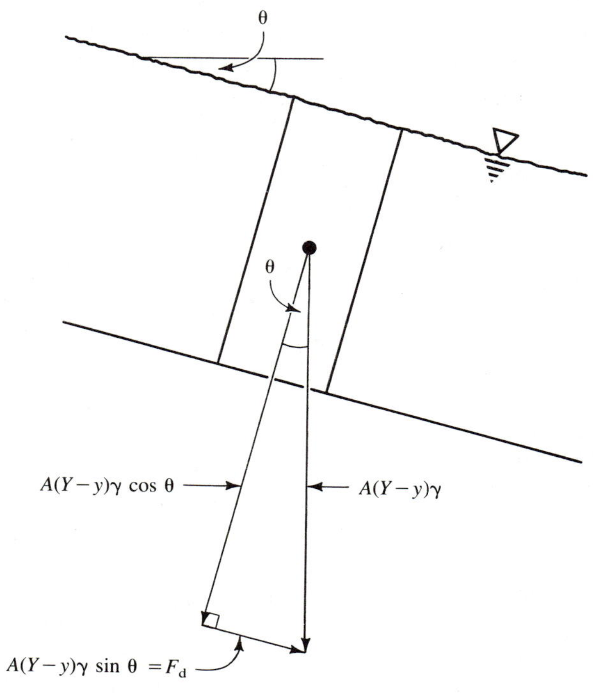

Figure 6.3
Vector diagram showing that $F_d = \gamma(Y - y)A \sin\theta$.

$$F_u = \tau A = \mu \frac{dv}{dy} A. \tag{6-8}$$

Equating F_d and F_u and eliminating A from both sides yields

$$\mu \frac{dv}{dy} = \gamma(Y - y) \sin \theta. \tag{6-9}$$

Equation 6-9 is a differential equation that can be readily solved by separating variables and integrating:

$$dv = \frac{\gamma \sin \theta}{\mu} (Y - y)\, dy,$$

$$\int dv = \frac{\gamma \sin \theta}{\mu} \int (Y - y)\, dy,$$

$$v = \frac{\gamma \sin \theta}{\mu} \left(Yy - \frac{y^2}{2}\right) + C, \tag{6-10}$$

where C is a constant of integration. To evaluate C, recall the no-slip condition, which states that when $y = 0$, $v = 0$. Substituting these values in Equation 6-10 shows that $C = 0$. Thus the vertical velocity distribution in steady laminar open-channel flow is a parabola (Figure 6.4) given by

$$v = \frac{\gamma \sin \theta}{\mu} \left(Yy - \frac{y^2}{2}\right). \tag{6-11a}$$

This formula can be written alternatively as

$$v = \frac{g \sin \theta}{\nu} \left(Yy - \frac{y^2}{2}\right). \tag{6-11b}$$

From Equations 6-8 and 6-9 we see that

$$\tau = \gamma(Y - y) \sin \theta. \tag{6-12}$$

Thus at the flow surface ($y = Y$) the shear stress is zero, while at the bottom ($y = 0$) the *boundary shear stress* τ_0 is

$$\tau_0 = \gamma Y \sin \theta. \tag{6-13}$$

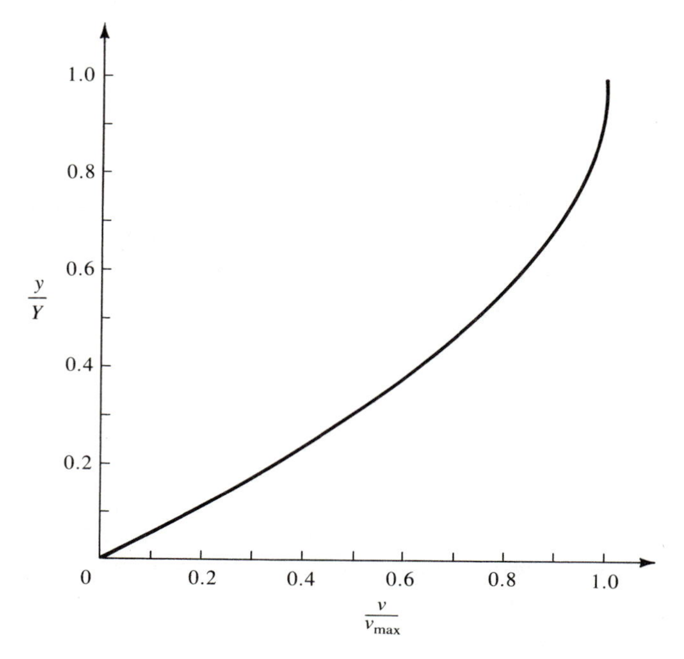

Figure 6.4
Dimensionless vertical velocity profile in a uniform laminar flow (Equation 6-11).

Average Velocity The definition of the mean value V of a variable v that is a continuous function of another variable y over the interval $y = 0$ to $y = Y$ is

$$V \equiv \frac{1}{Y} \int_0^Y v \, dy. \tag{6-14}$$

Substituting Equation 6-11a or 6-11b into 6-14 and integrating gives

$$V = \left(\frac{g \sin \theta}{3\nu}\right) Y^2 = \left(\frac{\gamma \sin \theta}{3\mu}\right) Y^2. \tag{6-15}$$

Thus we can find the average velocity of a steady laminar open-channel flow if we know the flow depth and the channel slope, along with appropriate fluid properties. Strictly speaking, Equation 6-15 applies to the average velocity at one location in a flow cross section. However, most laminar flows of hydrologic interest are wide sheet flows, in which the sides of the channel do not affect the velocity distribution significantly. Thus Equation 6-15 is commonly used to give the mean velocity in the entire cross section of a laminar flow.

For small values of θ,

$$\sin\theta \simeq \tan\theta = -\frac{dZ}{dx} = S_0, \tag{6-16}$$

where x is the horizontal direction (positive downstream) and Z is the elevation of the channel bottom above a horizontal datum. This fact has two consequences: (1) The slope tangent, often called simply the *slope* and denoted by S_0, can be substituted for $\sin\theta$ in Equations 6-11a, 6-11b, and 6-15; and (2) Equation 6-15 can be written as

$$V = -\left(\frac{Y^2}{3\nu}\right)\left(g\frac{dZ}{dx}\right) = \left(\frac{Y^2}{3\nu}\right)gS_0. \tag{6-17a}$$

Written this way, Equations 6-15 and 6-17a are clearly examples of equations of motion as discussed in Chapter 3. The slope times the gravitational acceleration provides the driving force of laminar flow, and the ''conductivity'' of the channel (the inverse of channel resistance) depends on the flow depth and viscosity. An alternative form of Equation 6-17a is

$$V = \frac{\gamma Y^2 S_0}{3\mu}. \tag{6-17b}$$

UNIFORM TURBULENT FLOW

Velocity Profile In turbulent flow the turbulent forces far outweigh viscous forces, but act to resist the flow in exactly the same way that viscosity does. This analogy led Ludwig Prandtl (1875–1953), a German hydraulician who has been called the father of modern fluid mechanics (Rouse and Ince, 1963), to formulate in 1926 an expression for turbulent flow exactly analogous to Equation 4-5 for laminar flow:

$$\tau = \varepsilon\frac{dv}{dy}. \tag{6-18}$$

In Equation 6-18, ε is the eddy viscosity, a friction within the flow that results from the vertical circulation of turbulent eddies rather than from the sliding of layers (see the discussion following Equation 5-25). Like dynamic viscosity, eddy viscosity expresses the vertical transfer of momentum, but by the mechanism discussed in Chapter 5: Slower-moving parcels of water are transferred to regions where the flow is faster, and vice versa.

We should also recall that in turbulent flows, the velocity at any point will fluctuate as a result of eddies, even if the overall flow is steady. Thus we must interpret v in Equation 6-18 and in subsequent discussions of turbulent flow as a time-averaged velocity at a point.

Although the mechanism of momentum transfer by eddies is readily grasped qualitatively, the major problem Prandtl faced in analyzing turbulent flows was that of evaluating ε. Unlike μ, it is not simply a property of the fluid, but instead depends on the characteristics of the flow and even varies from place to place within a given flow. Prandtl was the first to formulate an expression for ε that led to a satisfactory development of the turbulent velocity profile:

$$\varepsilon = \rho l^2 \frac{dv}{dy}, \tag{6-19}$$

where ρ is mass density and l is a "mixing length" characterizing the eddy size. (Compare Equation 5-26.) Prandtl then made two critical assumptions: (1) The mixing length is proportional to the distance y from the channel bottom,

$$l = ky, \tag{6-20}$$

where k is the constant of proportionality; and (2) the shear stress is constant throughout the flow and equal to the boundary shear stress given by Equation 6-13 for laminar flow. Combining Equations 6-13, 6-16, 6-18, 6-19, and 6-20 gives

$$\gamma Y S_0 = \left(\rho k^2 y^2 \frac{dv}{dy}\right) \frac{dv}{dy} = \rho k^2 y^2 \left(\frac{dv}{dy}\right)^2, \tag{6-21}$$

where S_0 is the water-surface and channel slope (sine or tangent for small slopes, but usually interpreted as the tangent). Proceeding from Equation 6-21 and noting that $\gamma/\rho = g$, we obtain

$$\frac{1}{k}(gYS_0)^{1/2} = y\frac{dv}{dy}. \tag{6-22}$$

Once again, this differential equation can be solved by separating variables and integrating:

$$dv = \frac{1}{k}(gYS_0)^{1/2}\frac{dy}{y};$$

$$\int dv = \frac{1}{k}(gYS_0)^{1/2}\int \frac{dy}{y};$$

$$v = \frac{1}{k}(gYS_0)^{1/2}\ln y + C. \tag{6-23}$$

Although the no-slip condition states that $v = 0$ when $y = 0$, this boundary condition can't be imposed on Equation 6-23 because ln 0 is not defined. Instead, we say that $v = 0$ when y is a small value y_0, which then gives

$$v = \frac{1}{k}(gYS_0)^{1/2} \ln\left(\frac{y}{y_0}\right). \tag{6-24}$$

Thus, by Prandtl's formulation, the velocity distribution in turbulent flow is logarithmic, as shown in Figure 6.5. Note that the velocity gradient is smaller throughout most of the flow than in laminar flow (Figure 6.4). This is because turbulence is a much more efficient distributor of momentum than is viscosity.

The factor k has become known as *von Karman's constant* after Theodore von Karman, another German hydraulician, who derived Equation 6-24 by a somewhat different approach in 1930. Although it is not strictly constant, a number of experiments have shown that $k \simeq 0.4$, and this value is generally assumed to be appropriate for open-channel and pipe flows. Further, it can be shown (Chow, 1959) that the most general formulation of Equation 6-24 uses the hydraulic radius R (defined in Equation 5-45) rather than the depth Y. The quantity $(gRS_0)^{1/2}$ has the units of velocity, and has become known as the *friction velocity*, or *shear velocity*, V^*:

$$V^* \equiv (gRS_0)^{1/2}. \tag{6-25}$$

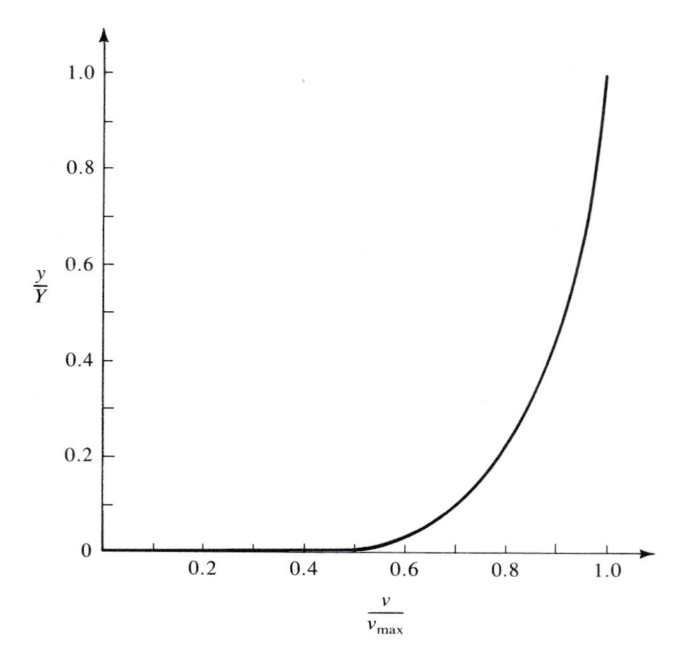

Figure 6.5
Dimensionless vertical velocity profile in a uniform turbulent flow (Equation 6-26).

Upon incorporation of this notation and with $k = 0.4$, Equation 6-24 becomes

$$v = 2.5V^* \ln\left(\frac{y}{y_0}\right), \tag{6-26}$$

and in this form is known as the *Prandtl–von Karman universal velocity-distribution law*. Numerous experiments, in both pipes and open channels, have shown that it is an excellent model for turbulent velocity profiles (see Chow, 1959, for references).

In order to apply Equation 6-26, we still have to evaluate y_0, the distance above the stream bed at which the Prandtl–von Karman analysis gives a velocity of zero. To appreciate the significance of this quantity it is helpful first to consider the physical characteristics of the flow above a smooth channel bed. We know from the no-slip property of water that the actual velocity at the bed is zero. The velocity increases with distance above the bed, so there must be a zone near the bed in which velocities and Reynolds numbers are low enough to be in the laminar range (see Figure 5.7). This zone is known as the *laminar sublayer,* and its thickness is designated y_l. Its upper boundary is indefinite, since the turbulent force gradually becomes more important, but experiments have shown that its average thickness can be computed as

$$y_l = \frac{4\nu}{V^*}. \tag{6-27}$$

As shown by Daily and Harleman (1966), the velocity profile in this zone is linear with distance above the bed, and is given by

$$v = \left(\frac{V^{*2}}{\nu}\right)y, \qquad y < y_l. \tag{6-28a}$$

From Equation 6-25, this profile can also be expressed as

$$v = \left(\frac{gRS_0}{\nu}\right)y, \qquad y < y_l, \tag{6-28b}$$

or, for sufficiently wide channels, as

$$v = \left(\frac{gYS_0}{\nu}\right)y, \qquad y < y_l. \tag{6-28c}$$

(Compare Equation 6-28c with Equation 6-11b.)

Above the laminar sublayer is a *buffer zone* in which the flow gradually becomes fully turbulent. Again, the boundaries of this zone are diffuse, but the mean position of its upper limit y_t is estimated as

$$y_t = \frac{50\nu}{V*} \tag{6-29}$$

(Daily and Harleman, 1966). The velocity profile in this zone is a gradual transition from Equation 6-28a to Equation 6-26, as indicated in Figure 6.6.

To appreciate the typical magnitudes of y_l and y_t, consider a flow with a hydraulic radius of 1 m and a slope of 0.001. From Equation 6-25, $V* = 0.099$ m s^{-1}. With a water temperature of 10°C, $\nu = 1.33 \times 10^{-6}$ m^2 s^{-1}. From Equations 6-27 and 6-29, $y_l = 5.25 \times 10^{-5}$ m and $y_t = 6.57 \times 10^{-4}$ m. Thus both the laminar sublayer and the buffer zone are typically extremely thin. Fully developed turbulence is present in well over 99% of the vertical flow profile in this example.

Experiments have shown that the value of y_0 for channels with smooth boundaries can be computed as

$$y_0 = \frac{0.11\nu}{V*} \tag{6-30}$$

(Chow, 1959). Clearly, y_0 is much smaller than y_l and falls well outside the zone for which Equation 6-26 describes the actual velocity profile. Thus we see that y_0 itself has no physical significance, but is simply a curve-fitting parameter that adjusts the position of the computed logarithmic profile in the zone above y_t. Figure 6.6 is a logarithmic plot of the dimensionless turbulent velocity profile in a smooth channel, summarizing the relationships among the physical flow zones defined by y_l and y_t, y_0, Equations 6-26 and 6-28, and the actual velocity profile.

Although the preceding discussion of velocity profiles above smooth channel beds is instructive, many natural channels are far from smooth. Instead, they are characterized by two scales of irregularities: (1) individual roughness elements such as particles of sand or gravel, which give rise to *skin friction;* and (2) general irregularities due to the presence of ripples, dunes, and areas of scour and deposition, which give rise to *form drag*. For the moment, we consider only the first type, such as would characterize a stream flowing over a generally flat bed of sand or gravel. (The second type is discussed in Chapter 8.)

Individual roughness elements on a stream bed can be described in terms of their average (or median) diameter, their shape, and their average spacing. All these factors influence the effect of the elements on the velocity profile, but it has not proved possible to treat each of them quantitatively. Since the most important factor is size, it is customary to use an *effective roughness height* k_s to characterize the roughness elements. This height is approximately equal to the diameter of the grains, with empirical adjustments to account for the effects of shape and spacing.

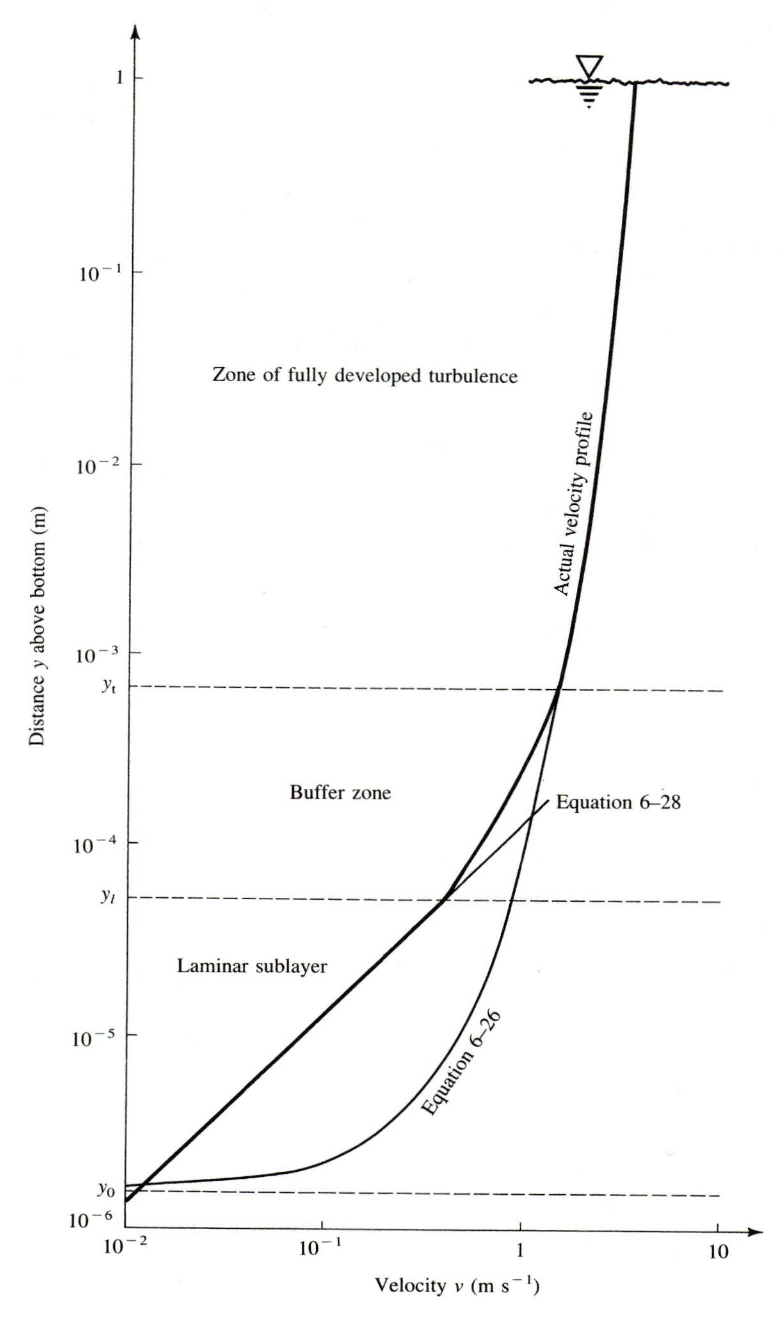

Figure 6.6
Vertical velocity profile in a smooth uniform turbulent flow with $Y = 1$ m, $S_0 = 0.001$, and $\nu = 1.3 \times 10^{-6}$ m^2 s^{-1}. The relations among the laminar sublayer, buffer zone, and zone of full turbulence are shown, along with the actual (heavy shading) and computed (by Equations 6-26 and 6-28) velocity distributions. Note logarithmic scales.

Experiments have shown that if k_s is less than the laminar sublayer thickness y_l, the laminar-flow zone near the bed is not significantly disrupted. Such flows are *hydraulically smooth* (Daily and Harleman, 1966). Figure 6.7 is a plot of Equation 6-27, which relates y_l to RS_0 for a fixed value of ν (1.3×10^{-6} m^2 s^{-1}). Approximately, if the effective roughness height for a given flow lies below the curve, the flow is hydraulically smooth and Equation 6-30 is used to compute y_0. If the roughness height lies above the curve, the flow is *hydraulically rough*. In this case, the laminar sublayer and the buffer zone effectively disappear and the fully turbulent flow profile exists at all levels above the tops of the roughness elements. For hydraulically rough flow y_0 is computed as

$$y_0 = 0.033k_s. \tag{6-31}$$

In summary, the Prandtl–von Karman turbulent velocity profile takes the following forms: For hydraulically smooth flow ($k_s < 4\nu/V^*$),

$$v = 2.5V^* \ln\left(\frac{9.00V^*y}{\nu}\right), \qquad y > y_t; \tag{6-32}$$

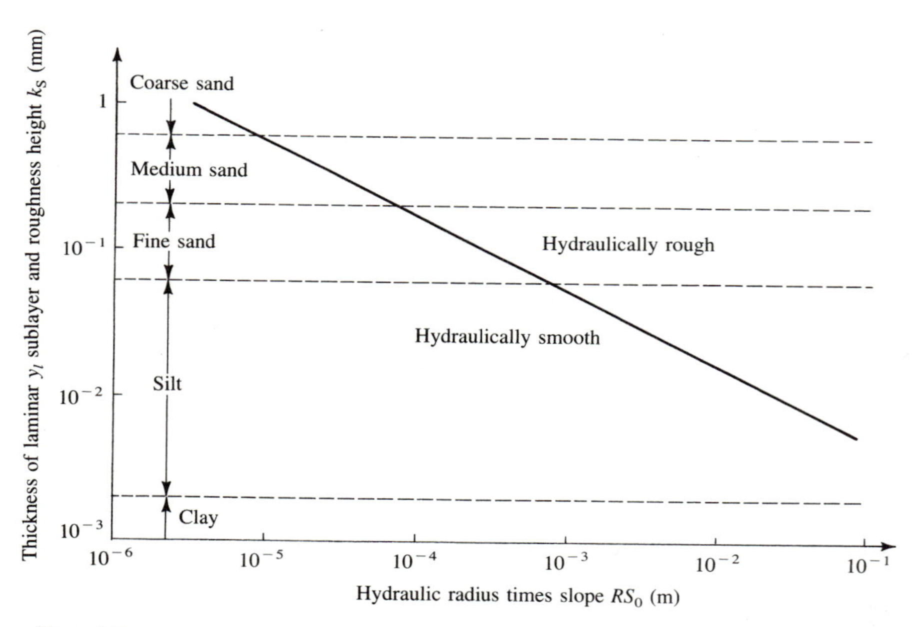

Figure 6.7
Plot of Equation 6-27. The diagonal line gives the thickness y_l of the laminar sublayer as a function of RS_0 for $\nu = 1.33 \times 10^{-6}$ m^2 s^{-1}. If the roughness height k_s exceeds y_l for a given flow, the flow is hydraulically rough.

and for hydraulically rough flow ($k_s > 4\nu/V^*$),

$$v = 2.5V^* \ln\left(\frac{30.0y}{k_s}\right), \qquad y > k_s. \tag{6-33}$$

Average Velocity at a Point in a Cross Section Just as for laminar flow, the mean velocity in a vertical profile can be computed by substituting the equation for the vertical variation of velocity (Equation 6-25) into Equation 6-14, except that the integration for turbulent flow must be done from $y = y_0$ to $y = Y$. (The deviation of the actual profile from the computed logarithmic profile, as illustrated in Figure 6.6, is ignored because it has little effect on the mean velocity.) Thus

$$V = \frac{1}{Y - y_0} \int_{y_0}^{Y} 2.5V^* \ln\left(\frac{y}{y_0}\right) dy. \tag{6-34}$$

Using the facts that

$$\ln\left(\frac{y}{y_0}\right) = \ln y - \ln y_0$$

and

$$\int \ln y \, dy = y \ln y - y,$$

we can evaluate Equation 6-34 as

$$V = \frac{2.5V^*}{Y - y_0}(Y \ln Y - Y + y_0 - Y \ln y_0). \tag{6-35}$$

With the reasonable assumption that y_0 is of negligible magnitude, Equation 6-35 becomes

$$V = 2.5V^*\left[\ln\left(\frac{Y}{y_0}\right) - 1\right]. \tag{6-36}$$

This equation for the mean velocity in a vertical profile can be used to solve a practical measurement problem. Much of the science of hydrology depends on the measurement of discharge at selected stream cross sections. Recall that discharge Q is given as

$$Q = VA = VwY, \tag{6-37}$$

where V is the mean velocity in the cross section, Y is the mean (hydraulic) depth in the cross section (Equation 5-45), and w is the width of the water surface at the cross section. The method generally used for determining discharge in natural streams is to measure w directly and to determine V and Y from a series of velocity and depth measurements across the stream. This is known as the *velocity–area method* (Figure 6.8). Detailed discussions of this and other stream-gaging methods are given in Corbett (1945), Boyer (1964), and Buchanan and Somers (1969). Since depth varies only with distance across the channel, Y can be calculated as simply the average of the several individual depth measurements. Velocity, however, varies both with distance across the channel and with depth at any point in the cross section, and so we have to find V by finding the average velocities in a series of vertical profiles across the stream and then taking the average of the vertical averages.

Figure 6.8
Stream gaging by the velocity–area method. The current meter is attached to the rod held by the observer and is adjusted to indicate the velocity at a distance above the bottom equal to 0.4 times the depth.

In order to determine the average velocity at each vertical profile, we take advantage of the fact that, at some level, the actual velocity must equal the average velocity V. If this average occurs at a level $y = bY$ above the bottom, then Equation 6-26 gives

$$V = 2.5V^* \ln\left(\frac{bY}{y_0}\right). \qquad (6\text{-}38)$$

Equation 6-38 can be equated to Equation 6-36 and solved to give

$$\ln\left(\frac{1}{b}\right) = 1,$$

$$b = \frac{1}{e} = 0.37. \qquad (6\text{-}39)$$

Thus the velocity measured at a distance of $0.37Y$ above the bottom is equal to the average velocity in the vertical profile. This conclusion is the basis for the U.S. Geological Survey's practice of making velocity measurements at "six-tenths depth," that is, 0.6 of the depth measured from the water surface downward, which is 0.4 ($\simeq$0.37) of the distance above the bottom.

Average Velocity in a Cross Section Equation 6-36 gives the mean velocity in a vertical profile for known values of V^*, Y, and y_0. If we were concerned only with sheet flows, which have widths many orders of magnitude greater than depths, that equation could be used to compute the mean velocity at any cross section. For streams, however, the frictional effects of the channel sides as well as of the bed are generally present throughout the cross section, and Equation 6-36 cannot be used. Ideally, we would like to formulate an equation accounting for the effects of the sides and giving the variation of velocity with width as well as with depth, and integrate this equation to give the average velocity for the cross section. This has not proved possible, however, because of the variable and complex shapes of natural channels. Thus it has been necessary to resort to empirical equations that relate the mean velocity for the cross section to the hydraulic radius of the channel, the frictional resistance that is transmitted from the channel via turbulent eddies (or the inverse of that resistance, the channel conductivity), and the slope of the water surface. These equations take the form

$$V = u\left(\frac{1}{\Omega}\right)R^{\alpha}S_0^{\beta}, \qquad (6\text{-}40)$$

where V is the mean velocity in the cross section ($[LT^{-1}]$), Ω is the channel resistance (treated as a dimensionless number), R is the hydraulic radius ($[L]$), and

S_0 is the slope ([1]). The constant u depends on the units in which V and R are measured, and is necessary because the equation is dimensionally inhomogeneous (see Equations 3-32 and 3-33).

The first equation of this type was developed by a French engineer, Antoine Chézy (1718–1798), in 1769. His development began with the consideration of a channel carrying a uniform turbulent flow (Figure 6.9). The basic assumption in his derivation was that the shear stress resisting the flow, τ_u, is proportional to the square of the mean velocity V:

$$\tau_u = K_T \rho V^2, \tag{6-41}$$

where K_T is the constant of proportionality (dimensionless) and ρ is mass density. The shear stress τ_u is due to the turbulent force, and Equation 6-41 looks very much like Equation 5-27. This stress acts over an area equal to the product of the wetted perimeter P and the reach length X. Thus the up-channel resistance force F_u is

$$F_u = \tau_u PX = K_T \rho V^2 PX. \tag{6-42}$$

The downstream component of the weight (force) of the flowing water, F_d, is

$$F_d = \gamma AX \sin \theta, \tag{6-43}$$

where A is the cross-sectional area. Since in uniform flow there is no acceleration, $F_u = F_d$, and

$$K_T \rho V^2 PX = \gamma AX \sin \theta. \tag{6-44}$$

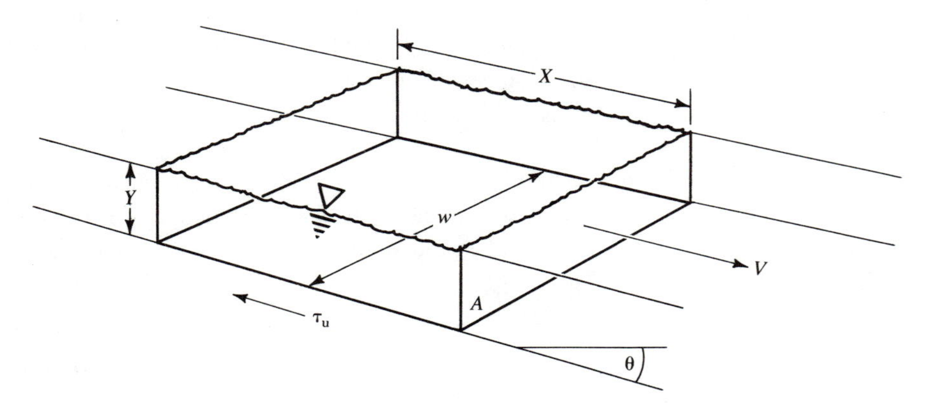

Figure 6.9
Definitions of the terms used in deriving the Chézy equation (Equation 6-48).

Writing S_0 for sin θ, we can solve Equation 6-44 for V:

$$V = \left(\frac{\gamma R S_0}{K_T \rho}\right)^{1/2} = \left(\frac{g R S_0}{K_T}\right)^{1/2}, \tag{6-45}$$

where R is the hydraulic radius A/P. Defining

$$C \equiv \left(\frac{g}{K_T}\right)^{1/2}, \tag{6-46}$$

we obtain from Equations 6-45 and 6-46

$$V = C R^{1/2} S_0^{1/2}, \tag{6-47}$$

which is called *Chézy's equation.*

From this development it can be seen that C (often called *Chézy's C*) expresses the inverse of the resistance of the channel to the flow and appears to have the dimensions $[L^{1/2}T^{-1}]$. In practice, however, C is usually treated as a dimensionless number characteristic of a particular channel type and size. Thus it is best to write the Chézy equation as

$$V = u_C C R^{1/2} S_0^{1/2}, \tag{6-48}$$

where u_C is a constant which has dimensions $[L^{1/2}T^{-1}]$ and which depends on the system of units in which V and R are measured. When V is in feet per second and R is in feet, $u_C = 1$; when V is in meters per second and R is in meters, $u_C = 0.552$. Chow (1959) gives ways for computing C for channels of various types.

The Chézy equation has a quasi-theoretical basis, as seen in the preceding development. However, a similar formulation based on experiments published in 1889 by the Irish engineer Robert Manning (1816–1897) is often used in its place:

$$V = u_M \left(\frac{1}{n}\right) R^{2/3} S_0^{1/2}. \tag{6-49}$$

Equation 6-49 is known as the *Manning equation.* In it the channel resistance is expressed by n, which is called the *Manning roughness coefficient,* or simply *Manning's n.* Like Chézy's C, n is usually considered to be dimensionless (however, see Chow, 1959, for a detailed discussion of the dimensions of n). When V is in feet per second and R is in feet, $u_M = 1.49$; when V is in meters per second and R is in meters, $u_M = 1$. Methods of estimating Manning's n for channels are discussed in detail in Chapter 8.

UNIFORM TRANSITIONAL OR MIXED FLOWS

As described earlier, when the Reynolds number $\mathbb{R}$ in open channels is between 500 and 2000, the flow is transitional between laminar and turbulent. When $\mathbb{R} < 500$, eddy disturbances generated by boundary irregularities or other means tend to be damped out by the effect of viscosity. When $\mathbb{R} > 2000$, such disturbances are unstable and tend to propagate throughout the flow, resulting in fully developed turbulence. In the transitional range, some eddy disturbances are damped and some tend to be propagated, that is, turbulence may be present but not fully developed.

Nominally laminar flows ($\mathbb{R} < 500$) are called *mixed flows* when either (1) eddy-generating boundary irregularities are large relative to the depth of flow, or (2) other disturbances, such as raindrops, generate eddies that are large relative to the flow depth. These conditions often exist in the sheet flows (or *overland flows*) that are sometimes sources of streamflow during rainstorms. These flows seldom exceed 1 cm in depth and 3 cm s^{-1} in velocity (Dunne and Leopold, 1978). Hence they usually have Reynolds numbers in the transitional or laminar range, and both viscosity and turbulence exert frictional resistance to the flow.

An equation for the mean velocity in a cross section for this type of uniform mixed flow has not been developed from first principles. Instead, an empirical approach has been taken by generalizing the equations for mean velocities of laminar flow (Equation 6-15) and turbulent flows (Equation 6-48, the Chézy equation; and Equation 6-49, the Manning equation). If the flow depth and hydraulic radius are taken to be equal, the general equation is

$$V = KR^{\alpha}S_0^{\beta}, \qquad (6\text{-}50)$$

where the values of K, α, and β appropriate for laminar and turbulent flows are summarized in Table 6.1. (The symbols are as defined earlier.) Thus, it might be expected that mixed overland flows would have values of $0.5 \le \alpha \le 2$ and $0.5 \le \beta \le 1$, with K depending to varying degrees on the viscosity and on the roughness of the boundary.

Most experiments (see Horton, 1945; Linsley et al., 1949) have found or assumed β to be 0.5 for this type of flow (although Butler, 1957, suggested using values of $0.6 \le \beta \le 1$), and most attention has been directed at determining α. In fact, experiments have yielded values of α over the expected range, and even as low as zero (Horton, 1945). However, α is commonly taken as equal to one (Eagleson, 1970), and a widely accepted equation for mixed overland flow is a modified form of the Manning equation, with $\alpha = 1$:

$$V = \left(\frac{u_{\mathrm{M}}}{n}\right)YS_0^{1/2}, \qquad (6\text{-}51)$$

where Y is the mean flow depth (equal to the hydraulic radius for wide sheet flows) and the other symbols are as previously defined. Values of n for overland flows are tabulated in Chapter 8.

Table 6.1
Terms for Generalized Uniform-Flow Equation (Equation 6-50)

Term	Laminar flow	Turbulent flow Chézy	Turbulent flow Manning
K	$\frac{\gamma}{3\mu}$	$u_C C$	$\frac{u_M}{n}$
α	2	$\frac{1}{2}$	$\frac{2}{3}$
β	1	$\frac{1}{2}$	$\frac{1}{2}$

As noted earlier, Horton (1945) reported values of α as low as zero, indicating that the mean velocity is independent of the flow depth under some conditions. He found such flows on densely vegetated slopes when the flow depth did not exceed the vegetation height, and he characterized them as ''superturbulent'' or ''subdivided.'' The latter term is probably more appropriate, since it is likely that such flows are described by Darcy's law of flow in porous media, which is discussed in detail in Chapter 12. This law has the form

$$V = K_D S_0, \tag{6-52}$$

where K_D depends on the size and shape of the spaces through which the flow occurs and on the viscosity and density of the fluid. It will be shown in Chapter 12 that flows through porous media can be treated as potential flows rather than as boundary-layer flows.

REEXAMINATION OF FORCE BALANCES, UNIFORM-FLOW FORMULAS, AND THE REYNOLDS AND FROUDE NUMBERS

The developments of formulas relating mean velocity to channel gradient and mean depth for laminar (Equation 6-17) and turbulent (Equation 6-47) uniform flows provide a basis for further insight into the expressions for turbulent, viscous, and gravitational forces developed in Chapter 5 (summarized in Table 5.1). These developments also lead to an instructive reexamination of the significance of the Reynolds and Froude numbers.

Actual Viscous Force In uniform flows the downslope component of gravity, $g \sin \theta$, is balanced by viscous or turbulent resisting forces. In laminar flow the viscous force dominates, and we saw in Equation 5-11 that this force is proportional to $\nu V/Y^2$. Writing k_v as the proportionality constant and equating the gravitational and viscous forces yields

$$g \sin \theta = \frac{k_v \nu V}{Y^2}. \qquad (6\text{-}53)$$

Assuming that $\sin \theta = S_0$, as before, and solving for V then gives

$$V = \left(\frac{1}{k_v \nu}\right) g S_0 Y^2. \qquad (6\text{-}54)$$

Comparing Equation 6-54 with Equation 6-17, which was derived from a microscopic rather than a macroscopic treatment of the balance of forces, we see that

$$k_v = 3. \qquad (6\text{-}55)$$

Thus the expression for the *actual* viscous force per unit mass, F_v/M, is

$$\frac{F_v}{M} = \frac{3\nu V}{Y^2}. \qquad (6\text{-}56)$$

Actual Turbulent Force An analogous development can be applied to the turbulent force. Using k_t for the proportionality factor, we obtain the turbulent-flow analog to Equation 6-53:

$$g S_0 = k_t \left(\frac{V^2}{Y}\right). \qquad (6\text{-}57)$$

Again solving for V yields

$$V = \left(\frac{g}{k_t}\right)^{1/2} Y^{1/2} S_0^{\,1/2}. \qquad (6\text{-}58)$$

Comparing this expression with the Chézy equation, Equation 6-48 (using mean depth rather than hydraulic radius) shows that

$$\left(\frac{g}{k_t}\right)^{1/2} = u_C C, \qquad (6\text{-}59)$$

where C is the Chézy resistance coefficient. Thus k_t is proportional to the channel reach resistance, and is in fact equal to K_T, defined in the development of the Chézy equation (Equation 6-46). Since Equation 6-59 can be rearranged to show that

$$k_t = \frac{g}{u_C^2 C^2}, \qquad (6\text{-}60)$$

it is clear from Equation 6-57 that the expression for the actual turbulent force per unit mass, F_t/M, is

$$\frac{F_t}{M} = \frac{g}{u_C^2 C^2}\left(\frac{V^2}{Y}\right). \qquad (6\text{-}61)$$

Since C is inversely proportional to the reach resistance, Equation 6-61 states that the turbulent force in a flow is proportional to the frictional resistance offered by the channel.

Reexamination of the Reynolds Number The Reynolds number is traditionally defined as being *proportional* to the ratio of turbulent to viscous forces (Equation 5-44). Equations 6-56 and 6-61 can now be used to formulate the *actual* ratio of turbulent to viscous force, which we designate $\mathbb{R}^*$.

$$\mathbb{R}^* \equiv \frac{F_t/M}{F_v/M} = \frac{Y^2 g V^2}{3\nu V u_C^2 C^2 Y} = \frac{gVY}{3u_C{}^2 \nu C^2}. \qquad (6\text{-}62)$$

The ratio $\mathbb{R}^*$ can be thought of as the "actual" Reynolds number for open-channel flows. Although it is not traditionally used in characterizations of such flows, it does clearly show that the variability in the critical Reynolds number for the laminar–turbulent transition is due largely to variation in channel resistance, which is reflected in the value of C.

As discussed in Chapter 5, experience has shown that the turbulent force becomes dominant at about $\mathbb{R} = 500$. We might therefore assume that at this value the turbulent force is some 500 times greater than the viscous force. Physical intuition, however, suggests rather that these forces should be more nearly equal at the transition from laminar to turbulent flow. The following reanalysis with the expressions incorporating the actual force magnitudes shows that this intuition is in fact correct.

From Equations 5-44 and 6-62,

$$\mathbb{R}^* = \left(\frac{g}{3u_C^2 C^2}\right)\mathbb{R}. \tag{6-63}$$

Substituting $\mathbb{R} = 500$ into Equation 6-63, we see that

$$\mathbb{R}^* = \frac{5370}{C^2}. \tag{6-64}$$

Values of C range between about 6 and 90 for natural channels. Thus the critical value of $\mathbb{R}^*$ ranges from 0.663 ($C = 90$) to 149 ($C = 6$), and equals exactly 1 when $C = 73.3$. Since C for many natural channels is close to this value, we see that a value of $\mathbb{R} = 500$ generally does in fact correspond to a near equality of the turbulent and viscous forces in open-channel flows.

Reexamination of the Froude Number We saw in Chapter 5 that one definition of the Froude number $\mathbb{F}$ is that it is *proportional* to the square root of the ratio of the turbulent to the gravitational force (Equations 5-48, 5-50). However, that formulation did not employ the expressions for the *actual* forces per unit mass, as written in Equation 6-57. Using that equation, we now define the "actual" Froude number $\mathbb{F}^*$ as

$$\mathbb{F}^* \equiv \left(\frac{k_t V^2}{g S_0 Y}\right)^{1/2} \tag{6-65}$$

or, using Equation 6-60, as

$$\mathbb{F}^* = \left(\frac{g V^2}{u_C^2 C^2 g S_0 Y}\right)^{1/2} = \left(\frac{V^2}{u_C{}^2 C^2 S_0 Y}\right)^{1/2}. \tag{6-66}$$

The relation between the actual and traditional Froude numbers is thus

$$\mathbb{F}^* = \left(\frac{g}{u_C^2 C^2 S_0}\right)^{1/2} \mathbb{F}. \tag{6-67}$$

(Note the similarity in the factors relating the starred and unstarred versions of $\mathbb{F}$ and $\mathbb{R}$ in Equations 6-63 and 6-67.)

Recall that when $\mathbb{F} = 1$, flow is critical, and the discussion in Chapter 5 implied that turbulent and gravity forces are balanced at that point. Now, however, we know that these forces are balanced in *all* uniform turbulent flows, regardless of Froude number. This means that $\mathbb{F}^* = 1$ for all such flows. If this is true, then Equation 6-67 can be rearranged to give an alternative expression for $\mathbb{F}$:

$$\mathbb{F} = \left(\frac{u_C^2 C^2 S_0}{g}\right)^{1/2}. \tag{6-68}$$

Thus to the extent that C and S_0 are constant for any channel reach, the Froude number for that reach is also fixed, and will not change as a function of discharge. Equation 6-67 thus implies that

$$\mathbb{F} = 1 \quad \text{when} \quad u_C^2 C^2 S_0 = g, \tag{6-69a}$$

$$\mathbb{F} < 1 \quad \text{when} \quad u_C^2 C^2 S_0 < g, \tag{6-69b}$$

$$\mathbb{F} > 1 \quad \text{when} \quad u_C^2 C^2 S_0 > g. \tag{6-69c}$$

Expressions 6-69a–c are in fact close to being true for most channels. In Chapter 8, however, we will see that C increases as the one-sixth power of the hydraulic radius, and since hydraulic radius increases with discharge, the Froude number will tend to increase slowly as discharge increases in a given reach (see Exercise 9-1). Of course, deviations from these general statements will occur if there are significant changes in C or S_0 due to changes in channel shape resulting from erosion and deposition accompanying discharge fluctuations.

Exercises

6-1. Suppose you want to measure the average velocity in a channel, but have no equipment except a watch and measuring tape. Using these items, you measure the time it takes for small sticks to float through a measured distance. What is the general relation between the surface velocity and the mean velocity? (*Hint:* Compare Equations 6-26 and 6-36.)

6-2. Show that in a uniform turbulent flow the friction velocity is equal to the actual velocity at a distance of $1.49y_0$ above the channel bottom.

6-3. When flow depths exceed 2 ft, the U.S. Geological Survey recommends determining the mean velocity by making two current-meter measurements at each position in the cross section, one at 0.2 and the other at 0.8 of the depth, measured upward from the bottom. These two values are then averaged to give the average for the vertical. Show that for a uniform turbulent flow this procedure is mathematically equivalent to a single measurement at 0.4 of the depth measured from the bottom.

6-4. Compute the thickness of the laminar sublayer for the following flows. The temperature of the water is 10°C.

	Flow				
	1	2	3	4	5
R (m)	0.10	0.10	1.00	10.00	2.00
S_0	0.00200	0.00800	0.00100	0.00100	0.00600

6-5. Determine whether the flow is rough or smooth for the following uniform turbulent flows in wide channels. The water temperature is 10°C.

	Flow				
	1	2	3	4	5
$R = Y$ (m)	1.00	2.00	0.30	0.10	5.00
S_0	0.00800	0.00400	0.0100	0.00100	0.00600
k_s (m)	0.01	0.05	0.05	0.01	0.50

6-6. Compute and plot the vertical velocity profiles for the flows of Exercise 6-5. Use log–log paper, and plot at least ten points, spaced so that the profile is well defined. Show the laminar sublayer and the top of the zone of transition to full turbulence. Compute the mean velocity in each profile and indicate the depth at which it occurs.

6-7. Using the Manning equation, compute the mean velocity in stream reaches with the following characteristics. Assume that the flow is uniform and turbulent, and determine whether it is subcritical or supercritical.

	Flow				
	1	2	3	4	5
R (m)	1.00	2.00	0.30	0.10	5.00
S_0	0.00800	0.00500	0.0100	0.00100	0.00600
n	0.070	0.028	0.040	0.400	0.030

6-8. A rating curve is a relationship between the elevation (Z) of the water surface above an arbitrary datum and the discharge at a cross section. For the cross section in Exercise 5-8, assume a constant slope, $S_0 = 0.00600$, and constant roughness, $n = 0.045$. Assuming uniform turbulent flow, compute a rating curve for this section with the datum (i.e., the level at which $Z = 0$) at the lowest point in the section. Compute at least eight well-spaced points up to bank-full gage height ($Z = 4.1$ m).

6-9. Write Equation 6-49 for discharge, and Equations 6-17 and 6-51 for discharge per unit width.

CHAPTER 7

Mechanical Energy in Open-Channel Flows

BASIC CONCEPTS

The analysis of hydrostatics in Chapter 5 led to Equation 5-12, an expression for the total potential energy in a fluid at equilibrium. We now want to modify this expression so that it applies to open-channel flows, and to develop concepts of total mechanical energy (i.e., potential energy plus kinetic energy) and specific energy in such flows.

Figure 7.1 shows a column of water extending from a sloping surface to a streamline at a depth y below the surface. For simplicity, the channel is considered to be rectangular, and has a width w and a downstream extent x. (Note that Figure 7.1 is identical to Figure 6.2 except that y is now measured in the opposite direction.) The expressions for gravitational potential energy E_{pg} and elevation head H_{pg} are the same as they would be for hydrostatic conditions:

$$E_{pg} = \gamma z w \, dx, \tag{7-1}$$

$$H_{pg} = z, \tag{7-2}$$

where dx is the dimension of the column in the downstream direction. (Compare Equations 5-7 and 5-13.)

From Figure 6.3 it is clear that the component of the weight of the column of water above the streamline that is normal to the cross section, F_n, is

$$F_n = \gamma y \cos\theta \, dx \, w, \tag{7-3}$$

where θ is the slope. Thus the pressure P_n normal to the streamline is

$$P_n = \frac{F_n}{w \, dx} = \gamma y \cos\theta. \tag{7-4}$$

By analogy with the development for hydrostatic conditions (Equations 5-10–5-13) the pressure head H_{pp} at the streamline is

$$H_{pp} = y \cos\theta, \tag{7-5}$$

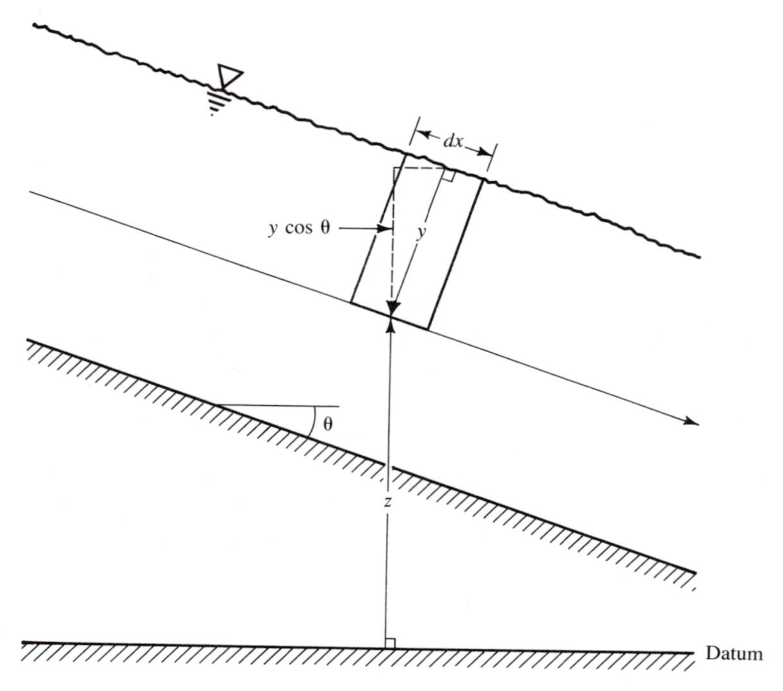

Figure 7.1
Potential head relations for a streamline in an open-channel flow.

which is the vertical projection of the depth y measured at right angles to the slope.

Adding Equations 7-2 and 7-5 gives the total potential head (or piezometric head) H_p at a streamline in an open-channel flow:

$$H_p = z + y \cos \theta, \tag{7-6}$$

where z is the vertical distance above a datum and y is the distance below the surface measured normal to the flow direction. Note that Equation 5-13 is a special case of Equation 7-6 in which $\theta = 0°$ and $\cos \theta = 1$.

An important implication of Equation 7-6 is that potential head does not change along a line drawn at right angles to the streamlines. However, from the discussion of equations of motion in Chapter 3 we know that flow occurs if and only if there is a gradient of potential energy, and that the flow is in the direction from higher to lower potential energy. This means that potential energy along a streamline must decrease downstream in any flow, and obviously does so in Figure 7.1. The first law of thermodynamics states that energy is conserved; it cannot be created or destroyed, but only converted to other forms. The downstream decrease in potential energy in a flow represents conversion of potential energy to the kinetic energy of the flow and any entrained sediment. This kinetic energy is ultimately transformed to heat energy due to the friction between parcels of water moving at different velocities, that is, between eddies in turbulent flows and between adjacent laminae in laminar flows.

To formulate the expression for kinetic energy, consider the motion of a parcel of water of volume B along a streamline from point x_1 to point x_2 in the steady flow shown in Figure 7.2. If we assume for purposes of this derivation a frictionless channel, the work dW expended in this motion over the infinitesimal distance dx is related to the net force F on the parcel as

$$dW = F\,dx \tag{7-7}$$

(recall that work equals energy equals force times distance). Thus the energy W expended between x_1 and x_2 is

$$W = \int_{x_1}^{x_2} F\,dx. \tag{7-8}$$

But

$$F = ma = \rho B a, \tag{7-9}$$

where a is the acceleration of the parcel. Total acceleration is given by Equation 6-5, but since the flow in Figure 7.2 is steady, the local acceleration is zero and

$$a = v\frac{dv}{dx}, \tag{7-10}$$

where v is the velocity on the streamline.

Equations 7-9 and 7-10 can now be substituted into Equation 7-8 to give

$$W = \int_{x_1}^{x_2} \rho B v \frac{dv}{dx}\, dx = \rho B \int_{v_1}^{v_2} v\, dv, \tag{7-11}$$

where v_1 represents the velocity at x_1 and v_2 the velocity at x_2. Integrating Equation 7-11, we obtain

$$W = \rho B\left(\frac{v_2^2}{2}\right) - \rho B\left(\frac{v_1^2}{2}\right). \tag{7-12}$$

Equation 7-12 represents the change in kinetic energy between x_1 and x_2; thus the

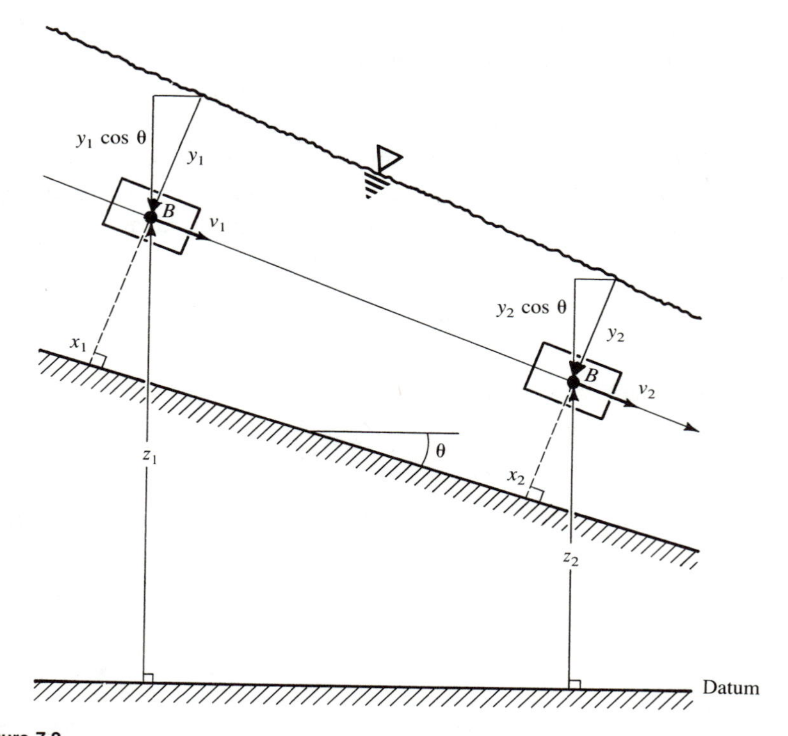

Figure 7.2
Definitions of terms for developing the expression for kinetic energy at a point on a streamline.

two terms on the right side must represent the kinetic energy at x_1 and x_2, respectively. Therefore we can assert that the kinetic energy E_v at any point in the flow is

$$E_v = \rho B\left(\frac{v^2}{2}\right). \tag{7-13}$$

A kinetic-energy head H_v can be defined analogously to potential head by dividing E_v by the weight of the parcel:

$$H_v \equiv \frac{E_v}{\gamma B} = \frac{v^2}{2g}. \tag{7-14}$$

The quantity H_v is commonly called the *velocity head,* and of course has the dimensions $[L]$. Combining the foregoing analyses, we see that the total mechanical energy at any point along a streamline is represented by the total head H_y, which is the sum of the potential head (Equation 7-6) and the velocity head (Equation 7-14):

$$H_y = z + y \cos \theta + \frac{v^2}{2g}. \tag{7-15}$$

Equation 7-15 is central to the analysis of fluid flow, and will be invoked repeatedly in subsequent chapters. It is often called the *Bernoulli equation* in honor of the Swiss mathematician and physicist Daniel Bernoulli (1700–1782). Interestingly, although Bernoulli introduced the fundamental concept of head, the equation that bears his name was actually first derived by his colleague Leonhard Euler (1707–1783; Rouse and Ince, 1963).

In practical problems, we would like to be able to characterize the total head over the entire flow depth, not for just a single streamline. Doing so for potential head presents no problem, for we have seen that $z + y \cos \theta$ is constant at right angles to the streamlines, and thus is constant over any cross section. Therefore the total potential energy at a cross section can be conveniently expressed by substituting the mean depth Y into Equation 7-15 and taking z equal to the elevation, designated Z, of the channel bottom above a datum. From Chapter 6, however, it is clear that velocity changes nonlinearly from zero at the bottom to a maximum value at the surface, so velocity head cannot be constant for all streamlines at a cross section. To account for this variation in velocity, a *kinetic-energy factor,* or *energy coefficient, α,* is introduced as a multiplier. With these modifications the final expression for the total head H at a cross section in an open-channel flow becomes

$$H = Z + Y \cos \theta + \frac{\alpha V^2}{2g}, \tag{7-16}$$

where V is the mean velocity at the cross section.

Chow (1959) reports $1.15 \leq \alpha \leq 1.50$ for natural streams, with an average value of 1.30, and discusses ways of estimating α. However, changes in velocity head are usually much smaller than changes in potential head in natural flows, so we will not be much concerned here with the evaluation of α.

DOWNSTREAM CHANGES IN ENERGY: THE ENERGY EQUATION

While the first law of thermodynamics is the principle of energy conservation, the second law states that any conversion of energy from one form to another inevitably includes an irreversible conversion of at least some of the energy to heat. Thus the conversion of potential energy to kinetic energy in a flow is accompanied by a downstream decrease in the total mechanical energy of the flow, and the total head must therefore decrease from an upstream section to one farther downstream. The difference in total head between any two sections represents mechanical energy that is converted to heat due to friction and "lost"; it is manifested as a minute temperature rise of the water, and cannot be reconverted to mechanical energy.

Equation 7-16 and the second law of thermodynamics combine to give

$$Z_1 + Y_1 \cos \theta + \frac{\alpha_1 V_1^2}{2g} = Z_2 + Y_2 \cos \theta + \frac{\alpha_2 V_2^2}{2g} + h_\Delta, \qquad (7\text{-}17)$$

where the subscripts 1 and 2 refer to the upstream and downstream sections, respectively, and h_Δ is the head loss between the two sections (see Figure 7.3). Equation 7-17 is the *one-dimensional energy equation,* a fundamental relation that is used to solve many practical problems. It applies to laminar, transitional, and turbulent flows and to flows that are subcritical, critical, and supercritical. In the derivation we assumed that the flow was steady, but allowed the depth and velocity to change downstream. It was tacitly assumed, however, that the channel slope θ did not change from one section to the next, and that all streamlines at a cross section were parallel. These assumptions will be strained if the flow is rapidly varied, so Equation 7-17 should be applied only to flows that can be considered steady and gradually varied.

The expression of energy in terms of head allows construction of diagrams like Figure 7.3, where each component of Equation 7-17 is represented by a vertical distance. In contrast to the hydrostatic conditions illustrated in Figure 5.6, the piezometric head line slopes downstream and lies some distance below the water surface. The slope of the line represents the gradient of potential energy that causes the flow to occur. It should be noted that the slope of most natural streams is less than 5°, so $\cos \theta$ will be larger than 0.996 and the piezometric head line will coincide with the water surface for all practical purposes.

The *energy grade line* shown in Figure 7.3 represents the total mechanical energy at each point in the flow. Its slope, the *energy slope* S_e, is proportional to the rate at which energy is dissipated as heat due to the frictional resistance to the flow. In uniform flows the energy slope is equal to the channel slope, which we denote by S_0, so that

$$S_0 \equiv \tan \theta = -\frac{dZ}{dx}, \tag{7-18}$$

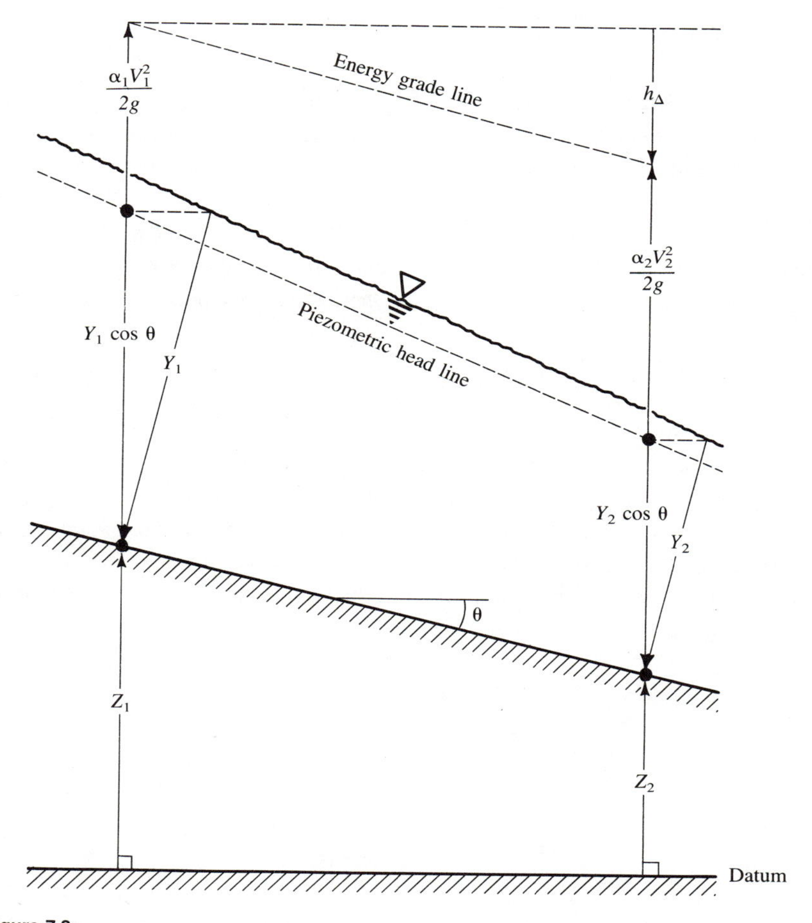

Figure 7.3
Graphical representation of terms of the one-dimensional energy equation for steady gradually varied flow (Equation 7-17).

where x is the distance downstream. Since the Manning equation applies to uniform flows, it can be solved to give S_e for such flows:

$$S_e = \frac{n^2 V^2}{u_M^2 R^{4/3}}. \tag{7-19}$$

For varied flow, $S_e \neq S_0$. If the flow is gradually varied, however, it is assumed that Equation 7-19 applies at any cross section. Extensive use of this approximation will be made in estimating water-surface profiles in Chapter 9.

SPECIFIC ENERGY

The concept of specific energy is the starting point for computing water-surface profiles in flows that can be considered steady and gradually varied. It also provides further insight into the distinction between subcritical, critical, and supercritical flows.

The specific energy of a flow is its mechanical energy measured with respect to the channel bottom rather than to a horizontal datum. Thus the elevation-head term of Equation 7-16 disappears and the *specific head* H_s at any point is given by

$$H_s \equiv Y \cos \theta + \frac{\alpha V^2}{2g}. \tag{7-20}$$

Note that because of the elimination of one component of total potential energy, Equation 7-20 is no longer an expression of the conservation of energy. Thus specific head may increase or decrease downstream, and the relative magnitudes of the two components of specific head can vary as we move downstream.

If we consider a cross section in a flow of constant discharge Q and width w we can use the fact that

$$Q = VwY \tag{7-21}$$

to rewrite Equation 7-20 as

$$H_s = Y \cos \theta + \frac{\alpha Q^2}{2gw^2 Y^2}. \tag{7-22}$$

With Q and w constant, Equation 7-22 shows that specific head depends only on flow depth. However, since it contains both Y and Y^2 it can be solved by two different values of Y. The lower limit for solutions is given by $Y \rightarrow 0$ (when velocity approaches a maximum for a given discharge and width), and the upper limit by $Y \rightarrow H_s$ (when velocity approaches zero). Furthermore, there is a particular value $Y = Y_c$ for which H_s reaches a minimum value. Thus a graph of Equation 7-22 looks like Figure 7.4: For all values of H_s greater than the minimum there are two values of Y corresponding to each value of H_s; the upper limb is asymptotic to the line $Y = H_s$, and the lower limb is asymptotic to the line $Y = 0$.

The significance of the two alternate depths will become apparent after we examine the value of Y_c. To do this, we find the value of Y when H_s takes on its minimum value, using the standard technique of calculus. (For simplicity, we assume that cos $\theta = 1$ and $\alpha = 1$ in this development.) When H_s is a minimum, dH_s/dY is zero, so

$$\frac{dH_s}{dY} = 1 - \frac{2Q^2}{2gw^2Y_c^3} = 0$$

and

$$Y_c^3 = \frac{Q^2}{gw^2}. \tag{7-23}$$

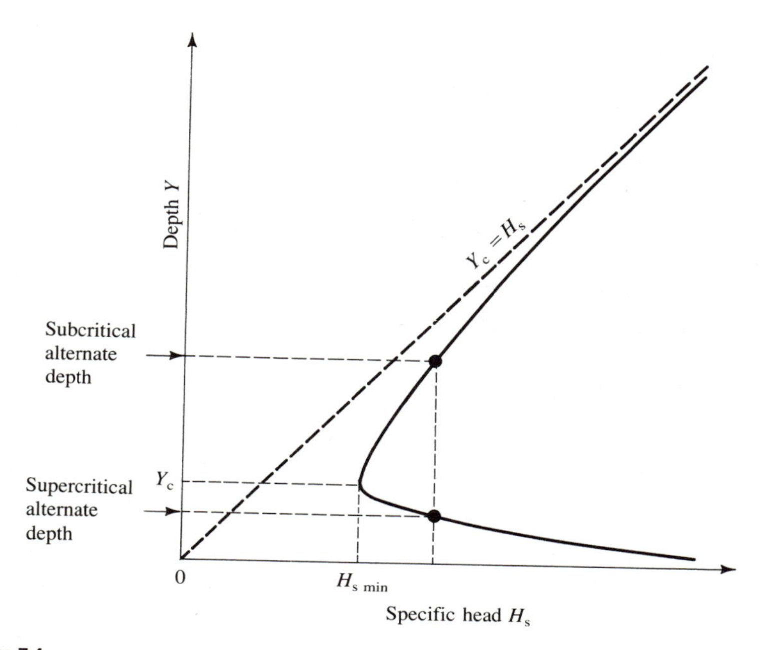

Figure 7.4
General form of a specific-head diagram for a particular discharge, showing one pair of alternate depths corresponding to a particular relation between velocity and depth.

Upon substitution of Equation 7-21, Equation 7-23 becomes

$$Y_c^3 = \frac{V^2 w^2 Y_c^2}{g w^2};$$

$$Y_c = \frac{V^2}{g}. \tag{7-24}$$

Equation 7-24 shows that when specific head is at its minimum value, the depth (pressure head) is twice the velocity head. Furthermore, recalling the definition of the Froude number $\mathbb{F}$ (Equation 5-50), we can manipulate Equation 7-24 to show that

$$1 = \frac{V^2}{gY_c} = \mathbb{F}^2. \tag{7-25}$$

This equation shows that when a flow has the minimum specific head, the Froude number equals one and the flow is therefore critical. The value of Y_c is thus called the *critical depth*. When specific head exceeds its minimum value, the depth $Y > Y_c$ is the depth of a subcritical flow, while the alternate depth $Y < Y_c$ is the depth of a supercritical flow.

From this and previous discussions, note that the Froude number has three interpretations: (1) It is proportional to the square root of the ratio of turbulent to gravitational forces in the flow (Equation 5-48); (2) it is the ratio of the mean flow velocity to the celerity of a gravity wave (Equation 5-50); and (3) it is the square root of twice the ratio of the velocity head to the pressure head.

Figure 7.4 represents the specific-head–depth relations for a particular discharge in a channel of a given width. Clearly there are an infinite number of pairs of alternate depths that can exist for each discharge. The pair that will apply for a given cross section is determined by the relation between velocity and depth (which can be expressed by the Manning equation) at that section. For example, consider a rectangular channel with a width of 10 m. With a discharge of 100 $m^3\ s^{-1}$ (and $\cos\theta = 1$ and $\alpha = 1$), Equation 7-22 becomes

$$H_s = Y + \frac{5.10}{Y^2}, \tag{7-26}$$

where Y and H_s are in meters. Suppose the channel slope is 0.005 and Manning's n is 0.03. Then from Manning's equation, assuming that $Y \simeq R$, we have

$$V = 2.36Y^{2/3}, \tag{7-27}$$

where V is in meters per second. Thus Equation 7-21 can be written in terms of Y as

$$Q = 23.6Y^{5/3}. \tag{7-28}$$

Substituting Equation 7-28 into Equation 7-22 gives

$$H_s = Y + 0.284Y^{4/3}. \tag{7-29}$$

Equation 7-28 can be solved for $Y = 2.38$ m when $Q = 100 \text{ m}^3 \text{ s}^{-1}$. Putting this value of Y into Equation 7-29 gives $H_s = 3.28$ m, which checks with the value found via Equation 7-26.

Suppose now that the same discharge occurs in a channel reach with the same width but a slope of 0.001 and a Manning's n of 0.04. In this case, the same sequence of calculations gives $Y = 4.58$ m and $H_s = 4.83$ m, which again satisfy Equation 7-26. The alternate depths for the two cases can be found by finding the other roots of Equation 7-22 for the appropriate values of H_s.

STREAM POWER

Definitions As shown in Table 2.3, power is the time rate of energy expenditure or, equivalently, the time rate of doing work. If we assume that the flow is uniform or gradually varied, then the slope of the channel, $|dZ/dx|$, represents the vertical distance that the flowing water falls while traveling a unit distance downstream (Figure 7.5). Furthermore,

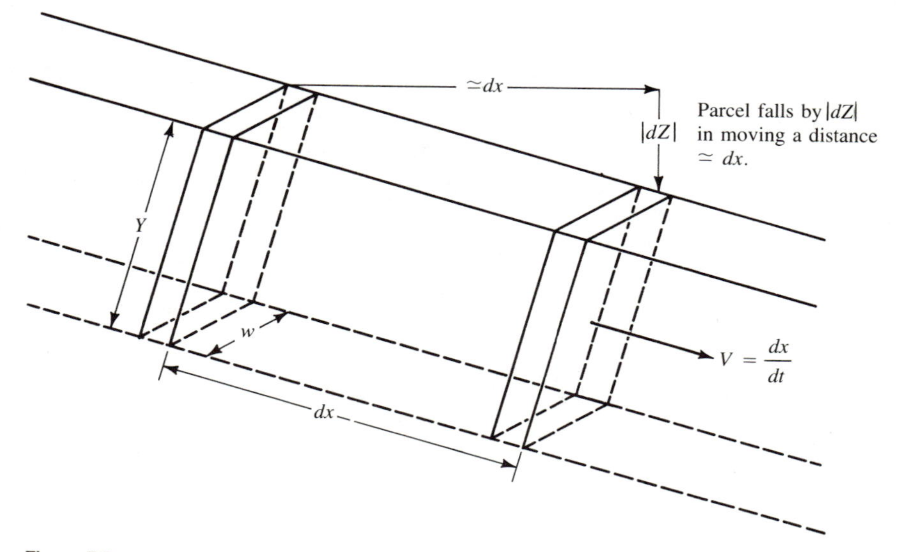

Figure 7.5
Definitions of the terms used to derive the expressions for stream power.

$$\left|\frac{dZ}{dt}\right| = \left|\frac{dZ}{dx}\right|\frac{dx}{dt}. \tag{7-30}$$

If the channel slope is small, dx/dt is the mean velocity V, and the time rate of fall of the water is thus

$$\left|\frac{dZ}{dt}\right| = \left|\frac{dZ}{dx}\right|V \tag{7-31}$$

or, with the alternative expression for channel slope (Equation 6-16),

$$\left|\frac{dZ}{dt}\right| = S_0V. \tag{7-32}$$

The weight of water per unit length of channel is γwY. Since energy is force times distance, the time rate of energy expenditure, or power, per unit length of channel is the product of the time rate of fall, S_0V, times the weight of water per unit length:

$$\Omega = \gamma wYVS_0, \tag{7-33a}$$

where Ω is the *stream power per unit channel length,* sometimes called simply the *stream power*. An equivalent expression is

$$\Omega = \gamma QS_0, \tag{7-33b}$$

where Q is discharge. Clearly, Ω has the dimensions $[FT^{-1}]$.

It has proved useful to define two additional expressions of stream power: (1) *Stream power* ω_A *per unit of bed area,*

$$\omega_A \equiv \frac{\Omega}{w} = \gamma YVS_0 = V\tau_0, \tag{7-34}$$

where w is stream width and τ_0 is boundary shear stress, defined as

$$\tau_0 \equiv \gamma YS_0 \tag{7-35}$$

(see Equation 6-13), and ω_A has the dimensions $[FL^{-1}T^{-1}]$; and (2) *stream power* ω_w *per unit weight of water,* where

$$\omega_w \equiv \frac{\Omega}{\gamma Y w} = VS_0. \tag{7-36}$$

The quantity ω_w is sometimes called simply the *unit stream power;* from Equation 7-32 it is clear that this is equal to the time rate of fall of the flowing water and the time rate of head loss, and has the dimensions $[LT^{-1}]$.

Stream power is of interest to hydrologists in three important areas of research: (1) exploration of the fundamental physical principles determining the shapes of streams in cross section, map view, and longitudinal profile; (2) prediction of sediment transport rates; and (3) prediction of bed forms and thus of flow resistance of alluvial streams. This chapter concludes with a brief discussion of the first area; the other two are discussed in Chapter 8.

Relation to Channel Shape Langbein and Leopold (1964) were the first to suggest that two basic tendencies underlie the behavior of streams and determine the shapes of their channels: (1) the tendency toward equal expenditure of energy on each unit area of the channel bed, which requires that ω_A be constant along the length of a river; and (2) the tendency toward minimization of the total energy expended over the whole length L of the river, which requires that $\int_0^L \Omega\, dx$ achieve its minimum possible value, where x is distance along the channel. They pointed out, however, that these two conditions cannot be simultaneously satisfied in rivers because of physical constraints, and therefore the shapes of longitudinal profiles and the downstream changes in width, depth, and velocity (discussed in the section on hydraulic geometry in Chapter 11) that are observed in nature are the result of "compromises" between the two opposing tendencies.

Later, Chang (1979) applied the principle of minimization of stream power (Ω) to the prediction of the forms of river channels as viewed on a map. These forms have been classified as straight, meandering (consisting of quasi-regular wavelike sinuosity, as shown in Figure 7.6), or braided (consisting of multiple joining and diverging channel segments separated by islands, as shown in Figure 7.7). He concluded that the form taken by a given river is determined by the tendency to minimize Ω under the constraints of discharge and sediment load imposed by conditions in the drainage basin.

Of particular note is that Chang (1979) postulated that the cause of meandering, which has been a persistent subject of debate in the literature of hydraulics, hydrology, and geomorphology, is the tendency toward minimization of stream power. This debate, however, is probably not over. Interestingly, among the many contributions to it are those of both Albert Einstein (1926) and his son, Hans Albert (Einstein and Shen, 1964). The son's contribution, a study of spiral flow patterns in straight channels, was far more substantial than that of his famous father, who attributed meandering to the Coriolis force (see the discussion of the relative magnitudes of forces in Chapter 5). Reviews of earlier attempts to explain the phenomenon have been published by Yang (1971) and Chitale (1973).

Figure 7.6
Aerial view of an intensely meandering stream in interior Alaska.

Figure 7.7
A braided glacial stream in interior Alaska.

Exercises

7-1. Derive the general expression for Y_c and $\mathbb{F}$ for cases when $\cos\theta \neq 1$ and $\alpha \neq 1$.

7-2. Show that for critical flow in a rectangular channel, the specific head is 1.5 times the flow depth, assuming that $\alpha = 1$ and $\cos\theta = 1$.

7-3. Show that for critical flow, the discharge is a maximum for a given specific energy.

7-4. Figure 7.8 is a diagram of an open-channel flow in a rectangular channel between two lakes. The accompanying table gives elevations at cross sections from 0 to 6, and the channel width is constant at 5 m. The discharge is 12 $m^3\ s^{-1}$. Assume that $\cos\theta = 1$ and $\alpha = 1$, and (a) compute the elevation, pressure, velocity, specific, and total heads at cross sections 0–6, using the elevation of the bed at cross section 6 as the datum; (b) plot the channel bottom, water-surface profile, and total head line on a graph, using a high degree of vertical exaggeration. Assume uniform flow except where the flow is rapidly varied, and use dashed lines to show the surface in regions of rapidly varied flow.

	Cross section						
Elevation†	0	1	2	3	4	5	6
Bed	400.00	399.80	397.80	397.21	395.63	395.21	393.61
Surface	403.00	402.00	399.78	398.01	396.38	396.26	395.91

†In meters above sea level.

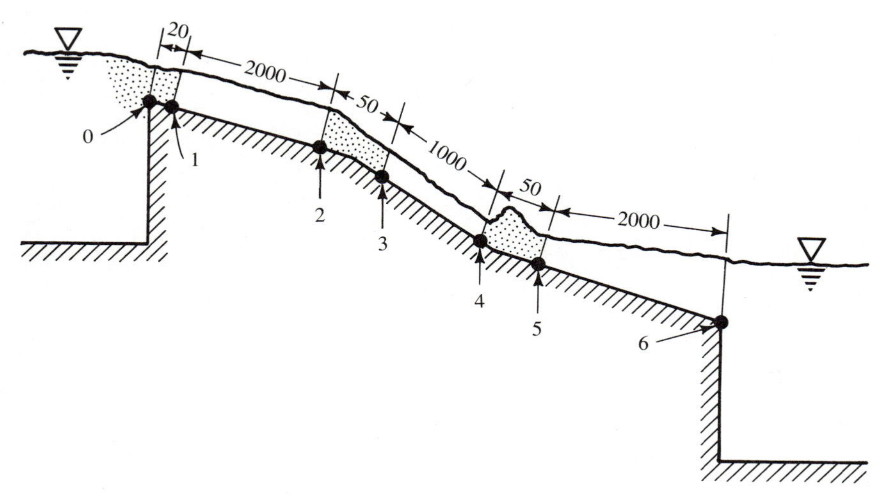

Figure 7.8
Relationships for Exercise 7-4.

7-5. Construct a specific-head diagram for the flow in Exercise 7-4. What is the critical depth?

7-6. Consider a rectangular channel with a width of 3 m and a maximum depth of 2 m. Assume that $\cos \theta = 1$ and $\alpha = 1$. (a) Construct on a single graph a family of specific-head diagrams for discharges of 2.00, 4.00, 6.00, 8.00, 10.00, and 12.00 $m^3 s^{-1}$. (b) Construct another graph showing critical depth as a function of discharge for this situation.

7-7. (a) Write out all the calculations in the development of Equations 7-26–7-29, and show that the values of Y and H_s computed for the two examples are correct. (b) Construct a specific-head diagram for a discharge of 100 $m^3 s^{-1}$ and a rectangular channel with a width of 10 m and show where the two examples discussed in the text fall on the diagram. (c) Determine whether the two examples are subcritical or supercritical flows. (d) Find the alternate depths for the two examples.

CHAPTER 8

Flow Resistance, Erosion, and Sediment Transport

As expressed in the Manning or Chézy coefficient, resistance to flow appears at first to be a straightforward physical parameter. As we will see in this chapter, however, any attempt to understand this concept fully leads quickly to the realization that it and all the other characteristics of a flow are so intimately interrelated that it is almost impossible to separate cause from effect. Thus almost any statement that we make in describing a flow—for example, giving its mean velocity, mean depth, and slope—is an implicit statement about flow resistance as well. In the next section of this chapter, the several terms that express flow resistance are examined and their interrelationships discussed. In addition, practical approaches to estimation of flow resistance are presented.

One of the factors that contributes to resistance is the bed forms that develop in alluvial streams (streams whose beds are composed of granular materials that can be transported by the flows in the channel). These forms arise when erosion and transportation of bed material occur. Furthermore, energy is expended in transporting these materials, and hence discussion of erosion and sediment transport is a natural extension of examining flow resistance.

FLOW RESISTANCE

In spite of the complexities associated with the concept of flow resistance, we can specify all the factors that contribute to it within a reach of channel. First, there is *skin friction,* which is contributed by the boundary material itself. For granular materials, this is the roughness due to the grains themselves, and it increases with grain size. It is also affected by grain spacing and shape. A second major contributor, called *form resistance* or *form drag,* is caused by the semiregular wavelike features (ripples, dunes, and antidunes, described more fully later) that often appear when the stream bed consists of sand. These forms cause small eddies, or zones of flow separation, that divert energy from the main flow. The magnitude of their effect depends on their shape, wavelength, and amplitude.

Flow resistance is also exerted by non-bed-material obstructions, such as living and dead vegetation, and by bends and abrupt changes in cross-section geometry. All these features cause eddies that are driven by energy from the main flow, and hence detract from the energy available to that flow. Energy is also lost by acceleration and deceleration as the water moves over local variations in slope, which may be caused by the bed forms mentioned earlier. Finally, energy is lost and flow resistance increased by breaking waves and hydraulic jumps (described in Chapter 10), which may occur when the flow is especially rapid, and which introduce additional eddies and turbulence (Simons and Richardson, 1966).

Local Resistance: Boundary Shear Stress and Friction Velocity As we saw in Chapter 6 the development of the equations for uniform flow in laminar and turbulent boundary layers is based on balancing the downslope component of gravitational forces with the forces resisting the flow. The origin of these resistance forces is the boundary. In laminar flow the frictional resistance of the boundary is transmitted throughout the fluid by the viscosity, while in turbulent flows this resistance is transmitted beyond the laminar sublayer by turbulent eddies. The basic equations describing these processes for the two flow regimes are Equations 4-5 and 6-18, respectively. Thus one way of describing the magnitude of the boundary resistance is in terms of the boundary shear stress τ_0, which for both states of flow is given by Equation 6-13:

$$\tau_0 = \gamma Y S_0, \tag{8-1}$$

where γ is the weight density of water, Y is the depth, and S_0 is the channel slope.

Equation 6-25 defined the friction velocity V^*, another term that reflects the magnitude of the boundary resistance at a point in the flow:

$$V^* = (gYS_0)^{1/2}, \tag{8-2}$$

where g is the gravitational acceleration. A comparison of Equations 8-1 and 8-2 shows immediately that

$$V^* = \left(\frac{\tau_0}{\rho}\right)^{1/2}, \tag{8-3}$$

where ρ is the mass density of water; and conversely,

$$\tau_0 = \rho V^{*2}. \tag{8-4}$$

As just defined, both the boundary shear stress and friction velocity arose in the development of vertical velocity profiles and reflect the skin friction (i.e., the direct effect of a point on the boundary on the overlying flow). We will see later that their meanings can be broadened.

Reach Resistance: Chézy's *C* and Manning's *n* Chézy's C and Manning's n are also clearly expressions of the channel resistance, although C is inversely related to resistance. A comparison of Equations 6-48 and 6-49 shows that

$$\frac{1}{C} = \frac{u_C n}{u_M R^{1/6}}, \tag{8-5}$$

where u_C and u_M depend on the system of units as described earlier.

Although τ_0 and V^*, and C and n, all reflect the resistance of the channel, there is a distinction between the first pair and the second. In contrast to the definitions of τ_0 and V^* given earlier, C and n reflect the skin friction plus all the other characteristics of a channel that contribute resistance over a stream reach, such as variations in cross-section shape and size, bends, and obstructions. Thus C and n are properly defined only for a finite stream reach rather than for a point. A later section of this chapter describes methods for estimating n from channel characteristics.

Other Measures of Reach Resistance: Energy Slope and Friction Factors As discussed in Chapter 7, the energy loss in a flow reflects the energy converted to heat in the process of overcoming the frictional resistance of the boundary. Thus the rate at which the total flow energy decreases with distance is a direct measure of the total flow resistance exerted by a stream reach and is expressed by the energy slope S_e. In a uniform flow, S_e is of course equal to the channel slope S_0.

The energy slope is often related to the characteristics of the flow by a dimensionless proportionality factor f_D:

$$S_e = f_D\left(\frac{V^2}{8gR}\right), \tag{8-6}$$

where f_D is called the *Darcy–Weisbach friction factor*. This designation acknowledges the early work on flow resistance of Henri Darcy (1803–1858) and Julius

Weisbach (1806–1871), who were French and German engineers, respectively. Rearranging Equation 8-6 expresses f_D in terms of other flow characteristics:

$$f_D = \frac{8gRS_e}{V^2}. \tag{8-7}$$

(The constant 8 appears in Equations 8-6 and 8-7 for geometric reasons, since the formulas were originally developed for flow in pipes.) Thus f_D is yet another measure of reach resistance. For pipes, it has been related by experiment to the Reynolds number and to the roughness of the pipe wall (see Chow, 1959; Daily and Harleman, 1966), and such relations are sometimes applied to open-channel flows.

A final expression of reach resistance can be developed from the consideration of actual turbulent forces at the end of Chapter 6. Equation 6-57 stated that the downslope forces of gravity are balanced by the resisting forces due to turbulence:

$$gS_0 = k_t\left(\frac{V^2}{Y}\right), \tag{8-8}$$

where k_t is also a factor proportional to reach resistance. Solving for k_t and using the more appropriate hydraulic radius instead of mean depth gives

$$k_t = \frac{gRS_0}{V^2}. \tag{8-9}$$

For uniform flow, when $S_0 = S_e$, we see that k_t is a "pure" dimensionless reach-resistance factor that is related to the Darcy–Weisbach friction factor as

$$k_t = \frac{f_D}{8}. \tag{8-10}$$

Relations among Resistance Terms As noted earlier, the origins of the several measures of flow resistance suggest that they can be classified into two groups. The first includes those that express the direct effect of the channel bed on the flow immediately above it (τ_0, V^*), and the second includes those that reflect the total resistance of a channel reach of finite length ($1/C$, n, S_e, f_D, k_t). It turns out, however, that τ_0 and V^* can be readily generalized into measures of reach resistance by simply redefining them as

$$\tau_0 = \gamma RS_0 \tag{8-11}$$

and

$$V^* = (gRS_0)^{1/2}, \tag{8-12}$$

that is, by replacing the flow depth with the hydraulic radius and by considering the slope to be measured over the reach rather than at a point or cross section.

With these definitions, all seven indices of flow resistance become essentially equivalent and interchangeable. This equivalence is emphasized in Table 8.1, which gives each measure in terms of the other six for uniform flows. A glance at this table might make you wonder why measures of flow resistance have proliferated so. The answer is simply that each has arisen in a different context, as shown in Chapters 6 and 7.

Table 8.1 also illustrates a fundamental point about open-channel flows: All the flow variables—slope, width, depth (or hydraulic radius), velocity, and flow resistance—are interdependent. None of them can be considered a truly independent variable. If we consider the discharge to be a given value, as is often done, the five quantities mutually adjust themselves to convey that discharge. However, it is not really possible to consider discharge as an independent variable either, since it is the product of width, depth, and velocity. If these variables adjust along with slope and resistance in one reach, they determine the discharge in the next reach, and so on downstream. This is why some scientists have attempted to discover independent underlying principles, such as the minimization of stream power as discussed in Chapter 7, that determine the mutual adjustment of the more commonly measured flow characteristics. The quest for these principles continues, as evidenced by the recent work of Yang et al. (1981).

Estimation of Flow Resistance In spite of the mutual variability of flow characteristics, the practicing hydrologist or hydraulic engineer needs techniques that make it possible to estimate velocity or discharge when width, depth, and slope are known. This need in turn requires a method for estimating flow resistance based on the nature of the channel. Most of these practical techniques are concerned specifically with estimating Manning's n, because the empirical Manning's equation has proved to be the most useful for estimating the characteristics of actual flows. Once n is estimated, and R and S_0 are known, we can compute V (and Q if width is known) from Equation 6-49. From column 4 of Table 8.1 it is clear that any of the other measures of flow resistance can be computed when n, V, and R are specified.

A procedure for estimating n that clearly shows the several factors that determine resistance in a reach was developed by Cowan (1956). As presented by Chow (1959), the basic equation of this procedure is

$$n = (n_0 + n_1 + n_2 + n_3 + n_4)m, \tag{8-13}$$

where n_0 is a basic value of n for a straight, uniform, smooth channel in various

Table 8.1

Relations among Resistance Parameters and Flow Characteristics for Uniform Flows

Symbols as defined in text. Expressions on diagonal are definitions of resistance parameters in terms of flow characteristics. Values of u_C and u_M in the three unit systems are $u_C = 0.552$ (SI), 5.52 (cgs), 1.00 (English); $u_M = 1.00$ (SI), 4.64 (cgs), 1.49 (English).

Parameter	τ_0	V^*	$1/C$	n	S_e	f_D	k_t
τ_0 $[FL^{-2}]$	$\gamma R S_0$	ρV^{*2}	$\frac{\gamma V^2}{u_C^2}\left(\frac{1}{C^2}\right)$	$\frac{\gamma V^2 n^2}{u_M^2 R^{1/3}}$	$\gamma R S_e$	$\frac{\rho V^2 f_D}{8}$	$\rho V^2 k_t$
V^* $[LT^{-1}]$	$\left(\frac{\tau_0}{\rho}\right)^{1/2}$	$(gRS_0)^{1/2}$	$\frac{g^{1/2}V}{u_C}\left(\frac{1}{C}\right)$	$\frac{g^{1/2}Vn}{u_M R^{1/6}}$	$(gRS_e)^{1/2}$	$\frac{V f_D^{1/2}}{2.38}$	$V k_t^{1/2}$
$1/C$ $[1]^\dagger$	$\frac{u_C \tau_0^{1/2}}{\gamma^{1/2} V}$	$\frac{u_C V^*}{g^{1/2} V}$	$\frac{u_C R^{1/2} S_0^{1/2}}{V}$	$\frac{u_C n}{u_M R^{1/6}}$	$\frac{u_C R^{1/2} S_e^{1/2}}{V}$	$\frac{u_C f_D^{1/2}}{2.38 g^{1/2}}$	$\frac{u_C k_t^{1/2}}{g^{1/2}}$
n $[1]^\dagger$	$\frac{u_M R^{1/6} \tau_0^{1/2}}{\gamma^{1/2} V}$	$\frac{u_M R^{1/6} V^*}{g^{1/2} V}$	$\frac{u_M R^{1/6}}{u_C}\left(\frac{1}{C}\right)$	$\frac{u_M R^{2/3} S_0^{1/2}}{V}$	$\frac{u_M R^{2/3} S_e^{1/2}}{V}$	$\frac{u_M R^{1/6} f_D^{1/2}}{2.38 g^{1/2}}$	$\frac{u_M R^{1/6} k_t^{1/2}}{g^{1/2}}$
S_e $[1]$	$\frac{\tau_0}{\gamma R}$	$\frac{V^{*2}}{gR}$	$\frac{V^2}{u_C^2 R}\left(\frac{1}{C^2}\right)$	$\frac{V^2 n^2}{u_M^2 R^{4/3}}$	$-\frac{dH}{dx} \simeq S_0$	$\frac{V^2 f_D}{8gR}$	$\frac{V^2 k_t}{gR}$
f_D $[1]$	$\frac{8\tau_0}{\rho V^2}$	$\frac{8V^{*2}}{V^2}$	$\frac{8g}{u_C^2}\left(\frac{1}{C^2}\right)$	$\frac{8gn^2}{u_M^2 R^{1/3}}$	$\frac{8gRS_e}{V^2}$	$\frac{8gRS_e}{V^2}$	$8k_t$
k_t $[1]$	$\frac{\tau_0}{\rho V^2}$	$\frac{V^{*2}}{V^2}$	$\frac{g}{u_C^2}\left(\frac{1}{C^2}\right)$	$\frac{gn^2}{u_M^2 R^{1/3}}$	$\frac{gRS_e}{V^2}$	$\frac{f_D}{8}$	$\frac{gRS_0}{V^2}$

†Usually treated as if dimensionless.

materials (Table 8.2). The values of n_1–n_4 take into account the flow resistance added by surface roughness; variations in slope and size of the channel cross section; the presence of flow obstructions such as boulders, logs, and roots; and vegetation of the surface, respectively. The value of m is determined by the degree of meandering of the channel. Chow (1959) provides further guidance for choosing values from Table 8.2, and notes that the method has not been tested for channels where R is greater than about 5 m.

Table 8.2
Values for Calculating Manning's n by Cowan's (1956) Method

From V. T. Chow, *Open Channel Hydraulics,* © 1959, Table 5.5. Reprinted by permission of McGraw-Hill Book Company, New York.

Channel conditions	Value
Material involved	n_0
Earth	0.020
Rock cut	0.025
Fine gravel	0.024
Coarse gravel	0.028
Degree of irregularity	n_1
Smooth	0.000
Minor	0.005
Moderate	0.010
Severe	0.020
Variations of channel cross section	n_2
Gradual	0.000
Alternating occasionally	0.005
Alternating frequently	0.010–0.015
Relative effect of obstructions	n_3
Negligible	0.000
Minor	0.010–0.015
Appreciable	0.020–0.030
Severe	0.040–0.060
Vegetation	n_4
Low	0.005–0.010
Medium	0.010–0.025
High	0.025–0.050
Very high	0.050–0.100
Degree of meandering	m
Minor	1.000
Appreciable	1.150
Severe	1.300

Table 8.3
Values of Manning's *n* for Natural Streams

From V. T. Chow, *Open Channel Hydraulics,* © 1959, Table 5.6. Reprinted by permission of McGraw-Hill Book Company, New York.

Type of channel and description	Minimum	Normal	Maximum
Minor streams (top width at flood stage <100 ft)			
Streams on plain			
1. Clean, straight, full stage, no riffles or deep pools	0.025	0.030	0.033
2. Same as above, but more stones and weeds	0.030	0.035	0.040
3. Clean, winding, some pools and shoals	0.033	0.040	0.045
4. Same as above, but some weeds and stones	0.035	0.045	0.050
5. Same as above, but lower stages, more ineffective slopes and sections	0.040	0.048	0.055
6. Same as 4, but more stones	0.045	0.050	0.060
7. Sluggish reaches, weedy, deep pools	0.050	0.070	0.080
8. Very weedy reaches, deep pools, or floodways with heavy stand of timber and underbrush	0.075	0.100	0.150
Mountain streams, no vegetation in channel, banks usually steep, trees and brush along banks submerged at high stages			
1. Bottom: gravels, cobbles, and few boulders	0.030	0.040	0.050
2. Bottom: cobbles with large boulders	0.040	0.050	0.070
Floodplains			
Pasture, no brush			
1. Short grass	0.025	0.030	0.035
2. High grass	0.030	0.035	0.050
Cultivated areas			
1. No crop	0.020	0.030	0.040
2. Mature row crops	0.025	0.035	0.045
3. Mature field crops	0.030	0.040	0.050
Brush			
1. Scattered brush, heavy weeds	0.035	0.050	0.070
2. Light brush and trees, in winter	0.035	0.050	0.060
3. Light brush and trees, in summer	0.040	0.060	0.080
4. Medium to dense brush, in winter	0.045	0.070	0.110
5. Medium to dense brush, in summer	0.070	0.100	0.160
Trees			
1. Dense willows, summer, straight	0.110	0.150	0.200
2. Cleared land with tree stumps, no sprouts	0.030	0.040	0.050
3. Same as above, but with heavy growth of sprouts	0.050	0.060	0.080
4. Heavy stand of timber, a few down trees, little undergrowth, flood stage below branches	0.080	0.100	0.120
5. Same as above, but with flood stage reaching branches	0.100	0.120	0.160
Major streams (top width at flood stage >100 ft)†			
Regular section with no boulders or brush	0.025	—	0.060
Irregular and rough section	0.035	—	0.100

†The *n* value is less than that for minor streams of similar description, because banks offer less effective resistance.

Table 8.3 gives n values for natural streams and their floodplains. Values for in-channel flows shown in the table range from 0.025 to 0.150. Barnes (1967) has presented a useful series of photographs and cross sections of streams to be used in making field estimates of n. His streams cover the range $0.024 \leq n \leq 0.075$ (Figure 8.1). A method is given in Barnes's report for calculating composite values when n varies over a cross section, as for flows on a floodplain.

Detailed studies of small streams show higher values of n than reported by Chow (1959) and Barnes (1967). For instance, Johnson (1964) found n values as high as 0.20 in mountain streams in Vermont, and Dingman (1971) found $0.089 \leq n \leq 0.421$, with an average value of 0.197, for a small, tortuous, and heavily vegetated Alaskan stream (Figure 8.2).

Finally, Table 8.4 shows n values for overland flows, to be used in Equation 6-51.

Changes of Resistance with Discharge Many studies have shown that Manning's n varies as discharge changes in a given reach. Three physical phenomena are largely responsible for these observations. First, depth increases as discharge increases (quantitative description of this relationship is discussed in the section on hydraulic geometry in Chapter 11). Thus the ratio of depth to the height of the roughness elements (such as sand or gravel particles on the stream bed) also increases with discharge, and the roughness elements become relatively less effectual in retarding the flow, so that n tends to decrease as discharge increases. This phenomenon is of course important only in flows that are hydraulically rough, which are typically those with gravel beds or those in which vegetation forms a significant part of the roughness elements.

The second effect arises principally in sand-bed streams, where a quasi-regular succession of bed forms (ripples, dunes, etc.) occurs as discharge changes. The size, shape, and spacing of these forms have a major influence on flow resistance, as discussed later in this chapter, but the effect is not monotonic with discharge.

Finally, if a stream overflows its banks, the flow resistance encountered in the floodplains is generally markedly higher than that in the channel, so the composite resistance typically increases when discharges reach flood stage. In this section we explore the first of these effects; the influence of bed forms is discussed later in the chapter, and ways of accounting for changes in resistance in overbank flows are discussed briefly in Chapter 9.

Equation 6-36 relates the mean velocity in a vertical profile to the depth. For rough flows, y_0 is computed from Equation 6-31, so Equation 6-36 becomes

$$V = 2.5V^*\left[\ln\left(\frac{30Y}{k_s}\right) - 1\right]. \tag{8-14}$$

Following the rules of logarithms, we can write Equation 8-14 as

$$V = 2.5V^*(\ln Y - \ln k_s + \ln 30 - 1),$$

(a) (b) (c) (d)

Figure 8.1
Photographs of river reaches covering a range of values of Manning's n. The values of n were computed from flow measurements made at the time the photographs were taken. (a) Columbia River at Vernita, WA: $n = 0.024$; (b) West Fork Bitterroot River near Conner, MT: $n = 0.036$; (c) Moyie River at Eastport, ID: $n = 0.038$; (d) Tobesofkee Creek near Macon, GA: $n = 0.041$; (e) Grande Ronde River at La Grande, OR: $n = 0.043$; (f) Clear Creek near Golden, CO: $n = 0.050$; (g) Haw River near Benaja, NC: $n = 0.059$; (h) Boundary Creek near Porthill, ID: $n = 0.073$. From Barnes (1967); photographs courtesy of U.S. Geological Survey.

and since $\ln 30 = 3.40$,

$$V = 2.5V^*\left[\ln\left(\frac{Y}{k_s}\right) + 2.40\right]. \tag{8-15}$$

If we now assume that the mean velocity in the reach is related to the mean depth by the same relationship as is expressed in Equation 8-14 for a vertical profile, we can equate Equation 8-15 to the Manning equation (Equation 6-49) and solve for n:

(e) (f) (g) (h)

$$n = \frac{u_M R^{2/3} S_0^{1/2}}{2.5V^*[\ln(Y/k_s) + 2.40]}. \quad (8\text{-}16)$$

With the definition of V^* (Equation 6-25),

$$n = \frac{u_M R^{1/6}}{2.5g^{1/2}[\ln(Y/k_s) + 2.40]}. \quad (8\text{-}17)$$

Theoretically, Equation 8-17 indicates the way in which Manning's n changes as mean depth or hydraulic radius varies for rough flows; the relationship is plotted for a range of k_s values in Figure 8.3 (with Y assumed equal to R). Because of the way in which it was developed, it does not account for other factors that influence n in actual reaches. However, Bray (1979) found that an equation of the form of Equation 8-17 was the most acceptable for estimating n from k_s for gravel-bed rivers.

Other studies discussing the variation of n with discharge are those of Leopold and Maddock (1953), Leopold et al. (1964), Gregory and Walling (1973), and Dunne and Leopold (1978). Johnson (1964) and Dingman (1971) found especially rapid decreases of n with increasing discharge in the very small streams they studied. In both these studies, vegetative obstructions were an important part of flow resistance.

Figure 8.2
This small, heavily vegetated stream in central Alaska has Manning's n values ranging from 0.089 at high flows to 0.42 at very low flows (Dingman, 1971).

Table 8.4
Values of n for Overland Flows (Equation 6-51)
From Crawford and Linsley (1966).

Surface	n
Smooth asphalt	0.012
Asphalt or concrete paving	0.014
Packed clay	0.03
Light turf	0.20
Dense turf	0.35
Dense shrubbery and forest litter	0.40

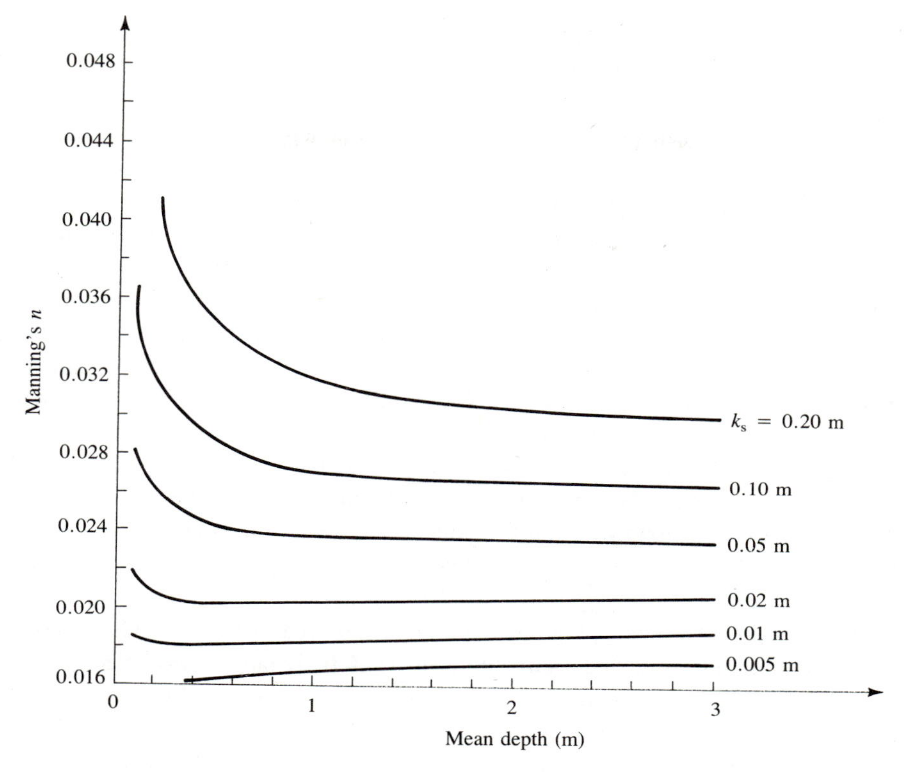

Figure 8.3
Theoretical variation of Manning's n with mean flow depth, computed for various roughness heights (k_s) via Equation 8-17.

CHANNEL EROSION AND SEDIMENT TRANSPORT

In natural rivers studied by hydrologists, as well as in artificial canals designed by hydraulic engineers, two important questions arise: (1) What size particles is the stream or canal capable of eroding and transporting under varying conditions? and (2) What is the total load of sediment (i.e., *sediment discharge*) it will carry under varying conditions? The first question concerns the *competence* of the stream; the second concerns its *capacity*.

Answers to these questions have been difficult to come by, first because of the previously discussed interactions among all flow characteristics and the impossibility of identifying true independent variables, and second because of the great difficulties in obtaining measurements with which to validate predictions of sediment transport. In recent years, however, significant progress has been made. Because a complete development of the physics of the subject would require book-length treatment, we will touch on only those aspects of the subject that have an especially close relationship to the other subjects discussed in this book. You can explore this field in greater depth via the references cited.

We begin with some important definitions. Sediment moving in a stream is classified according to its origin as either bed-material load or wash load. *Bed-material load* is eroded directly from the channel bed, whereas *wash load* is typically considerably finer-grained material derived from sources other than the stream bed (e.g., overland flow, tributaries, or caving of banks). Bed-material load is further subdivided into *bed load,* which is carried within a few grain diameters of the bed and moves essentially parallel to it; and *suspended load,* which is transported farther above the bed by turbulent eddies and moves downstream in an irregular path. In general, the bed load consists of larger particles than are present in suspension. Experiments have shown, however, that a given particle may alternate fairly rapidly between bed load and suspended load (Kikkawa and Ishikawa, 1978). Virtually all attempts at predicting sediment transport consider only the bed-material load, since the amount of wash load bears little relation to the characteristics of the stream. Some approaches estimate suspended load, some bed load, and some the total bed-material load.

Fall Velocity The concept of fall velocity, or *settling velocity,* is fundamental to any discussion of erosion and sediment transport. A body falling through a vacuum in response to gravity continuously accelerates. If the body is falling through a fluid, however, forces arise that resist the motion. Since these resisting forces strengthen as velocity increases (see Equations 5-24 and 5-28), the body eventually achieves a speed at which the gravitational force is balanced, and thereafter it falls at a constant, or terminal, velocity. The terminal velocity of a sediment particle is its fall velocity, or settling velocity. In water this velocity is reached very quickly, and the brief period of acceleration can be ignored for purposes of analysis.

It should not be surprising that the forces resisting gravity are those due to viscosity and turbulence. In 1851 the English physicist G. G. Stokes (1819–1903) first formulated the expression for the viscous forces F_v resisting the fall of a spherical particle in a fluid:

$$F_v = 6\pi\mu r v_f, \tag{8-18}$$

where μ is the dynamic viscosity, r is the radius of the sphere, and v_f is its fall velocity. The numerical constants arise from the geometry of the situation; a full derivation of this equation is given by Daily and Harleman (1966).

The effective mass density of a particle in water is equal to the difference between the density of the particle, ρ_s, and the density of the water, ρ. Thus we can readily write the expression for the gravitational force F_g on the particle as

$$F_g = (\rho_s - \rho)g\tfrac{4}{3}\pi r^3, \tag{8-19}$$

where g is the gravitational acceleration and $\frac{4}{3}\pi r^3$ is the volume of the sphere. At the terminal velocity, $F_v = F_g$; equating Equations 8-18 and 8-19 and solving for v_f gives

$$v_f = \frac{2(\rho_s - \rho)gr^2}{9\mu}. \tag{8-20a}$$

Since sediments are generally described in terms of their diameter d, it is convenient to express v_f as

$$v_f = \frac{(\rho_s - \rho)gd^2}{18\mu}. \tag{8-20b}$$

Equation 8-20 is usually called *Stokes's law*. It applies only to cases in which turbulent force can be ignored and only viscous force resists the motion. We can formulate a *particle Reynolds number* $\mathbb{R}_p$ to provide an indication of the relative magnitudes of the two forces:

$$\mathbb{R}_p \equiv \frac{v_f d\rho}{\mu}. \tag{8-21}$$

Experiments have shown deviations from Stokes's law when $\mathbb{R}_p \simeq 1$. By substituting Equation 8-20b into 8-21 and taking an appropriate value for μ from Table 4.2 and a typical value of ρ_s, 2650 kg m^{-3}, we can show that the critical value of $\mathbb{R}_p$ occurs when $d \simeq 0.1$ mm. Thus Stokes's law gives the fall velocity of typical sediment particles larger than 2×10^{-4} mm (where Brownian motion begins to become important and no settling occurs) and smaller than about 0.1 mm.

Particles larger than 0.1 mm are affected by turbulent as well as viscous forces. Consistent with discussions in Chapters 5 and 6, the turbulent force per unit area is proportional to $\rho v_f^2/2$ (which is equal to $\gamma v_f^2/2g$). Since the area of the particle as it is encountered by the fluid is πr^2, the turbulent force F_t acting on the falling sphere can be expressed as

$$F_t = \left(\frac{C_D}{2}\right)\rho v_f^2 \pi r^2, \tag{8-22}$$

where C_D is a dimensionless *drag coefficient*.

The American geologist W. W. Rubey formulated an expression (Rubey, 1933) for fall velocity for larger particles by including both turbulent and viscous resisting forces as

$$F_g = F_t + F_v. \tag{8-23}$$

(Note that this is identical to Equation 5-41, which was formulated for open-channel flows.) Substituting Equations 8-18, 8-19, and 8-22 into Equation 8-23 gives

$$\frac{4}{3}\pi r^3(\rho_s - \rho)g = \left(\frac{C_D}{2}\right)\rho v_f^2 \pi r^2 + 6\pi\mu r v_f.$$

This is a quadratic equation in v_f that can be reduced to

$$\left(\frac{C_D}{4}\right)\rho r v_f^2 + 3\mu v_f - \frac{2}{3}r^2(\rho_s - \rho)g = 0$$

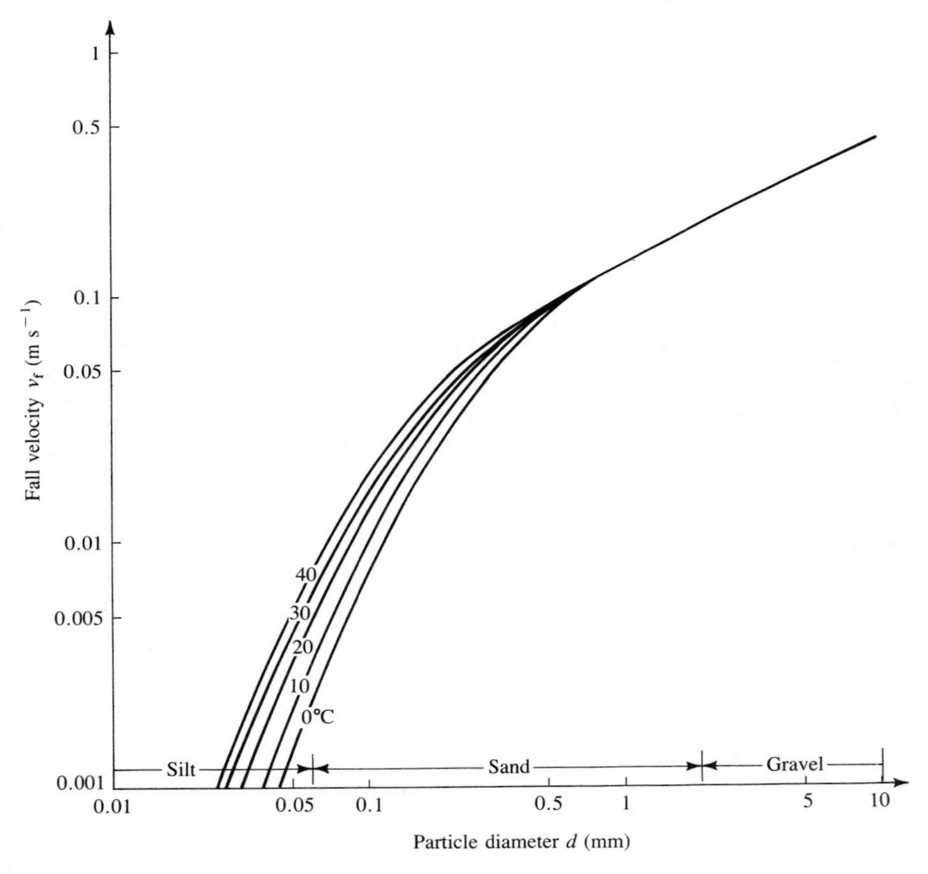

Figure 8.4
Fall velocity as a function of diameter for spherical particles with a density of 2650 kg m^{-3}, computed via the Rubey–Watson law (Equation 8-25b) for temperatures of 0, 10, 20, 30, and 40°C. The curves are essentially identical to those obtained with Stokes's law (Equation 8-20b) for particles less than about 0.07 mm diameter.

and solved to give

$$v_f = \frac{[9\mu^2 + \frac{2}{3}C_D\rho g(\rho_S - \rho)r^3]^{1/2} - 3\mu}{(C_D/2)\rho r}. \tag{8-24}$$

Rubey (1933) used $C_D = 2$, and the resulting form of Equation 8-24 has become known as *Rubey's law*. However, Watson (1969) found that Rubey's law gave fall velocities that were somewhat slower than those measured in settling tanks. Thus he used the experimental data to compute modified drag coefficients for both the turbulent and viscous terms in Equation 8-24, and arrived at

$$v_f = \frac{[3.48\mu^2 + 0.707\rho g(\rho_s - \rho)r^3]^{1/2} - 1.87\mu}{0.531\rho r} \tag{8-25a}$$

or, using diameter instead of radius,

$$v_f = \frac{[3.48\mu^2 + 0.0884\rho g(\rho_s - \rho)d^3]^{1/2} - 1.87\mu}{0.265\rho d}. \tag{8-25b}$$

Equation 8-25 will be referred to as the *Rubey–Watson law;* it can be used for computing fall velocity for particles of silt size and larger (Figure 8.4).

In many treatments of sediment transport and deposition, the sediment size is discussed in terms of fall velocities rather than diameters because of the one-to-one relation between those variables expressed in Equations 8-20 and 8-25. It should be noted, however, that particle shape can affect the value of the drag coefficient and hence the relation between fall velocity and particle diameter.

Stokes's law and the Rubey–Watson law can be solved for diameter as a function of fall velocity. In this form, they provide a method for measuring particle-size distributions of soils and sediments. This method is routinely used for silts and finer materials; the technique is described in detail in Hillel (1980). Because of the dependence of viscosity on temperature, you must specify the latter when computing fall velocities.

Competence: Critical Tractive Force and Critical Velocity The competence of a flow is expressed as the diameter of the largest particle that the flow can erode from its bed. It is of course of considerable practical interest to determine this value, and we do so here by relating first the tractive force and then the mean velocity to the size of particles that can be eroded.

Figure 8.5 is a simplified view of particles on the bed of a stream. Following the approach of Engelund and Hansen (1967), we note that each such particle is subject to three principal forces: (1) the submerged weight of the particle, acting vertically downward; (2) a downstream-directed *drag force* due to the skin friction of the grain and to a difference in pressure between its upstream and downstream sides, caused by induced eddies; and (3) an upward-directed *lift force* due to induced deviations from hydrostatic pressure at the top and sides of the grain. If the particle is in the silt or clay size range (<0.06 mm in diameter), an important fourth force, due to electrostatic attractions among the particles, acts along with the particle weight to resist movement. Earth materials in which this force is present are characterized as *cohesive soils*. Our analysis will ignore this last force; hence what follows applies only to *noncohesive soils* (sands and gravels).

The magnitude of the downward weight component F_g is

$$F_g = K_1(\rho_s - \rho)gd^3, \tag{8-26}$$

where K_1 is a proportionality factor. (As in Equation 8-19, $K_1 = \pi/6$ if the particle is perfectly spherical.) It turns out that both the drag and the lift forces are proportional to $\rho V^{*2}d^2/2$ (see Equation 8-22 for an analogous expression), so

$$F_e = K_2\left(\frac{\rho V^{*2}}{2}\right)d^2, \tag{8-27}$$

where F_e is the total erosive force, equal to the vector sum of the lift and drag forces, V^* is the friction velocity, and K_2 is the constant of proportionality.

Reasoning that movement of the particle will depend on the relative magnitudes of F_g and F_e, we can ignore the constants in Equations 8-26 and 8-27 and define the dimensionless ratio θ_e:

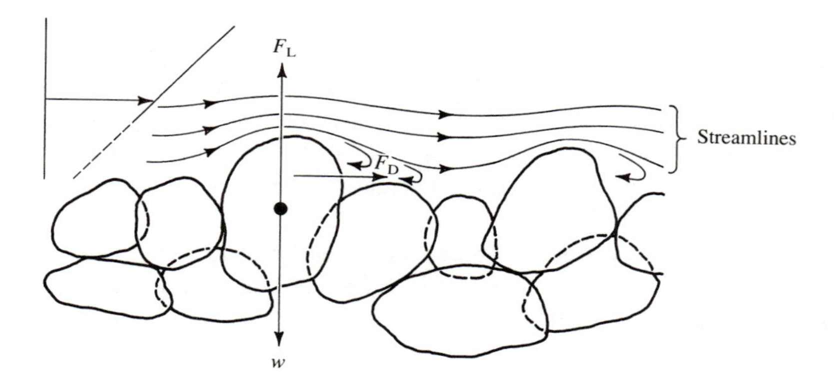

Figure 8.5
Forces on a particle in a stream bed: w, weight (gravitational force); F_L, lift force; F_D, drag force. Velocity gradient shown at left.

$$\theta_e = \frac{\rho V^{*2} d^2}{(\rho_s - \rho) g d^3} = \frac{\rho V^{*2}}{(\gamma_s - \gamma) d}, \tag{8-28}$$

where γ_s is the weight density of the particle and γ is the weight density of water. An equivalent form of Equation 8-28 uses the relation between V^* and the boundary shear stress τ_0 (Equation 8-4) to give

$$\theta_e = \frac{\tau_0}{(\gamma_s - \gamma) d}. \tag{8-29}$$

Thus θ_e can be thought of as a *dimensionless shear stress* (check the dimensions in Equation 8-29). In the context of erosion, the term *tractive force* is often used as a synonym for shear stress.

Equation 8-29 was first formulated in 1936 by the German hydraulician A. Shields, who also conducted experiments to determine the critical value of θ_e at which motion of the particle (erosion) begins. He found that this critical value depended on another dimensionless parameter, the *erosive Reynolds number* $\mathbb{R}_e$, where

$$\mathbb{R}_e \equiv \frac{V^* d}{\nu} \tag{8-30}$$

and ν is the kinematic viscosity. By reexamining Shields's and subsequent experimental data, Yalin and Karahan (1979) developed the curves relating θ_{ec} and $\mathbb{R}_e$ that are shown in Figure 8.6, where θ_{ec} is the critical value of θ_e.

Rearranging Equation 8-29 gives

$$\tau_{0c} = \theta_{ec}(\gamma_s - \gamma) d \tag{8-31}$$

where τ_{0c} is the critical value of tractive force and θ_{ec} is determined from Figure 8.6. Where defined, the curve for laminar flows lies above that for turbulent flows, but the two coincide at very low values of $\mathbb{R}_e$. This is because low values of $\mathbb{R}_e$ correspond to high values of the thickness of the laminar sublayer (see Equation 6-27). Turbulent flows become hydraulically rough beginning at about $\mathbb{R}_e = 4$ and are fully rough for $\mathbb{R}_e \geq 70$. The curve of Figure 8.6 shows that $\theta_{ec} = 0.044$ for fully rough flows.

Equations 8-30 and 8-31 and Figure 8.6 can be used to construct a useful graph that directly expresses stream competence in terms of tractive force. Figure 8.6 shows that for flows with $\mathbb{R}_e \geq 70$, θ_{ec} has the constant value 0.044. Since for most common sediments the weight density is close to 26000 N m^{-3}, the critical tractive force is found from Equation 8-31 as

$$\tau_{0c} = 0.044 \cdot 16200 \cdot d = 713d, \tag{8-32}$$

where d is in meters and τ_{0c} is in newtons per square meter. The actual tractive force is given by Equation 8-11 as

$$\tau_0 = 9800RS_0, \tag{8-33}$$

where R is in meters and τ_0 is in newtons per square meter. The critical point for erosion to occur is when the actual shear equals the critical shear, so

$$713d_c = 9800RS_0,$$

$$d_c = 13.7RS_0, \tag{8-34}$$

where d_c is the critical sediment diameter that can just be eroded by the flow. Equation 8-34 holds for $RS_0 \geq 1.5 \times 10^{-4}$ m; for smaller values we have to use a trial-and-error procedure to determine d_c. The results are shown in Figure 8.7, where stream competence is related to RS_0 over the range of values commonly encountered in natural streams.

It is also possible to express competence in terms of a critical velocity for erosion. One way to develop this relation makes use of Figure 8.7 and the equation for mean velocity that was derived from the expression for a vertical velocity profile in

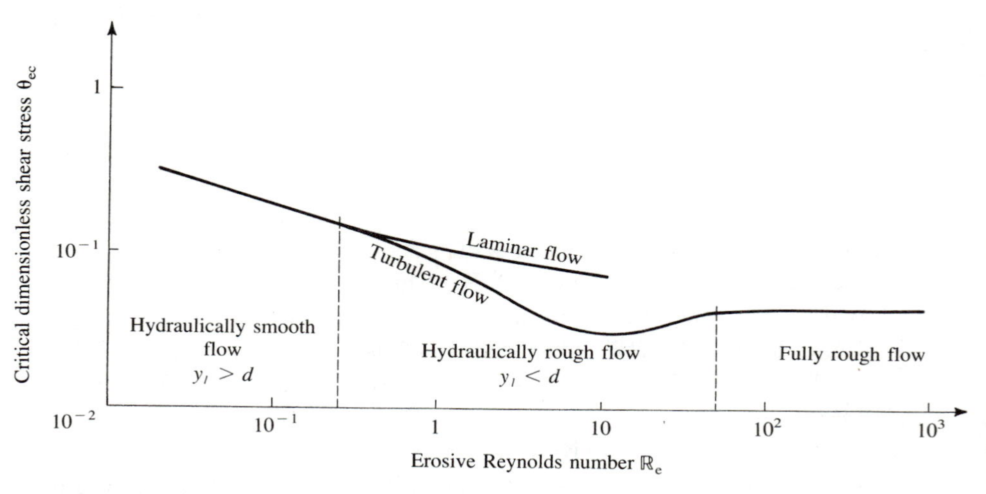

Figure 8.6
Relation between critical dimensionless shear stress θ_{ec} (Equation 8-29) and erosive Reynolds number $\mathbb{R}_e$ (Equation 8-30) for laminar and turbulent flows. After Yalin and Karahan (1979).

turbulent flow (Equation 6-36):

$$V = 2.5V^*\left[\ln\left(\frac{Y}{y_0}\right) - 1\right], \tag{8-35}$$

where $y_0 = 0.11\nu/V^*$ for smooth flow ($y_l \geq k_s$), $y_0 = 0.033k_s$ for rough flow ($y_l < k_s$), y_l is the thickness of the laminar sublayer ($y_l = 4\nu/V^*$), and k_s is the roughness height.

In order to pursue this approach we must specify a particular hydraulic radius (considered equal to mean depth) and slope separately, instead of using their product, RS_0, as a single independent variable. Once a hydraulic radius and slope are specified, V^* can be computed and Figure 8.7 can be used to find the corresponding d_c. We then consider the case when the flow is over grains of diameter d_c, and thus assume that $k_s = d_c$. With this information we can decide whether the flow is rough or smooth, compute y_0 accordingly, and then calculate the critical erosion velocity from Equation 8-35. The results of this analysis are shown in Figure 8.8 for five

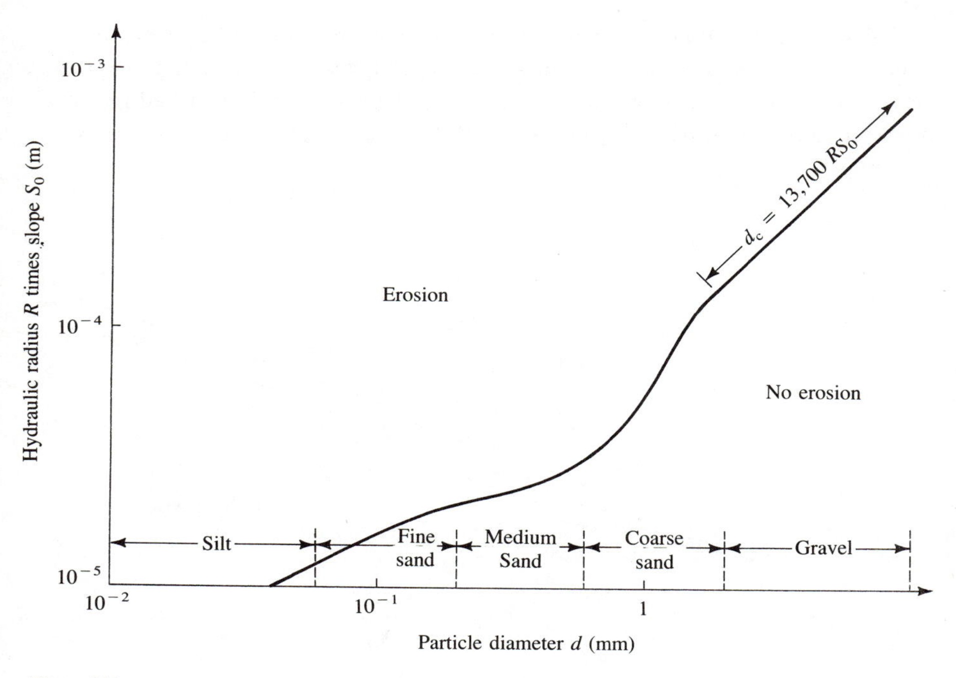

Figure 8.7

Relation between particle diameter d and critical value of hydraulic radius R times slope S_0 computed via Equations 8-30 and 8-31 and Figure 8.6. Multiply RS_0 by 9800 to get the critical tractive force in newtons per square meter.

different hydraulic radii. Electrostatic forces become important for particles smaller than about 0.1 mm in diameter, so the curves are not extended to diameters smaller than that.

Bed Forms in Alluvial Channels Observations of rivers and experiments in flumes have revealed that in flows over sand beds, there is a typical sequence of bed forms that occurs as discharge changes. These forms are intimately related to processes of erosion that begin when the critical tractive force or velocity is reached, and in turn they strongly influence the velocity and tractive force because of their effects on flow resistance.

Table 8.5 describes and Figures 8.9–8.12 illustrate the sequence of forms. The *lower flow regime* includes all stages up to and including dunes; flows in this range have generally high resistances and low bed-material transport rates. The transition to the *upper flow regime* is characterized by a recurrence of a plane bed, and flows in this regime have relatively low resistances and relatively high bed-material discharges. It should be noted that different parts of a given stream reach may show different bed forms at the same time as a result of variations in flow conditions (e.g., higher velocities and depths in the center of straight reaches or in the outer parts of bends).

Considerable research effort has been expended on predicting the flow conditions under which the various forms occur and in developing means for estimating their effects on resistance. Initial attempts considered tractive force and bed-material diameter to be the principal determinants of bed form, but an extensive series of

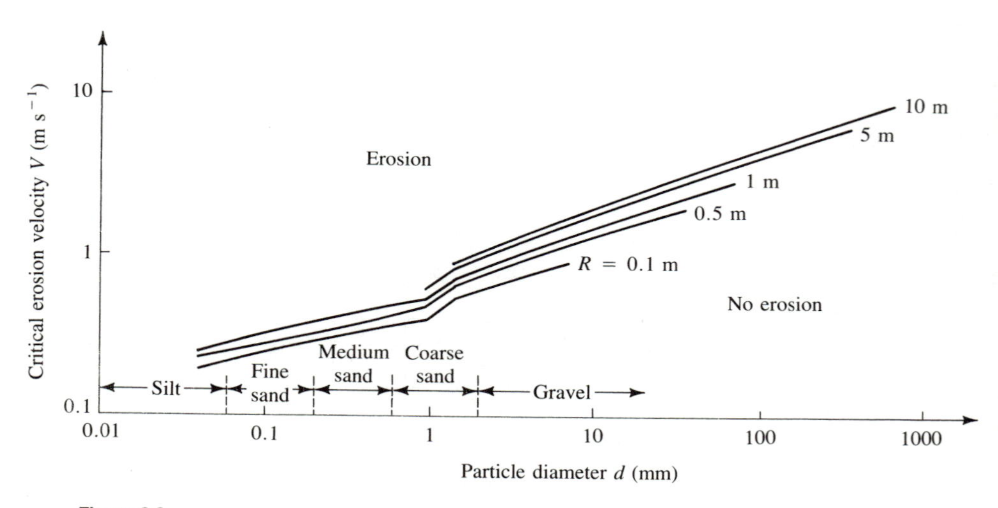

Figure 8.8
Relation between particle diameter d and critical mean velocity for erosion V computed via Equation 8-35 and Figure 8.7 for selected values of hydraulic radius R.

Table 8.5
Nomenclature of Bed Forms in Sand-Bed Streams
After Task Force on Bed Forms in Alluvial Channels (1966).

	Lower flow regime
Plane bed	Generally flat bed, often with irregularities due to deposition; occurs where stream power is insufficient to cause erosion of the bed.
Ripples	Small wavelike bed forms with amplitudes <0.03 m and wavelength <0.3 m. May be triangular to sinusoidal in longitudinal cross section. Crests may be short and irregular to long, parallel, regular ridges transverse to the flow. Typically move downstream at velocities much lower than the stream velocity. May occur on upslope portions of dunes.
Dunes	Larger wavelike forms, generally triangular in longitudinal cross section with gentle upstream slopes and steep downstream slopes. Crest lengths approximately same magnitude as wavelength. Move downstream at velocities much lower than stream velocity.
	Upper flow regime
Transitional plane bed	Occurs often with heterogeneous, irregular bed forms, a mixture of flat areas and low-amplitude ripples and/or dunes.
Antidunes	Large wavelike forms with triangular to sinusoidal longitudinal cross sections that are in phase with gravity waves in the water. Crest lengths approximately equal to wavelengths. May move up- or downstream or remain stationary.
Chutes and pools	Large mounds of sediment that form steep chutes in which flow is supercritical, separated by pools in which the flow may be either supercritical or subcritical. Hydraulic jumps (see Chapter 10) form at transitions from supercritical to subcritical flow. Chutes move slowly upstream.

flume studies by Simons and Richardson (1966) showed that stream power per unit bed area and bed-material diameter gave better results. Recall from Equation 7-34 that this measure of stream power, ω_A, is computed as

$$\omega_A = V\tau_0. \tag{8-36}$$

Instead of the actual median diameter of the bed material, the median *fall diameter* was used as a measure of grain size. This quantity is defined as the diameter of a spherical particle with a mass density of 2650 N m^{-3} that would have the same fall velocity as the actual grains in water at a temperature of 24°C. It is thus a standardized effective diameter that reflects the relative sedimentation behavior of the bed material more realistically than actual diameter.

A plot of ω_A versus the median fall diameter can be divided into "fields" that

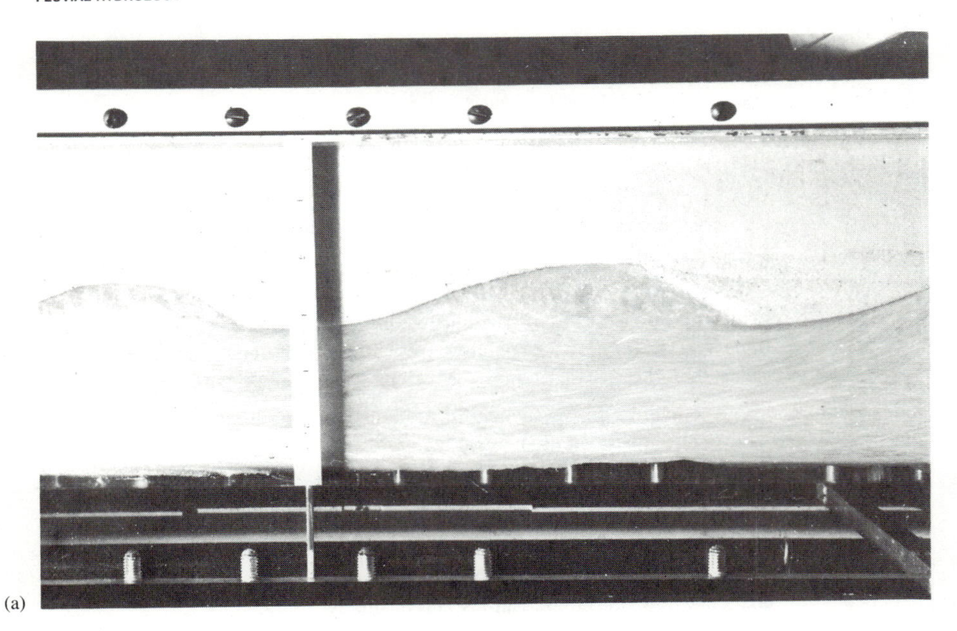

(a)

(b)

Figure 8.9

(a) Side view of ripples in a laboratory flume. The flow is from right to left at a mean depth of 0.064 m and a mean velocity of 0.43 m s^{-1}. Aluminum powder was added to the water to make the streamlines visible. Note that the water surface is unaffected by the ripples. Photograph courtesy of A. V. Jopling, University of Toronto. (b) Ripples on the bed of the Delta River in central Alaska. Flow was from left to right.

(a)

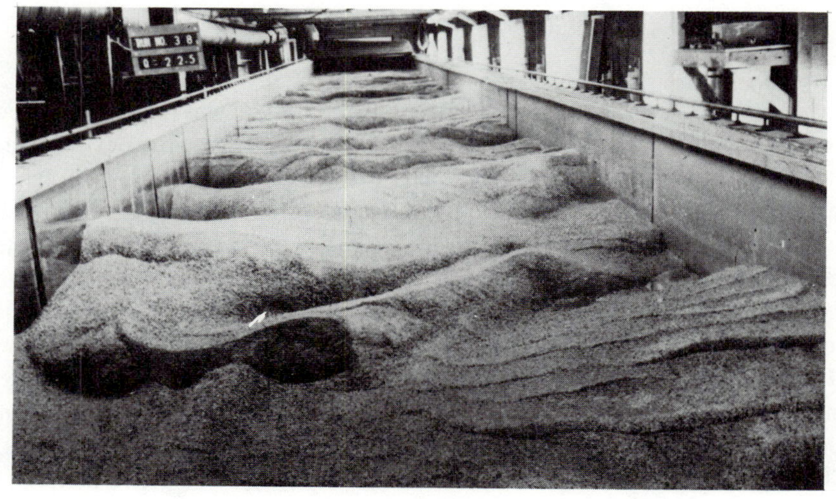

(b)

Figure 8.10

(a) Side view of dunes in a laboratory flume. The flow is from right to left at a mean depth of 0.064 m and a mean velocity of 0.67 m s^{-1}. Aluminum powder was added to the water to make the streamlines visible. The water-surface wave is out of phase with the bed form, and there is a region of flow separation in the lee of the dune. Photograph courtesy of A. V. Jopling, University of Toronto. (b) Sand dunes in a laboratory flume. The flow was toward the observer at a mean depth of 0.31 m and a mean velocity of 0.85 m s^{-1}. Note ripples superimposed on some dunes. Photograph courtesy of D. B. Simons, Colorado State University.

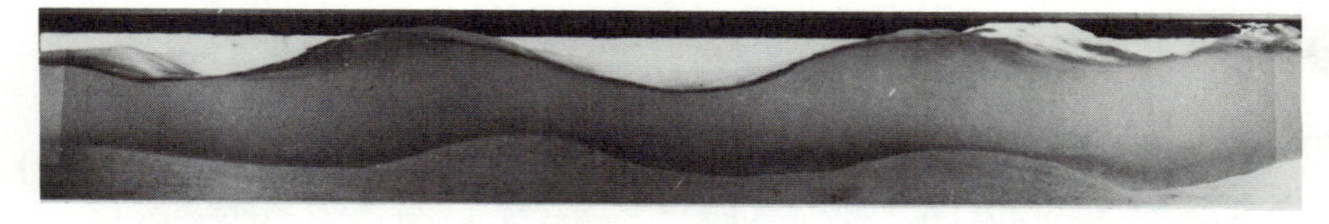

Figure 8.11
Side view of antidunes in a laboratory flume. The flow is from left to right at a mean depth of 0.11 m and a mean velocity of 0.79 m s^{-1}. Note that the surface waves are approximately in phase with the bed forms, which are also migrating to the right. Photograph courtesy of J. F. Kennedy, University of Iowa.

Figure 8.12
Side view of chutes and pools in a laboratory flume. The flow is from right to left at a mean depth of 0.07 m and a mean velocity of 1.0 m s^{-1}. The Froude number is 1.21; note the highly irregular water surface. Photograph courtesy of J. F. Kennedy, University of Iowa.

delineate the conditions under which the various bed forms occur (Figure 8.13). For a given median fall diameter, bed-load movement begins when ω_A reaches its critical value, represented by the solid line in Figure 8.13. The location of this line can be computed by multiplying values of critical tractive force and critical velocity obtained for each value of d_c from Figures 8.7 and 8.8, respectively. If the diameter is less than 0.6 mm, ripples will form initially, and dunes follow at higher values of ω_A. With larger sediment the ripple phase is bypassed and dunes are the initial form. According to Engelund and Hansen (1967), the occurrence of ripples or dunes is determined by whether the flow is hydraulically smooth (ripples) or rough (dunes). As ω_A increases still further, the dunes tend to become ''washed out,'' and a nearly plane bed typically occurs, marking the limit of the lower flow regime.

The upper flow regime is characterized by a highly irregular water surface that is an approximate replica of the stream bed. In the case of antidunes, both the stream and its bed have wavy surfaces; when chutes and pools form, the stream bed consists of alternating steep and less steep reaches.

The effect of bed forms on flow resistance is profound, and several attempts have been made to describe it quantitatively. To provide a conceptual basis for understanding these attempts, note that if we assume that the Chézy equation applies to a flow,

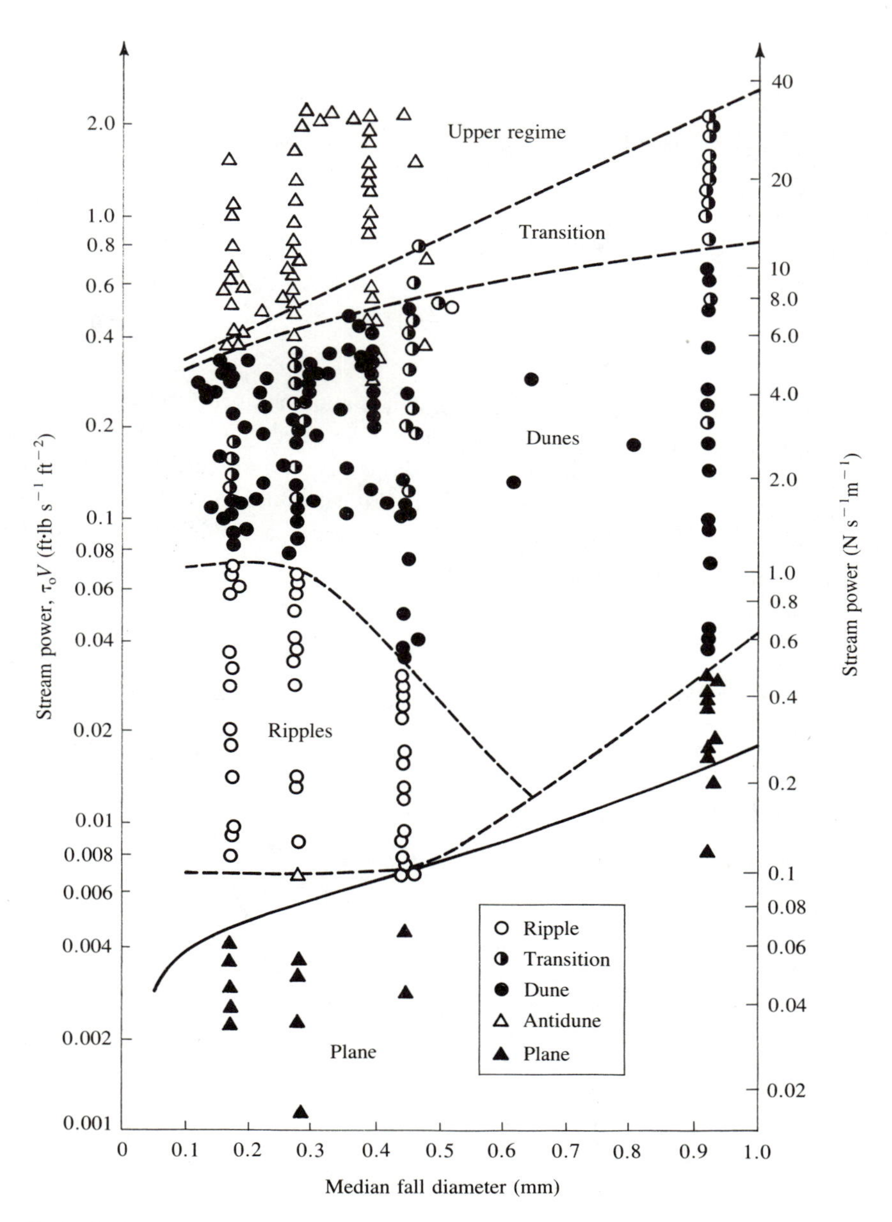

Figure 8.13

Occurrence of various bed forms as a function of unit stream power and median fall diameter, as determined by studies of Simons and Richardson (1966). The solid line is the critical value of unit stream power $\tau_0 V$ determined from Figures 8.7 and 8.8. Modified from Simons and Richardson (1966).

$$V = u_C C R^{1/2} S_0^{1/2}$$

and

$$RS_0 = \frac{V^2}{u_C^2 C^2}. \tag{8-37}$$

Now Equation 8-37 can be substituted into the definition of tractive force to give

$$\tau_0 = \left(\frac{\gamma}{u_C^2 C^2}\right) V^2. \tag{8-38}$$

Consider a straight reach in which only skin friction and bed form contribute to flow resistance. As long as the bed is flat and C depends only on skin friction, the relation between τ_0 and V would follow the dashed curve in Figure 8.14, reflecting Equation 8-38. When stream power reaches its critical value, however, the sequence of bed forms begins and changes the relation between τ_0 and V in a manner similar to that indicated by the solid line in Figure 8.14.

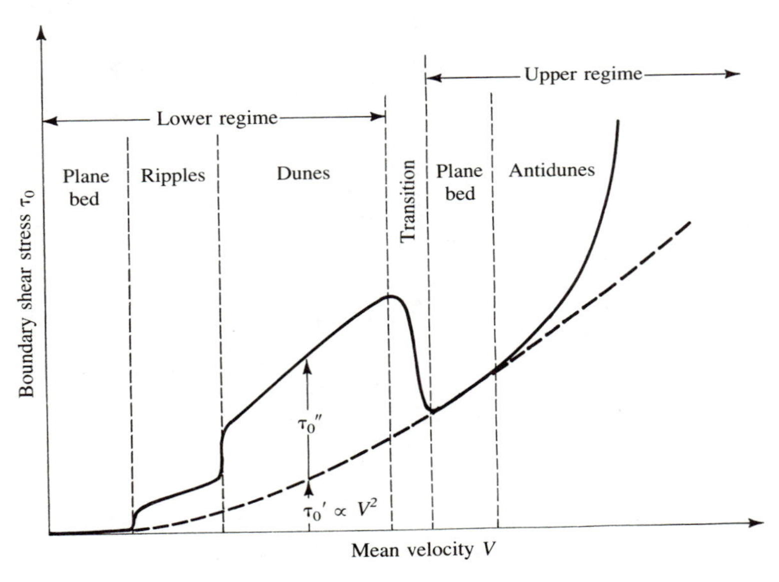

Figure 8.14
Effect of bed forms on the relation between the boundary shear stress τ_0 and the mean velocity V. The dashed line shows the relation τ_0' due to skin friction alone, while τ_0'' represents the effect of bed forms. See Equation 8-39. From Engelund and Hansen (1967).

Engelund and Hansen (1967) thus postulated that

$$\tau_0 = \tau_0' + \tau_0'', \tag{8-39}$$

where τ_0' and τ_0'' are the contributions of skin friction and form drag, respectively, to total shear stress. They then expressed these shear stresses in dimensionless form and transformed Equation 8-39 to

$$\theta_e = \theta' + \theta'', \tag{8-40}$$

where θ_e is defined by Equation 8-29,

$$\theta' = 0.02 S_0^{1/2} \left[\frac{V}{(gd)^{1/2}} \right]^{8/5}, \tag{8-41}$$

and

$$\theta'' = \frac{\alpha \mathbb{F}^2 \gamma h^2}{2(\gamma_s - \gamma) d\lambda} \tag{8-42}$$

(h is the height and λ the wavelength of the bed forms, α is the kinetic energy coefficient, and $\mathbb{F}$ is the Froude number of the flow). Equations 8-41 and 8-42 were derived from dimensional analysis.

After examination of experimental evidence, Engelund and Hansen (1967) combined Equations 8-29, 8-41, and 8-42 to arrive at an equation of the Manning form that accounts for the effects of dunes on flow resistance:

$$V = \left(\frac{10.9}{d^{3/4}} \right) R^{5/4} S_0^{9/8}, \tag{8-43}$$

where d and R are in meters and V is in meters per second (a sediment density of 2650 kg m^{-3} is assumed). Comparing Equation 8-43 with the usual form of the Manning equation gives an indication of the dependence of n on particle size, hydraulic radius, and slope (compare with Equation 8-17):

$$n = \frac{d^{3/4}}{10.9 R^{7/12} S_0^{5/8}}. \tag{8-44}$$

It must be emphasized that Equations 8-43 and 8-44 are empirical, and strictly speaking apply only when dunes are present; Engelund and Hansen (1967) state that Equation 8-43 has satisfactorily predicted velocity in actual rivers under those conditions.

The most intensive and best-known study of the effects of bed form on flow resistance is that of Simons and Richardson (1966). In addition to establishing the relations between bed form, grain size, and stream power shown in Figure 8.13, they developed methods for estimating flow resistance that cover a wider range of conditions than those to which Engelund and Hansen's (1967) equations apply. Conceptually, these methods involve representing the boundary shear, or tractive force, as

$$\tau_0 = \gamma R(S_0' + S_0'') \qquad (8\text{-}45a)$$

or

$$\tau_0 = \gamma S_0(R' + R''). \qquad (8\text{-}45b)$$

In these equations, the primed quantities are the values of slope or hydraulic radius that would exist in a particular flow if only skin friction were present, and the double-primed quantities are the incremental values required to make the equation correct for the actual flow. Thus the latter represent the contribution of the bed forms. Simon and Richardson's (1966) report provides an extensive series of graphs for finding R'' or S_0'' as a function of various flow characteristics, and describes the trial-and-error procedure required to arrive at a complete prediction of the flow characteristics.

Sediment Transport Capacity One of the most intractable problems for hydraulic engineers has been that of predicting the sediment discharge of streams for a given set of flow and bed-material characteristics. Many approaches have been devised over the last 100 years, but to describe even a few of these in detail is beyond the scope of this book. We will instead use some of the concepts developed earlier to show how the problem has been approached analytically, and then suggest works that treat the subject in more depth and detail. Initially, we will consider only the suspended load; subsequent discussions will deal briefly with estimating bed load and total bed-material load.

The first task is to compute the vertical distribution of the concentration of suspended sediment; the product of concentration times the volume flow rate of water at each level will give the vertical distribution of sediment-transport rate, and the integration of this distribution over the entire flow cross section will give the total sediment discharge, or capacity.

Consider a plane parallel to the bottom in a steady uniform flow that is transporting sediment of a uniform size. Under equilibrium conditions, the rate of sediment transport upward through the plane, Q_u, is equal to the rate of transport downward, Q_d, so

$$Q_u = Q_d, \qquad (8\text{-}46)$$

where Q_u and Q_d are expressed as mass per unit time per unit area ($[MT^{-1}L^{-2}]$). The downward rate can be expressed as the product of the sediment concentration c ($[ML^{-3}]$), and the fall velocity v_f:

$$Q_d = cv_f. \tag{8-47}$$

The upward rate is expressed by considering it to be a diffusion process. Referring to Equation 3-22, we can write

$$Q_u = -D_s \frac{dc}{dy}, \tag{8-48}$$

where D_s is the vertical diffusivity of sediment in turbulent flows ($[L^2T^{-1}]$) and y is normal distance measured upward from the bottom.

Substituting Equations 8-47 and 8-48 into 8-46 then gives

$$-D_s \frac{dc}{dy} = cv_f,$$

which can be rearranged to

$$\frac{dc}{c} = -\frac{v_f}{D_s} dy. \tag{8-49}$$

It seems reasonable to assume that v_f is not a function of y. If we also assume for the time being that D_s is constant, Equation 8-49 can be directly integrated:

$$\ln c = -\frac{v_f y}{D_s} + C_1,$$

$$c = C_2 \exp\left(-\frac{v_f y}{D_s}\right), \tag{8-50}$$

where C_1 is a constant of integration and C_2 its antilog. The constants are evaluated by choosing a reference level, y_a at which the concentration c_a can be determined. Then Equation 8-50 becomes

$$c_a = C_2 \exp\left(-\frac{v_f y_a}{D_s}\right),$$

so that

$$C_2 = \frac{c_a}{\exp(-v_f y_a / D_s)}. \tag{8-51}$$

When this equation is substituted into Equation 8-50, we have

$$\frac{c}{c_a} = \exp\left[-\frac{v_f}{D_s}(y - y_a)\right] \tag{8-52}$$

as the expression for the relative concentration of sediment as a function of distance above the bottom.

In order for Equation 8-52 to be useful, the concentration c_a at level y_a and the value of D_s would have to be independently estimated or measured, and the value of D_s must be effectively constant throughout the flow. Putting aside the problem of determining c_a for the moment, we can exploit aspects of the analysis of turbulent velocity profiles in Chapter 6 to gain some insight into D_s and its variability. We begin by assuming that the diffusivity of sediment is identical to the diffusivity of momentum D_m. The latter is evaluated by recalling Equation 6-18 relating the shear stress τ to the velocity gradient in a turbulent flow:

$$\tau = \varepsilon \frac{dv}{dy}, \tag{8-53}$$

where ε is the eddy viscosity, analogous to dynamic viscosity. If ε is divided by the mass density, we have a quantity that is analogous to kinematic viscosity. From the discussion of Equation 3-23 we know that kinematic viscosity represents the diffusivity of momentum. Therefore we can define

$$D_m \equiv \frac{\varepsilon}{\rho}, \tag{8-54}$$

which gives the diffusivity of momentum for a turbulent flow. To continue the analysis, we now assume that

$$D_s = D_m. \tag{8-55}$$

With Equations 8-53–8-55, we can write

$$D_m = \frac{\tau}{\rho(dv/dy)}. \tag{8-56}$$

This expression is evaluated by recalling that in the analysis of turbulent velocity profiles, τ was assumed to have a constant value,

$$\tau = \tau_0 = \gamma Y S_0, \tag{8-57}$$

where τ_0 is the boundary shear stress, Y is the flow depth, and S_0 is the slope (Equation 6-21). Also from that analysis, we found that

$$\frac{dv}{dy} = \frac{(gYS_0)^{1/2}}{ky}. \tag{8-58}$$

Combining Equations 8-55–8-58 yields

$$D_s = (gYS_0)^{1/2}ky = V^*ky, \tag{8-59}$$

where V^* is the friction velocity (see Equation 6-25).

Equation 8-59 states that D_s is a function of y, whereas the analysis leading to Equation 8-52 required that D_s be constant. We will examine the implications of these conflicting requirements, and ultimately compare them with measured values. To obtain a constant value of D_s we can assume that the variation is linear with depth, as Equation 8-59 indicates, and take the average value:

$$D_s = \frac{V^*kY}{2}. \tag{8-60}$$

Using this expression in Equation 8-52 gives

$$\frac{c}{c_a} = \exp\left[-\frac{2v_f}{V^*kY}(y - y_a)\right]. \tag{8-61}$$

An alternative approach is to accept that D_s varies with y as indicated by Equation 8-59, and introduce this relationship into the original differential equation, Equation 8-49:

$$\frac{dc}{c} = -\frac{v_f}{V^*k}\left(\frac{dy}{y}\right). \tag{8-62}$$

Integration of this equation and evaluation at a reference level y_a as before leads to the following alternative expression for the relative concentration of suspended sediment as a function of distance above the bottom:

$$\frac{c}{c_a} = \left(\frac{y_a}{y}\right)^{v_f/kV^*}. \tag{8-63}$$

For comparison with measured values it is convenient to take the reference level as the value measured at half the depth, that is, $y_a = Y/2$. Then Equations 8-61 and 8-63 become

$$\frac{c}{c_{md}} = \exp\left[-\frac{2v_f}{kV*}\left(\frac{y}{Y} - \frac{1}{2}\right)\right] \quad (8\text{-}64)$$

and

$$\frac{c}{c_{md}} = \left(\frac{Y}{2y}\right)^{v_f/kV*}, \quad (8\text{-}65)$$

respectively, where c_{md} is the concentration at mid-depth.

Figure 8.15 compares Equations 8-64 and 8-65 with sediment concentrations measured in a flume. It shows that Equation 8-65, incorporating the linear variation of D_s with y, gives somewhat better predictions, but that neither does particularly well.

Disappointing results such as these have led to a further reexamination of how D_s is related to y. Reasonable success has been obtained by assuming that

$$D_s = V*k\left(y - \frac{y^2}{Y}\right), \quad (8\text{-}66)$$

which represents D_s as a parabola with a value of 0 at the top and bottom of the flow and a maximum value of $V*kY/4$ at mid-depth. When Equation 8-66 is substituted into the original differential equation (Equation 8-49), integration leads to

$$\frac{c}{c_a} = \left[\left(\frac{Y - y}{y}\right)\left(\frac{y_a}{Y - y_a}\right)\right]^{v_f/kV*}, \quad (8\text{-}67)$$

and when $y_a = Y/2$,

$$\frac{c}{c_{md}} = \left(\frac{Y}{y} - 1\right)^{v_f/kV*}. \quad (8\text{-}68)$$

Equation 8-67 was first published by Rouse (1937).

Figure 8.15 shows that Equation 8-68 does quite well at predicting the measured sediment concentrations through most of the flow, although it underestimates near the bed. The discrepancies between the measured and predicted values are probably largely the result of the high concentrations near the bed, which (1) cause the actual fall velocity to differ from the assumed value, which applies strictly only to isolated sediment particles; and (2) cause the apparent fluid density to be higher than the constant values assumed for clear water. Other possible reasons are examined by Garde and Ranga Raju (1978).

It is interesting to note that the experimental verification of Equation 8-67 implies that Equation 8-66 is essentially correct, and hence that the diffusivities of momentum and suspended sediment are in fact not equal, as was assumed in the development of Equation 8-59.

Two cautions about Equation 8-67 should be noted. First, it predicts a zero concentration at the surface and an infinite concentration at the bed, neither of which occurs in nature. Thus that equation should not be considered reliable at the extremes of depth. Second, the equation was developed for steady uniform flow and a single sediment size, and it should not be expected to provide accurate predictions when natural conditions depart markedly from those assumed.

In spite of these cautions, it appears reasonable to accept the essential validity of Equation 8-67, and to use it to further our understanding of the sediment-transport process. For convenience, we denote the exponent in that equation by Z:

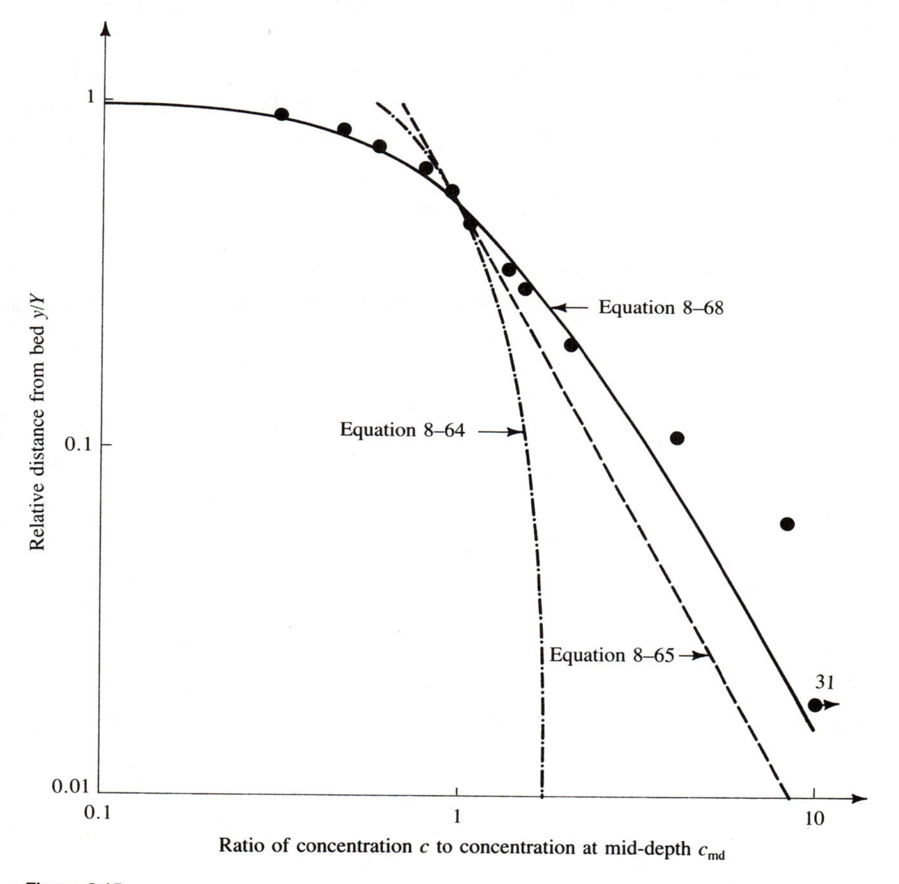

Figure 8.15
Comparison of vertical distribution of suspended sediment concentration predicted by Equations 8-64, 8-65, and 8-68 with values (closed circles) measured by Barton and Lin (1955). For this case, $d = 0.18$ mm and $V^* = 0.055$ m s^{-1}.

$$Z \equiv \frac{v_f}{kV*}. \tag{8-69}$$

Figure 8.16 shows that at small values of Z the sediment distribution is nearly uniform, whereas at high values the concentrations near the bed are many orders of magnitude greater than those near the surface. Recalling the definition of $V*$ (Equation 6-25), we see that the value of the exponent increases with increasing sediment size (v_f) and decreases with increasing hydraulic radius (which approximately equals the mean depth) and slope. Thus for given flow conditions the distribution of sediment is more uniform for finer-grained bed materials. When Z exceeds about 5, virtually all the transport is very close to the bed, where velocities are low, and hence the transport of suspended sediment is negligible (Garde and Ranga Raju, 1978).

Although the analysis leading to Equation 8-67 has provided considerable insight into the process of sediment transport, the problem of determining the reference concentration c_a has not been completely resolved. One possible approach is to take

$$y_a = 2d, \tag{8-70}$$

where d is the particle diameter (Einstein, 1950), and to use the value $c_a = 480$ kg m^{-3} for that level. This value is suggested by experimental results (Garde and Ranga

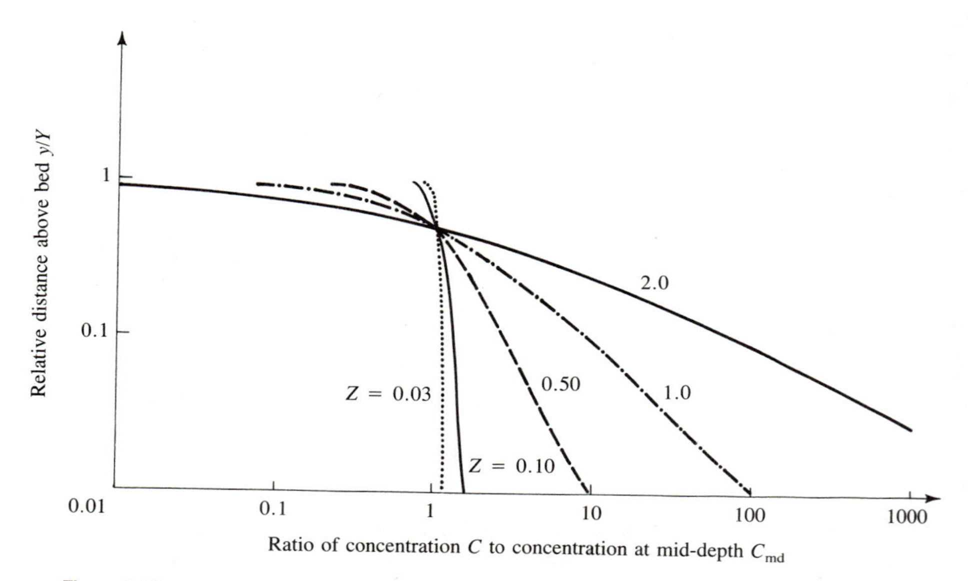

Figure 8.16
Effect of the exponent Z on the vertical distribution of the suspended sediment.

Raju, 1978), but probably applies only to sand-bed streams. With it, Equation 8-67 becomes

$$c = 480\left[\left(\frac{Y - y}{y}\right)\left(\frac{2d}{Y - 2d}\right)\right]^Z$$

or, since $2d$ would typically be a very small number,

$$c = 480\left[\left(\frac{Y - y}{y}\right)\frac{2d}{Y}\right]^Z, \qquad (8\text{-}71)$$

where c is in kilograms per cubic meter. Other approaches to this problem are discussed in the texts by Graf (1971), Vanoni (1975), Raudkivi (1976), and Garde and Ranga Raju (1978).

As noted earlier, computation of the total sediment discharge q_s per unit width is accomplished by integration of the product of concentration and velocity over the flow depth:

$$q_s = \int_{y_a}^{Y} cv\, dy. \qquad (8\text{-}72)$$

To do this, c is found from Equation 8-67:

$$c = c_a\left[\left(\frac{Y - y}{y}\right)\left(\frac{y_a}{Y - y_a}\right)\right]^Z; \qquad (8\text{-}73)$$

and v from the turbulent velocity profile (Equation 6-26):

$$v = \frac{V^*}{k}\ln\left(\frac{y}{y_0}\right). \qquad (8\text{-}74)$$

The result is a lengthy expression that includes integrals that must be evaluated numerically by means of tables or graphs. This approach was the basis for the pioneering efforts by H. A. Einstein (1950) to develop a practical method for estimating sediment transport, and his graphs have been reprinted many times (e.g., Graf, 1971; Raudkivi, 1976).

Since Einstein's (1950) work, a number of researchers have addressed the problem of estimating sediment discharge. The Task Committee for Preparation of Sediment Manual (1971) of the American Society of Civil Engineers undertook a thorough evaluation of the various approaches, and concluded that three of them could be recommended for engineering application. The first of these was Colby's (1964a,b) modification of Einstein's (1950) procedure, mentioned earlier. A second approach developed by Engelund and Hansen (1967) from dimensional considerations and experimental data was found to give satisfactory results for sand-bed streams flowing over dune beds. Finally, a complex method devised by Toffaleti (1969) also predicted actual sediment discharge with acceptable accuracy. Details of these methods and comparisons of measured and predicted values can be found in the committee's report as well as in the original papers.

The preceding discussion indicates that general approaches to computing suspended-load discharge are well established and can provide useful estimates if sufficient information about the flow and sediment is available. However, estimation of bed-load transport has proved considerably more difficult, and validation of methods is uncertain because of the great difficulties of measuring bed load. Most of the approaches to this problem are based on one of the following general relations (Garde and Ranga Raju, 1978):

$$q_b \propto f(\tau_0 - \tau_{0c}); \tag{8-75}$$

$$q_b \propto f(q - q_c); \tag{8-76}$$

$$q_b \propto f(V - V_c); \tag{8-77}$$

or

$$q_b \propto f(\omega_A - \omega_{Ac}). \tag{8-78}$$

In these equations, q_b is bed-load discharge per unit width, τ_0 is tractive force, q is water discharge per unit width, V is mean flow velocity, and ω_A is stream power per unit bed area (Equation 7-34). The subscript c indicates critical values for the initiation of motion, and f indicates a functional relationship.

A comprehensive examination of approaches to this problem would be too lengthy for this text; useful reviews can be found in the several works devoted to sediment transport, including those of Graf (1971), Yalin (1972), Vanoni (1975), Raudkivi (1976), Simons and Senturk (1977), and Garde and Ranga Raju (1978). Nevertheless, specific mention should be made of the pioneering work of the French engineer du Boys (1879), who based his analysis on the tractive force at the bed (Equation 8-75). Refinements of this approach have been attempted by many researchers, the best known of whom are Shields (1936), Kalinske (1947), Meyer-Peter and Müller (1948), and Garde and Albertson (1961). A method based on Equation 8-76 by a German engineer, A. Schoklitsch, was brought to the attention of the English-speaking world by Shulits (1935). Einstein's (1950) intensive examination of sediment transport included a consideration of bed load; his treatment introduced probability concepts. Finally, recent work by Bagnold (1977) suggests that stream power (Equation 8-78) may provide a useful independent variable for estimating bed-load transport.

Not surprisingly, many of the approaches to computing suspended load and bed load have also been applied to the computation of total bed-material load, and descriptions of these methods can be found in the sediment-transport texts cited earlier. Prominent among these approaches are those of H. A. Einstein (1950) and its modifications by Colby and Hembree (1955), Bagnold's (1956, 1966) method based on stream power, the studies of Engelund and Hansen (1967) and Vittal et al. (1973) using tractive force, and the equations developed by Bishop et al. (1965) based on dimensional analysis and empirical data.

More recently considerable success in predicting total bed-material concentration has been claimed by Yang and Stall (1976). After reviewing previous literature, they concluded that "the rate of [bed-material] transport in an alluvial channel is dominated by the rate of potential energy expenditure per unit weight of water, i.e., the unit stream power [ω_w]" (Yang and Stall, 1976, p. 559). Using this principle, they formulated a dimensionless equation of the form

$$C_t = J\left(\frac{\omega_w}{v_f} - \frac{\omega_{wc}}{v_f}\right)^K, \tag{8-79}$$

where C_t is the concentration of bed-material sediment in the flow (weight of sediment per weight of water), ω_w is unit stream power (Equation 7-36), ω_{wc} is a critical, or threshold, value of ω_w below which there is no sediment transport, v_f is the median fall velocity of the sediment particles, and J and K are dimensionless empirical factors that depend on the characteristics of the flow and sediments (see Equations 8-83 and 8-84). Note that the unit stream power is made dimensionless by division by v_f, which seems physically reasonable, since v_f can be considered an index of a particle's "reluctance" to be transported.

In flow that is hydraulically rough, experiments have shown that the critical erosion velocity depends only on the fall velocity and not on other characteristics of the flow. Recall from Equation 6-33 that hydraulically rough flows exist when the roughness height exceeds $4\nu/V^*$. If we replace roughness height by grain diameter d, this criterion is

$$d > \frac{4\nu}{V^*},$$

and rearranging yields

$$\frac{V^*d}{\nu} > 4$$

or, from the definition of erosive Reynolds number (Equation 8-30),

$$\mathbb{R}_e > 4.$$

It turns out, however, that for present purposes flow is not considered fully rough until

$$\mathbb{R}_e \geq 70 \tag{8-80}$$

(see Figure 8.6 and the related discussion). For these conditions, Yang and Stall (1976) found empirically that the critical velocity for erosion V_c was related to fall velocity as

$$V_c = 2.05 v_f. \tag{8-81}$$

This relation can be expressed nondimensionally and put in terms of unit stream power simply by multiplying by the slope S_0 and rearranging: If $\mathbb{R}_e \geq 70$, then

$$\frac{\omega_{wc}}{v_f} = 2.05 S_0. \tag{8-82a}$$

For flows that are not fully rough, the relation between V_c and v_f should be a function of $\mathbb{R}_e$. From experimental data, Yang and Stall (1976) found that this relation could be expressed as

$$V_c = \left(\frac{2.5}{0.434 \ln \mathbb{R}_e - 0.06} + 0.66\right) v_f,$$

and making the same transformations as before gave the expression for critical stream power in flows that are not fully rough: If $\mathbb{R}_e < 70$, then

$$\frac{\omega_{wc}}{v_f} = \left(\frac{2.5}{0.434 \ln \mathbb{R}_e - 0.06} + 0.66\right) S_0. \tag{8-82b}$$

To complete their analysis, Yang and Stall (1976) used dimensional considerations to reason that J and K in Equation 8-79 should depend only on the particle Reynolds number $\mathbb{R}_p$ (Equation 8-21) and the ratio of friction velocity to fall velocity. Multiple regression analysis using over 450 individual measurements from flumes and natural rivers resulted in the following empirical relations:

$$J = \frac{272{,}000}{\mathbb{R}_p^{0.286}\left(\dfrac{V*}{v_f}\right)^{0.457}}; \tag{8-83}$$

$$K = 1.799 - 0.178 \ln \mathbb{R}_p - 0.136 \ln\left(\frac{V^*}{v_f}\right). \quad (8\text{-}84)$$

When Equations 8-82–8-84 are used in Equation 8-79, C_t is in parts per million. To compute the total sediment discharge or capacity Q_s, the following equation is used:

$$Q_s = 10^6 C_t \gamma_s Q, \quad (8\text{-}85)$$

where Q_s is in units of weight per unit time ($[FT^{-1}]$), γ_s is the weight density of the sediment (usually taken as 2.65γ), and Q is the water discharge. Figure 8.17 compares measured sediment discharges with those computed by using Equations 8-79 and 8-82–8-85. The agreement indicates that this approach based on unit stream power provides very satisfactory predictions over a wide range of conditions.

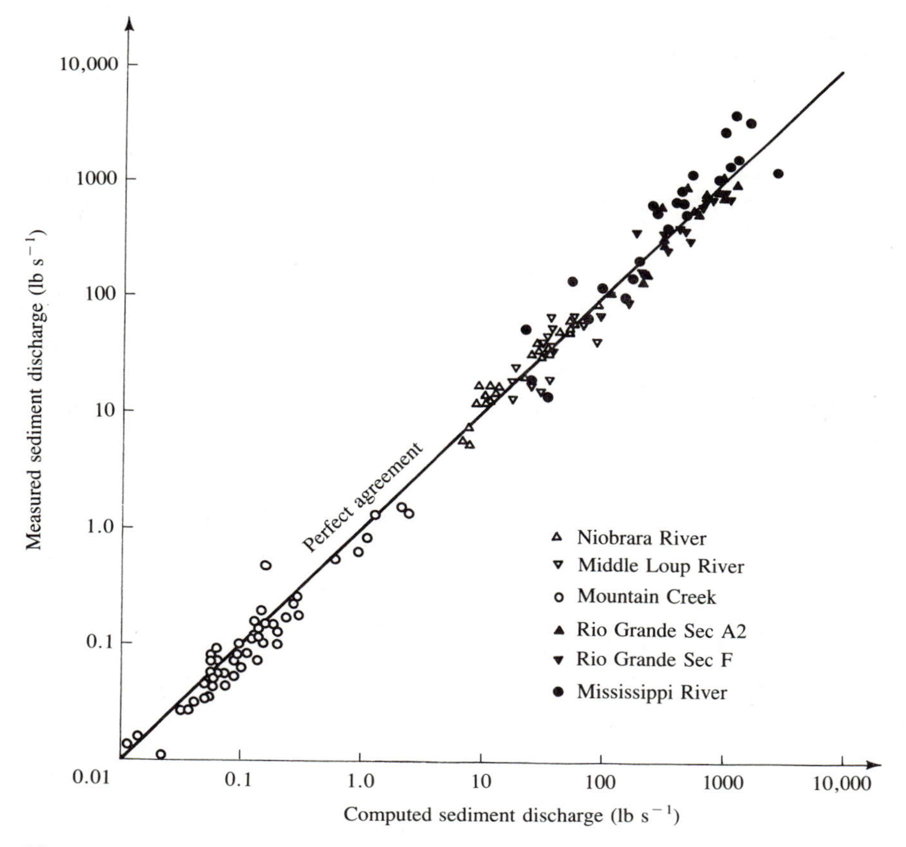

Figure 8.17
Comparison between sediment discharge measured at six river stations with that predicted by using Equations 8-79 and 8-82–8-84. From Yang and Stall (1976).

Exercises

8-1. Locate a reasonably uniform reach of a small stream and use Cowan's (1956) procedure (Equation 8-13) to estimate Manning's n. Then, with a transit or hand level, tape, and stadia rod, measure the slope of the water surface through the reach. Use standard U.S. Geological Survey techniques to compute the discharge and the mean velocity by the velocity–area method (Corbett, 1945; Boyer, 1964; Buchanan and Somers, 1969). Then calculate Manning's n from your data and compare it with your initial estimate.

8-2. Show that for hydraulically rough flow, the friction factor k_t is related to the ratio of the depth Y to the roughness height k_s as

$$k_t = \frac{1}{6.25[\ln(30Y/k_s) - 1]^2}.$$

(*Hint:* Note that $k_t = V^{*2}/V^2$ (Table 8.1) and use Equation 8-35.) Then plot k_t as a function of Y/k_s on arithmetic graph paper for $5 \leq (Y/k_s) \leq 100$.

8-3. Using Equation 8-21 and Figure 8.4, show that the critical value of the particle Reynolds number, $\mathbb{R}_p = 1$, occurs when $d \simeq 0.1$ mm. Assume a particle density of 2650 kg m^{-3} and a temperature of 10°C.

8-4. The following data were measured in actual streams by Leopold and Wolman (1957). Decide which would be competent to erode bed material of the given mean diameter, using first the critical-tractive-force criterion (Figure 8.7) and then the critical-velocity criterion (Figure 8.8). Assume that mean depth equals hydraulic radius.

Stream	At or near	Width (m)	Mean depth (m)	Mean velocity (m s^{-1})	Slope	Median particle diameter (mm)
a. North Platte River	North Platte, NE	165	0.52	0.73	0.0014	0.18
b. North Platte River	Douglas, WY	95	0.52	0.91	0.00095	6.1
c. Little Popo Agie River	Lander, WY	12	0.34	0.82	0.0049	33.5
d. Cottonwood Creek	Daniel, WY	13	0.24	0.55	0.0041	18.3
e. North Platte River	Sutherland, NE	104	0.28	0.52	0.0011	0.15
f. North Platte River	Lisco, NE	122	0.41	0.64	0.0013	0.49
g. Cheyenne River	Edgemont, SD	27	0.24	0.58	0.0013	0.05
h. Lance Creek	Spencer, WY	13	0.16	0.43	0.00073	3.0

8-5. For the flows in Exercise 8-4 that are transporting sediment, use Figure 8.13 to determine the bed forms that should be present.

8-6. Show the steps by which Equation 8-62 leads to Equation 8-63.

8-7. To obtain a visual representation of the vertical distribution of suspended sediment, compute c/c_{md} as a function of y/Y by using Equation 8-68 and plot the results on arithmetic graph paper. Compare this plot with Figure 8.15.

8-8. Compute the value of Z (Equation 8-69) for the flows in Exercise 8-4. Assume that mean depth equals hydraulic radius. Use Figure 8.4 to find the fall velocity (at 10°C) where possible, and use v_f values of 0.83 m s^{-1} and 0.60 m s^{-1} for flows c and d, respectively. Compare the values of Z with your conclusions in Exercise 8-4. Does the criterion of $Z = 5$ as a limit of significant erosion seem compatible with the critical tractive force and critical velocity criteria?

8-9. Select one of the flows in Exercise 8-8 for which $Z < 5$. Using Equation 8-71, plot the actual sediment concentration as a function of depth on log–log and arithmetic paper, with depth on the vertical axis.

8-10. For the flows in Exercise 8-4 that are transporting sediment, compute the total bed-material concentration by using Equations 8-79 and 8-82–8-84. Assume a temperature of 10°C and use Figure 8.4 to estimate fall velocity.

8-11. Your computations in Exercise 8-4 should have shown that flow h (Lance Creek, Spencer, WY) was not competent to transport sediment of the median size found on its bed (3.0 mm). It turns out, however, that it is competent to do so at slightly higher flows. Measurements show that at higher flows the stream would have the following characteristics:

Discharge (m^3 s^{-1})	Width (m)	Mean depth (m)	Mean velocity (m s^{-1})
5	16.5	0.41	0.77
10	18.1	0.59	0.98
50	22.3	1.38	1.69
100	24.4	2.00	2.14

Assuming that the slope remains constant at 0.00073, use Equations 8-79 and 8-82–8-84 to compute the bed-material concentrations for these higher flows. Then compute the sediment discharges in metric tons per day by means of Equation 8-85 (1 metric ton = 10^3 kg). Plot both C_t and Q_s versus Q on three-cycle log–log paper, fit a straight line through the points by eye, and determine the equations for the lines.

CHAPTER 9

Gradually Varied Steady Flows

It has become widely accepted among water-resource planners that the most cost-effective way to reduce future flood damages is to restrict development in areas that are subject to flooding (Figure 9.1). The process of identifying and mapping such areas involves, first, the selection of a *design flood*. The design flood is usually specified in terms of the probability that it will be exceeded in any given year, and federal regulations in the United States state that this probability should be 0.01 (i.e., there is one chance in a hundred that the design flood will be exceeded in any year). The second step in this process consists of a hydrologic analysis that provides estimates of the discharge of this design flood at points along the stream system. The next step is a hydraulic analysis in which the elevation of the water surface associated with the design flood is computed. Finally, the computed elevations are used in conjunction with topographic maps to identify those areas lying below the elevation of the design flood.

Our concern in this chapter is the third step in the foregoing sequence—the hydraulic analysis that provides estimates of the water-surface elevation associated with a given discharge in a given stream reach. This analysis is based on the one-dimensional energy equation for steady flows, as developed in Chapter 7 (Equation 7-17). In order to apply this equation, we assume that each cross section of the channel is characterized by a single velocity (the mean velocity, defined as the design discharge divided by the cross-sectional area), depth (the hydraulic mean depth, defined as the cross-sectional area divided by the water-surface width), slope, and resistance.[1] We also assume that the flow is *gradually varied,* that is, that the downstream changes in these flow characteristics are sufficiently gradual that the flow can be considered uniform.

If a flow is gradually varied, we can apply uniform-flow equations like the Chézy equation (Equation 6-48) or the Manning equation (Equation 6-49) to describe the relations among velocity, depth, slope, and channel resistance. Recall from Chapter 7 that in a uniform flow the channel slope S_0 equals the energy slope S_e. Since there is no downstream change in depth in a uniform flow, it is also true that

$$S_s = S_e = S_0, \tag{9-1}$$

where S_s is the slope of the water surface.

If the Manning equation is written in terms of discharge (Exercise 6-9), it can be

Figure 9.1
Computation of the profiles of gradually varied flows is used to identify areas that will be inundated by floods of a specified discharge. Development in such areas should be restricted in order to prevent occurrences like the one in this photograph.

[1]Sometimes a single cross section is divided into a small number of subsections, each with its own mean velocity, depth, and roughness. Methods for dealing with this situation are described later in this chapter.

rearranged to give

$$Y_n = \left(\frac{nQ}{u_M S_0^{1/2} w}\right)^{3/5}, \tag{9-2}$$

where Y_n is called the *normal depth*. Equation 9-2 is actually an equation of definition: For a given channel (S_0, w, n) and discharge (Q), Y_n is the depth that would exist if the flow were uniform. It is assumed in the derivation of Equation 9-2 and throughout this chapter that the channel is wide enough for the hydraulic radius to be approximated by the mean depth.

The critical depth Y_c was defined in terms of velocity in Equation 7-24. Making use of that relation and Equation 7-21, we obtain

$$Y_c = \left(\frac{Q}{wg^{1/2}}\right)^{2/3}, \tag{9-3}$$

which states that for a channel of constant width, the critical depth is proportional to the two-thirds power of the discharge and is independent of the channel slope and roughness. Thus a log–log plot of Y_c versus Q should be a straight line with a slope of 0.667, slightly steeper than the relation between Y_n and Q in Equation 9-2, in which the slope equals 0.600. Therefore we can reason that the two lines will cross at some critical value of discharge Q_c; when $Q < Q_c$, flow in the channel will be subcritical, and when $Q > Q_c$ it will be supercritical (assuming uniform flow).

CLASSIFICATION OF FLOW PROFILES IN NATURAL CHANNELS

Consider a reach of natural channel with a particular width, slope, and roughness, and transmitting a particular discharge. The flow is not necessarily uniform. Depths Y_n and Y_c can be computed from Equations 9-2 and 9-3 and shown as lines parallel to the channel bottom (Figure 9.2). If $Y_n > Y_c$, the channel slope is said to be *mild* (uniform flow would be subcritical); if $Y_n < Y_c$, the channel slope is *steep* (uniform flow would be supercritical). Although it is possible for Y_n to equal Y_c (*critical slope*), this precise condition is unlikely and we will not consider it in this chapter. [See Chow (1959) for a discussion of other profiles, including those for horizontal and adverse (upstream) slopes.]

Flow profiles for a given discharge are classified according to two criteria: (1) whether the normal depth at that discharge is greater or less than the critical depth; and (2) the relative magnitudes of Y_n, Y_c, and the actual depth Y. As indicated in Table 9.1, when $Y_n > Y_c$, the profile is an M (i.e., mild) profile; when $Y_n < Y_c$, it is an S (i.e., steep) profile. In either case, if the actual depth is higher than Y_n and Y_c, the profile is designated M1 or S1; if Y is between Y_n and Y_c, it is designated M2

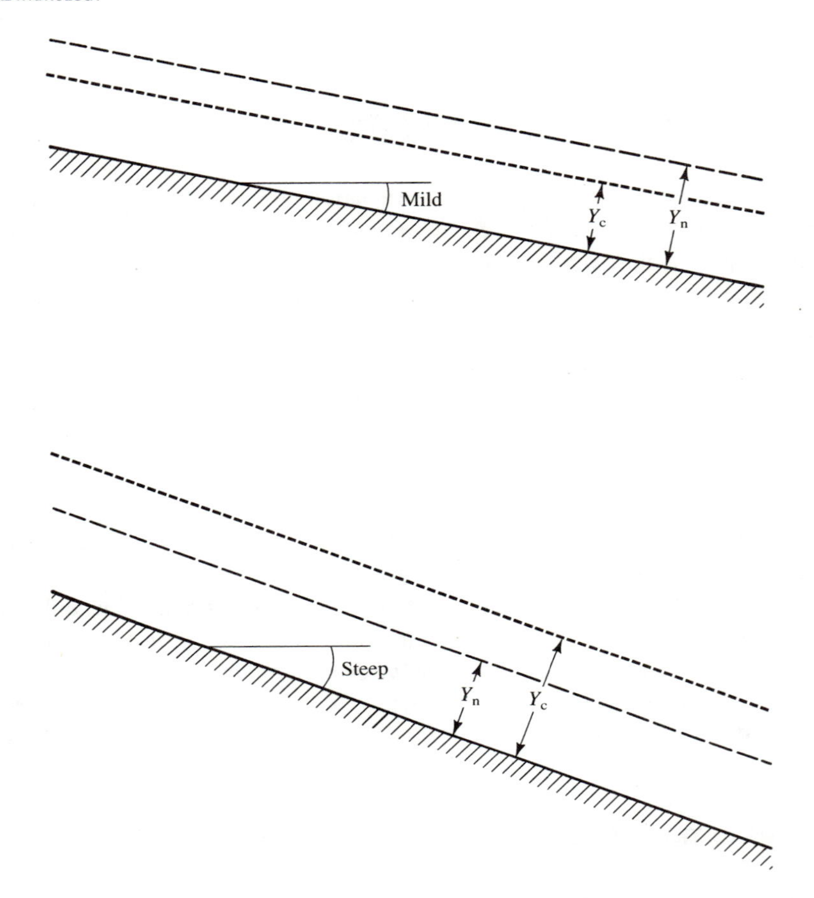

Figure 9.2
Relations between normal depth Y_n and critical depth Y_c for uniform flows on mild and steep slopes.

Table 9.1
Classification of Flow Profiles in Natural Channels

Mild Slopes ($Y_n > Y_c$)	Steep Slopes ($Y_n < Y_c$)
M1 backwater profile $Y > Y_n > Y_c$ (Figure 9.3a), subcritical flow	S1 backwater profile $Y > Y_c > Y_n$ (Figure 9.3d), subcritical flow
M2 draw-down profile $Y_n > Y > Y_c$ (Figure 9.3b), subcritical flow	S2 draw-down profile $Y_c > Y > Y_n$ (Figure 9.3e), supercritical flow
M3 backwater profile $Y_n > Y_c > Y$ (Figure 9.3c), supercritical flow	S3 backwater profile $Y_c > Y_n > Y$ (Figure 9.3f), supercritical flow

or S2; and if Y is less than Y_n and Y_c, it is designated M3 or S3. Typical situations giving rise to the six types of profiles found in natural channels are shown in Figure 9.3. Additional mathematical criteria for these profile types will be developed later in this chapter.

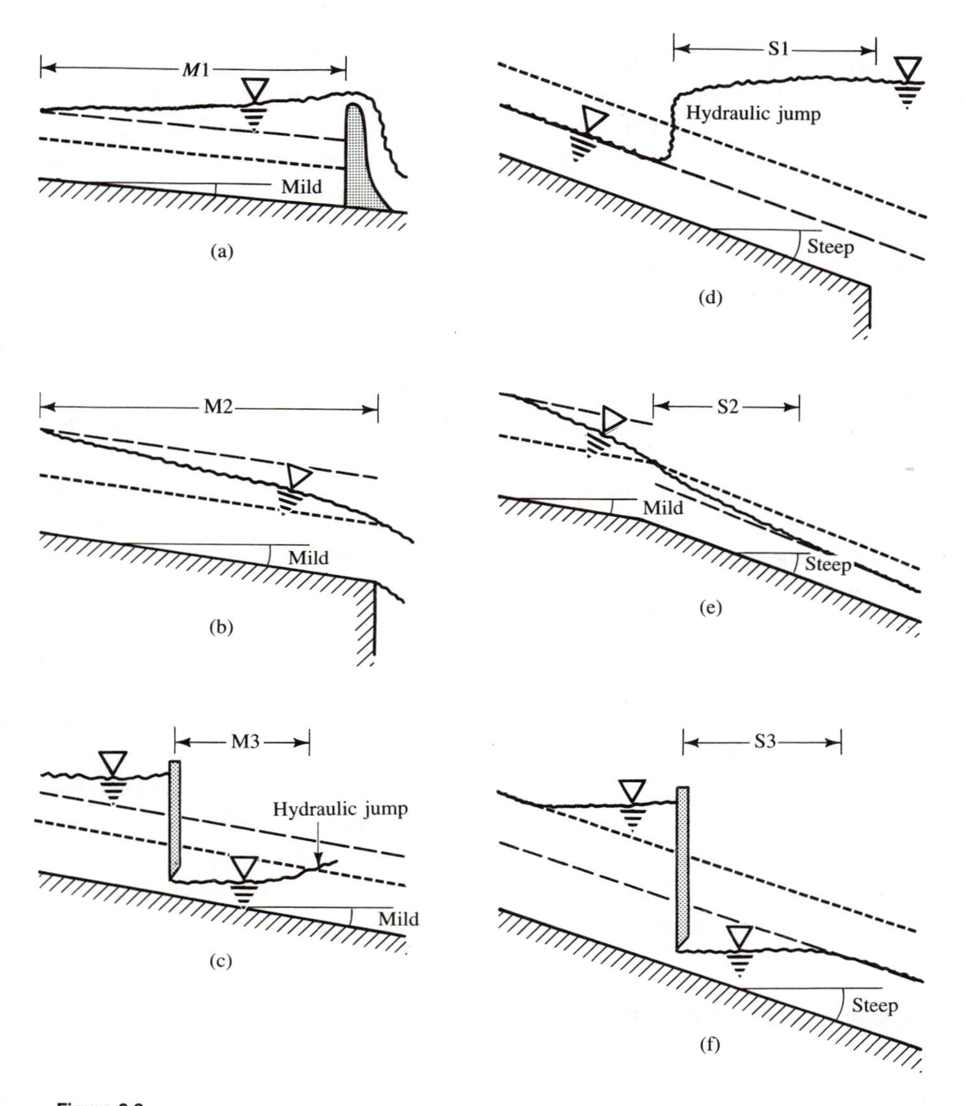

Figure 9.3
Typical situations associated with the most common types of water-surface profiles. Long-dashed lines represent normal depth; short-dashed lines represent critical depth. After Daily and Harleman (1966).

CONTROLS

The portion of the channel that determines the flow profile for some distance upstream (in subcritical flow) or downstream (in supercritical flow) is called a *control*. The control section is characterized by a unique relationship between the water-surface elevation (*stage*) and discharge, and is unaffected by conditions upstream or downstream. The weirs and flumes described in Chapter 10 are artificial controls used for stream gaging. Natural controls are generally sections where there is an abrupt change in width or slope. Generally a control is effective only up to a particular maximum discharge, above which it is effectively submerged and the stage-discharge relation at that point is determined by another control section upstream or downstream.

If the actual flow is subcritical for the discharge of concern (M1, M2, and S1 profiles), the flow profile is controlled by downstream conditions; for supercritical flow (M3, S2, and S3 profiles) the profile is controlled at its upstream end. This is consistent with the fact that a gravity-wave disturbance cannot be propagated upstream when $\mathbb{F} \geq 1$ (see Equations 5-49 and 5-50).

Figure 9.4 illustrates flow profiles over the types of slope changes that occur naturally. The break in slope acts as the control in these cases, although changes in width or depth causing acceleration or deceleration would have identical effects. In Figure 9.4a–c the flow upstream of the control is subcritical and the control therefore determines the profile to the next control upstream. In Figures 9.4d–f the upstream flow is supercritical, so the control determines the profile downstream. In Figure 9.4c the flow changes from subcritical to supercritical, so the control determines the profile both upstream and downstream. Note that the flow passes through the critical point at the control.

Figure 9.3c,d and Figure 9.4f illustrate the hydraulic jump that occurs when a flow goes from supercritical to subcritical. This phenomenon is a highly turbulent standing wave whose exact position and form are determined by the Froude number of the supercritical flow entering the jump. The profile of the hydraulic jump is irregular, and since it is a phenomenon of rapidly varied flow, its profile cannot be determined by the principles of gradually varied flow that are being applied in this chapter. Hydraulic jumps are discussed in some detail in Chapter 10.

COMPUTATION OF WATER-SURFACE PROFILES

In a regular channel with constant slope, width, and roughness a given steady discharge will tend to conform to the equations of uniform flow. The flow profile in such a case can thus be readily calculated from the Manning equation when the slope, roughness, and cross-sectional shape of the channel are known, as in Exercise 7-4.

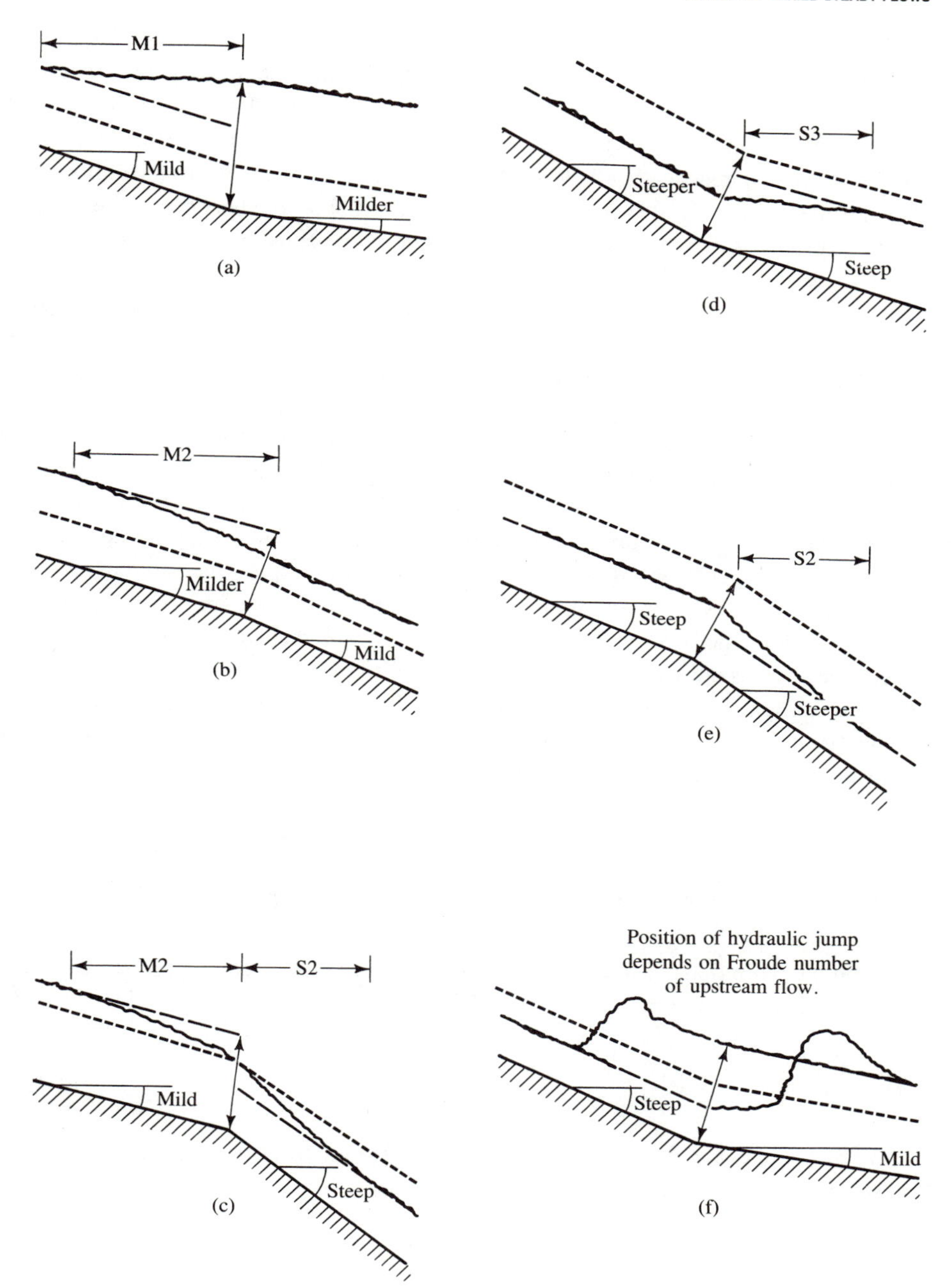

Figure 9.4
Water-surface profiles associated with controls exerted by changes in slope. Long-dashed lines represent normal depth; short-dashed lines represent critical depth. Vertical arrows indicate control sections.

As noted, Figure 9.4a–f shows profiles in transitions between uniform flows with different depths caused by a change in slope. (Changes in width or roughness would produce the same effects.) If the channels were regular, the flow profiles at some distance above and below the transitions could be calculated by the method for uniform flows. It appears that the transitional profiles are asymptotic to the uniform-flow profiles, but it is not clear at present just where the profile becomes uniform or what the profile is like in the transition. Furthermore, most natural channels provide even more complicated situations, since their slopes, widths, and roughnesses change more or less continuously, and we cannot be sure if uniform flow exists anywhere.

In the next sections we develop the basic quantitative approach to the estimation of water-surface profiles in gradually varied flows. Two approaches are presented, the first a theoretical development that provides some quantitative insight to the classification of profiles shown in Table 9.1 and Figure 9.3, and the second a more practical approach that is the basic method most commonly used for hydrologic studies. There is really no fundamental difference between the two approaches; both begin with the specific-head equation (Equation 7-20), employ the same assumptions, and end up with an expression for the change of depth over some distance. Because both methods compute the change of depth with distance, rather than depth itself, they both require that computations of profiles begin at a point where the flow depth is known for the particular discharge of interest.[2] If flow is subcritical, the computations proceed from that point in an upstream direction; if the flow is supercritical, they proceed downstream.

THEORETICAL FOUNDATION FOR PROFILE COMPUTATION

From the definitions of total head H (Equation 7-16) and specific head H_s (Equation 7-20), we know that

$$H = H_s + Z. \tag{9-4}$$

We now take the derivative of Equation 9-4 with respect to distance in the downstream (x) direction:

[2] Or, as will be explained later, if the depth is not known at any point in the reach of interest, we can estimate a starting depth at some point downstream, and compute the profile in an upstream direction. If we start far enough downstream, the calculated profile will converge to the true profile by the time we get to the reach of interest.

$$\frac{dH}{dx} = \frac{dH_s}{dx} + \frac{dZ}{dx}. \qquad (9\text{-}5)$$

In Equation 9-5 dH/dx represents the slope S_e of the energy grade line (refer to Figure 7.3) and dZ/dx represents the slope S_0 of the channel bottom. In natural channels, however, dH/dx and dZ/dx are always negative, whereas S_e and S_0 are treated as positive numbers. Thus $S_e = -dH/dx$ and $S_0 = -dZ/dx$, and Equation 9-5 can be rewritten as

$$-S_e = \frac{dH_s}{dx} - S_0. \qquad (9\text{-}6)$$

Rearranging this equation yields

$$\frac{dH_s}{dx} = S_0 - S_e. \qquad (9\text{-}7)$$

We now take advantage of the fact that

$$\frac{dH_s}{dx} = \frac{dH_s}{dY}\frac{dY}{dx} \qquad (9\text{-}8)$$

and substitute Equation 9-8 into Equation 9-7:

$$\frac{dY}{dx}\frac{dH_s}{dY} = S_0 - S_e;$$

$$\frac{dY}{dx} = \frac{S_0 - S_e}{dH_s/dY}. \qquad (9\text{-}9)$$

Thus Equation 9-9 states that the rate of change of depth equals the difference between the channel slope and the energy slope divided by the rate of change of specific head with respect to depth. The latter term was shown in the derivation of Equation 7-23 to be

$$\frac{dH_s}{dY} = 1 - \frac{Q^2}{gw^2Y^3} \qquad (9\text{-}10)$$

if we consider Q and w to be fixed. Substituting Equation 9-10 into Equation 9-9 gives

$$\frac{dY}{dx} = \frac{S_0 - S_e}{1 - Q^2/gw^2Y^3}. \qquad (9\text{-}11)$$

This expression can be simplified by noting Equation 9-3. If both sides of that definition of the critical depth are cubed, then

$$Y_c^3 = \frac{Q^2}{gw^2}, \tag{9-12}$$

and substituting Equation 9-12 into Equation 9-11 gives

$$\frac{dY}{dx} = \frac{S_0 - S_e}{1 - Y_c^3/Y^3} = \frac{S_0 - S_e}{1 - (Y_c/Y)^3}. \tag{9-13}$$

The numerator of Equation 9-13 can be put into a form analogous to the denominator by first writing it as

$$S_0 - S_e = S_0\left(1 - \frac{S_e}{S_0}\right). \tag{9-14}$$

Next, the Manning equation (Equation 9-2) can be rearranged to relate the channel slope S_0 to the normal depth Y_n:

$$S_0 = \left(\frac{nQ}{u_M Y_n^{5/3} w}\right)^2. \tag{9-15}$$

The next-to-last step is to make the assumption that the rate of energy loss S_e at a cross section in a gradually varied nonuniform flow is the same as for a uniform flow with the same velocity and depth. With this assumption, we can write an equation analogous to Equation 9-15, relating S_e and actual depth Y:

$$S_e = \left(\frac{nQ}{u_M Y^{5/3} w}\right)^2. \tag{9-16}$$

Now, from Equations 9-15 and 9-16,

$$\frac{S_e}{S_0} = \frac{Y_n^{10/3}}{Y^{10/3}}. \tag{9-17}$$

Finally, Equation 9-17 can be substituted into Equation 9-14, and the result into Equation 9-13, to obtain

$$\frac{dY}{dx} = S_0\left[\frac{1 - (Y_n/Y)^{10/3}}{1 - (Y_c/Y)^3}\right]. \tag{9-18}$$

Equation 9-18 is the form of the gradually varied flow equation we have been seeking. In application, we would calculate Y_n and Y_c from Equations 9-2 and 9-3, respectively, using the appropriate values of discharge, width, channel slope, and roughness. Then beginning at a point x where the depth Y is known for that discharge, we would compute dY/dx by using Equation 9-18. We then assume that

$$\frac{\Delta Y}{\Delta x} = \frac{dY}{dx}, \tag{9-19a}$$

where ΔY is the change in depth that occurs over a short distance, Δx. The value of ΔY is then found as

$$\Delta Y = \frac{dY}{dx} \Delta x \tag{9-19b}$$

and the new depth at $x + \Delta x$ is $Y + \Delta Y$.

Alternatively, Equation 9-18 can be directly integrated as described by Chow (1959) and Daily and Harleman (1966), although we still have to begin computations at a point where the depth is known or estimated for the discharge of interest. In all cases, computations proceed in a downstream direction if $Y < Y_c$ (supercritical flow) and in an upstream direction if $Y > Y_c$ (subcritical flow).

In practice, flow profiles are usually computed by a "step method" that will be described later rather than via Equation 9-18. Nevertheless, Equation 9-18 is of interest because it provides further insight into the classification of flow profiles described earlier and shown in Table 9.1. To see this, we define

$$N \equiv 1 - \left(\frac{Y_n}{Y}\right)^{10/3} \tag{9-20a}$$

and

$$D \equiv 1 - \left(\frac{Y_c}{Y}\right)^3, \tag{9-20b}$$

so that Equation 9-18 becomes

$$\frac{dY}{dx} = S_0\left(\frac{N}{D}\right). \tag{9-21}$$

Because S_0 is always positive, we see from Equations 9-20 and 9-21 that when $Y > Y_n > Y_c$,

$$1 > N > 0, \qquad 1 > D > 0, \qquad \frac{N}{D} > 0;$$

therefore,

$$\frac{dY}{dx} > 0$$

and depth increases downstream. Similarly, in conformance with Table 9.1, when $Y_n > Y > Y_c$,

$$N < 0, \qquad 1 > D > 0, \qquad \frac{N}{D} < 0;$$

therefore

$$\frac{dY}{dx} < 0$$

and depth decreases downstream. When $Y_n > Y_c > Y$,

$$N < 0, \qquad D < 0, \qquad \frac{N}{D} > 0;$$

consequently,

$$\frac{dY}{dx} > 0$$

and depth increases downstream. When $Y > Y_c > Y_n$,

$$1 > N > 0, \qquad 1 > D > 0, \qquad \frac{N}{D} > 0;$$

thus

$$\frac{dY}{dx} > 0$$

and depth increases downstream. When $Y_c > Y > Y_n$,

$$1 > N > 0, \qquad D < 0, \qquad \frac{N}{D} < 0;$$

so

$$\frac{dY}{dx} < 0$$

and depth decreases downstream. Finally, when $Y_c > Y_n > Y$,

$$N < 0, \qquad D < 0, \qquad \frac{N}{D} > 0;$$

therefore

$$\frac{dY}{dx} > 0$$

and depth increases downstream.

Two other implications of Equation 9-18 are of interest. When $Y = Y_n$, $dY/dx = 0$, which simply reflects that in uniform flow the water surface is parallel to the channel bed and depth does not change. When $Y = Y_c$, however, Equation 9-18 gives an infinite slope, that is, a vertical water surface. Although this is not what happens in real flows, the latter are unstable when they are near the critical depth, as is suggested by the specific-head diagram developed in Chapter 7 (Figure 7.4). That diagram shows that when the flow is near the critical state a small change in the energy of the flow will lead to a relatively large change in depth. In a natural channel there are ubiquitous small variations in slope, width, and roughness that affect the energy of the flow, so that when the flow is near critical, the surface is wavy and irregular (Figure 9.5). Under these conditions the flow is rapidly, rather than gradually, varied; and the assumptions that led to the development of the energy equation (Equation 7-17) and of Equation 9-18 are no longer valid.

THE STEP METHOD FOR PROFILE COMPUTATION

Basic Method Total mechanical energy H at a cross section can be written as

$$H = H_p + H_v, \tag{9-22}$$

where H_p is the potential head and H_v the velocity head. Note that in Figure 9.6 H_p represents the elevation of the water surface above a datum (assuming, as is done

Figure 9.5
A high flow in a small New England stream. The extremely uneven surface is characteristic of flows that are close to critical.

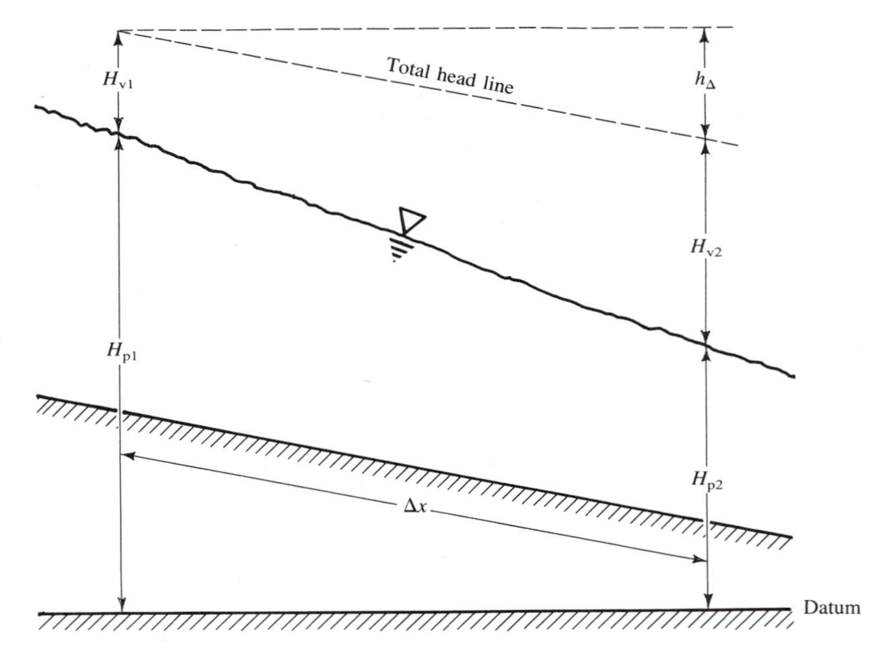

Figure 9.6
Definitions of terms for computation of water-surface profiles (Equation 9-23).

throughout this section, that $\cos\theta = 1$). In these terms, the energy equation (Equation 7-17) becomes

$$H_{p1} + H_{v1} = H_{p2} + H_{v2} + h_{\Delta}, \tag{9-23}$$

where h_{Δ} is the head loss between an upstream section (subscript 1) and a downstream section (subscript 2). For subcritical flow we compute in the upstream direction, so the working form of Equation 9-23 is

$$H_{p1} = H_{p2} + H_{v2} + h_{\Delta} - H_{v1}. \tag{9-24a}$$

For supercritical flow, we solve for the downstream water-surface elevation H_{p2}:

$$H_{p2} = H_{p1} + H_{v1} - H_{v2} - h_{\Delta}. \tag{9-24b}$$

Since subcritical flow is by far the more common in nature, subsequent developments herein will use only Equation 9-24a.

The total head losses h_{Δ} between the two sections are usually treated in two parts:

$$h_{\Delta} = h_f + h_s, \tag{9-25}$$

where h_f is the normal friction loss due to the roughness and channel geometry and h_s represents the eddy losses (sometimes called *shock losses*). The latter are included when the channel becomes wider in the downstream direction, giving rise to eddies that reduce the effective cross-sectional area of the channel (Figure 9.7). The friction-loss terms are computed from the average of the energy slopes computed from the Manning equation at the upstream (S_{e1}) and downstream (S_{e2}) sections:

$$\frac{h_f}{\Delta x} = \frac{S_{e1} + S_{e2}}{2},$$

$$h_f = \Delta x \frac{S_{e1} + S_{e2}}{2}, \tag{9-26}$$

where Δx is the distance between the two sections. The two energy slopes are computed by solving the discharge form of the Manning equation (Exercise 6-9):

$$S_{e1} = \left(\frac{n_1 Q_1}{u_M A_1 R_1^{2/3}}\right)^2; \tag{9-27}$$

$$S_{e2} = \left(\frac{n_2 Q_2}{u_M A_2 R_2^{2/3}}\right)^2. \tag{9-28}$$

The other component of the total head loss is computed from the following empirical formulas: When $w_2 > w_1$,

$$h_s = 0.5(H_{v1} - H_{v2}); \quad (9\text{-}29a)$$

and when $w_2 \leq w_1$,

$$h_s = 0 \quad (9\text{-}29b)$$

(w_1 and w_2 are the water-surface widths at the upstream and downstream sections, respectively).

Combining Equation 9-25 with Equation 9-24a gives

$$H_{p1} = H_{p2} + H_{v2} - H_{v1} + h_f + h_s. \quad (9\text{-}30)$$

We can now look at Equations 9-26–9-30 to get an overview of how profile computations proceed. The discharge Q is known, each cross section and the channel slope have been surveyed, and the roughness has been estimated in the vicinity of each section. The first step is to compute the normal and critical depths at each section. As the actual depths are computed, they can be compared with Y_c to determine whether computations should proceed upstream or downstream. The calculated depths can also be related to Y_n and Y_c to check, by means of Equation 9-18, whether the depth should be increasing or decreasing. The great majority of natural flows will be subcritical with profiles of the M1 and M2 type (Figure 9.3a,b).

As mentioned previously, computations must begin where the depth (or, equivalently, the water-surface elevation) is known. Thus if we assume a subcritical flow,

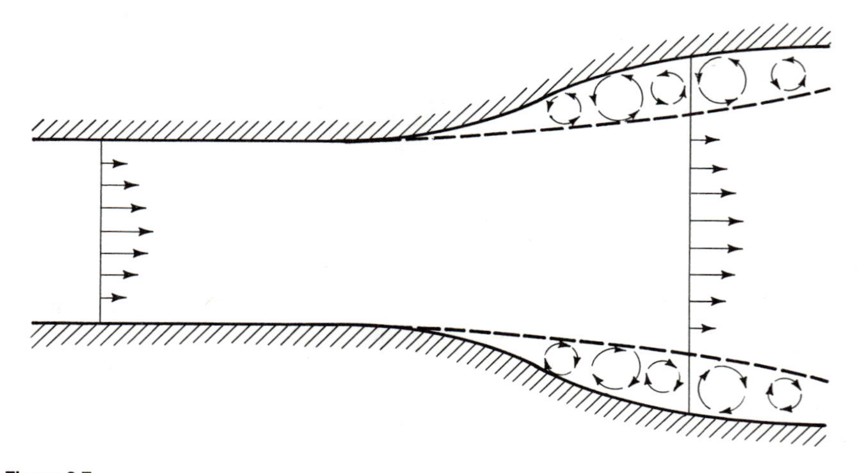

Figure 9.7
Map view of an expanding channel showing the formation of eddies that reduce the effective cross section and increase energy losses. Arrows represent velocity vectors; dashed lines are the boundaries of the zones of flow separation.

we will know the value of H_2 for the first step of the calculations. From the surveyed cross section, the cross-sectional area A_2 at that section can be found and the velocity head H_{v2} can then be computed from

$$V_2 = \frac{Q_2}{A_2}, \tag{9-31}$$

and

$$H_{v2} = \frac{\alpha_2 V_2^2}{2g}. \tag{9-32}$$

The remaining terms in Equation 9-30 must be computed on a trial-and-error basis. The procedure can best be conveyed by means of an example. To simplify the calculations, consider the rectangular channel of constant width ($w = 25$ ft) and roughness ($n = 0.022$) shown in Figure 9.8. The details of the computation are listed in Table 9.2. Columns 1 and 2 give the profile of the bed over a distance of 300 ft upstream of a control where the flow becomes critical as a result of an abrupt transition from a mild to a steep slope, as in Figure 9.4c. The discharge of concern is 700 $ft^3\ s^{-1}$. Because the flow becomes critical at the control, the depth at that point for that discharge is found by using Equation 9-3:

$$Y_c = \left[\frac{700}{(32.2)^{1/2}(25)}\right]^{2/3} = 2.91 \text{ ft}.$$

With this value, we can compute the information in columns 4–11 of the first row of Table 9.2. The mean velocity is calculated as Q/A, and we will assume that $\alpha = 1.3$ for all computations of velocity head. The energy slope S_e is calculated via Equation 9-28. Since this row is the starting point for the calculations, there are no entries in columns 12–15. The water-surface elevation in column 16 is simply the bed elevation (column 2) plus the depth (column 3).

Because it is known that the water surface has a relatively high curvature as it approaches the critical depth, the second cross section was surveyed only a short distance (10 ft) upstream; subsequent sections are located at increasing distances. Computations in the second and subsequent rows proceed as follows.

Column 3: Critical depth is the starting point for this computation. Subsequently, an initial depth is *assumed*. Calculations of Y_c and Y_n and experience guide the choice of the assumed value.

Columns 4–7: Computations of these hydraulic factors are made by using the depth assumed in column 3.

Column 8: Velocity is computed as Q/A.

Column 9: This value is computed from the value in column 8, here assuming that $\alpha = 1.3$.

Column 10: The assumed energy grade line elevation is the sum of the bed elevation (column 2), the assumed depth (column 3), and the assumed velocity head (column 9). From the previous discussions of energy in flowing water, we know that the energy grade line elevation must decrease downstream, so the initially calculated value should be checked against the value calculated in the preceding row. If it is less than the preceding value, return to column 3 and select a new (larger) assumed depth. In the worked example, the requirement was not satisfied with the initially assumed depth of 3.10 ft, but was satisfied with 3.35 and 3.40 ft.

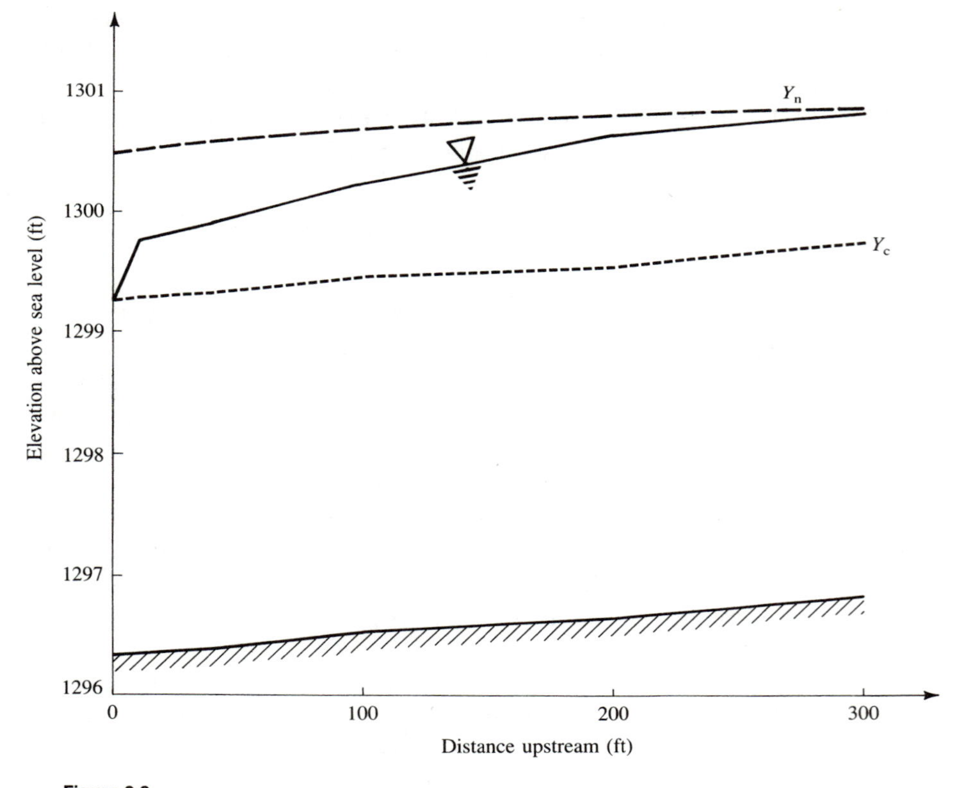

Figure 9.8
Water-surface profile as computed for the example worked in the text (Table 9.2); Y_n is the normal depth and Y_c the critical depth.

Column 11: The slope of the energy gradient at this point in the channel is estimated via Equation 9-28 by using the appropriate assumed values from the preceding columns.

Column 12: The average energy gradient over the reach between this station and the preceding station $\overline{S}_e$ is calculated as the arithmetic average of S_e at this station (column 11) and the preceding station (row 1, column 11).

Column 13: This value is Δx, the distance between this cross section and the preceding one, determined before the computations begin.

Column 14: The head loss h_f between this section and the downstream section is the product of the slope of the energy grade line between the two sections (column 12) and the distance between these (column 13), that is, $h_f = S_e \, \Delta x$. In this example, the channel width is constant, so there is no additional loss due to eddies.

Column 15: The calculated depth at this station is found from Equation 9-30. In practice, this is done by taking the elevation of the energy grade line at the preceding station (this is equal to H_{p2} plus H_{v2} and is found in column 10 of the preceding row), subtracting the velocity head at the present station (H_{v1}, column 9) and adding the computed h_f (column 14). This gives an estimate of H_{p1}. The depth Y_1 is now calculated by subtracting the bed elevation at this station (column 2), and the result is compared with the initially assumed depth. If the two differ by less than a predetermined amount, we accept the depth originally assumed in this row and proceed to column 16. If the depth in column 15 is not close enough to the assumed depth, we return to column 3, select a new assumed depth that is closer to the value computed in column 15, and repeat the previous steps. In the example of Table 9.2, the criterion for accepting the depth calculated in column 15 is that it differs from the assumed value in column 3 by no more than 0.05 ft. The second assumed depth, 3.35 ft, did not satisfy this criterion, but the third one of 3.40 ft did. A second depth had to be assumed at station 200 also.

Column 16: The water-surface elevation at this point (H_{p1} in Equation 9-30) is the bed elevation (column 2) plus the final assumed depth (column 3).

Practical Details of Profile Computation Although the exercises at the end of this chapter provide additional practice in computing profiles, you should consider the material presented in this chapter as only an introduction to this important

Table 9.2

Computation of Water-Surface Profile in Figure 9.8

Discharge (Q) = 700 $ft^3\ s^{-1}$; width (w) = 25 ft; roughness (n) = 0.022. BE is the bed elevation (ft above sea level, a.s.l.), Y the mean depth (ft), A the cross-sectional area (ft^2), P the wetted perimeter (ft), R the hydraulic radius (ft), V the mean velocity ($ft\ s^{-1}$), g the gravitational acceleration (32 $ft\ s^{-2}$), EGL the energy grade line elevation (ft a.s.l.), S_e the energy slope at point, $\bar{S}_e$ the mean energy slope in the reach, Δx the distance between points (ft), h_f the head loss due to friction (ft), WSE the water surface elevation (ft a.s.l.). The underlined values were established by survey.

1	2	3	4	5	6	7	8	9	10	11	12	13	14	15	16
Distance from initial point (ft)	BE	Assumed Y	A	P	R	$R^{2/3}$	V	$\frac{1.3V^2}{2g}$	EGL	S_e	$\bar{S}_e$	Δx	h_f	Calculated Y	WSE
0	1296.35	2.91	72.75	30.82	2.36	1.78	9.62	1.87	1301.13	0.0064					1299.26
10	1296.37	3.10	77.50	31.20	2.48	1.84	9.03	1.65	1301.12†						
10	1296.37	3.35	83.75	31.70	2.64	1.92	8.36	1.41	1301.13	0.0041	0.0053	10	0.0527	3.41‡	
10	1296.37	3.40	85.00	31.80	2.67	1.93	8.24	1.37	1301.14	0.0040	0.0052	10	0.0519	3.44	1299.77
40	1296.41	3.50	87.50	32.00	2.73	1.96	8.00	1.29	1301.20	0.0036	0.0038	30	0.114	3.55	1299.91
100	1296.55	3.70	92.50	32.40	2.85	2.02	7.57	1.16	1301.41	0.0031	0.0033	60	0.198	3.69	1300.25
200	1296.65	3.90	97.50	32.80	2.97	2.07	7.18	1.04	1301.59	0.0026	0.0029	100	0.286	4.01‡	
200	1296.65	4.00	100.00	33.00	3.03	2.10	7.00	0.99	1301.64	0.0024	0.0028	100	0.276	4.05	1300.65
300	1296.85	4.00	100.00	33.00	3.03	2.10	7.00	0.99	1301.84	0.0024	0.0024	100	0.242	4.04	1300.85

†This value is lower than the previous value, which is physically impossible. Return to column 3 and assume a higher value of Y.

‡This value differs from the assumed value by more than 0.05 ft. Return to column 3 and assume a new value of Y.

technique. In practice, considerable engineering judgment is required, particularly in the estimation of roughness values and in the location of surveyed cross sections. Nevertheless, some of the basic considerations can be outlined.

Methods for estimating roughness were discussed in Chapter 8. When overbank flow exists, or when any marked variation of roughness or depth occurs at a cross section, that section should be subdivided into the appropriate number of more homogeneous subsections. Computations then proceed by our first assuming an initial water-surface elevation (H_{p2}) and then using the cross-section survey to compute the average depth, area, wetted perimeter, and hydraulic radius in each subsection. Chow (1959) gives methods for computing the velocity-distribution coefficients in each subsection. The slope is considered the same in all subsections, and is computed by first defining the *conveyance* K as

$$K \equiv \frac{u_M R^{2/3} A}{n}, \qquad (9\text{-}33)$$

where u_M is the unit-conversion factor in the Manning equation, R is the hydraulic radius, A is the cross-sectional area, and n is the roughness. Thus the Manning equation takes the form

$$Q = K S_e^{1/2}. \qquad (9\text{-}34)$$

The value of K is computed for each subsection from the appropriate values of R, A, and n; and the individual K values are added to obtain the total conveyance ΣK for the section. The energy slope is then found by solving Equation 9-34:

$$S_e = \left(\frac{Q}{\Sigma K}\right)^2, \qquad (9\text{-}35)$$

which is equivalent to Equation 9-27 or 9-28. This value is then entered in column 11 of the computation table (Table 9.2).

Location of cross sections is important for the accuracy of the computed profile. The following extract from a U.S. Army Corps of Engineers (1969) manual on profile computation succinctly presents the principal guidelines.

> Cross-sections used in the computation of water surface profiles should not necessarily be restricted to the actual surveyed cross-sections that are available. The location of cross-sections is more important than the exact shape and area of the cross-section for properly defining the energy loss. For large rivers where the cross-sections are fairly uniform and slopes are approximately 1-foot per mile (about 2 meters per kilometer), cross-sections may be spaced up to one mile (1.6 kilometers) apart. For small tributaries on very steep slopes, 5 or more cross-sections per mile may be required. Additional cross-sections should be added when the cross-sectional area changes appreciably, when a change in roughness occurs, or when a marked change in bottom slope occurs.

Instructions for accounting for tributary inflows are given in Chow (1959). Computing profiles at bridges is a matter of special concern, and methods for doing this are described in the U.S. Army Corps of Engineers (1969) manual.

As noted earlier, profile computations should begin whenever possible at a control section where the depth is known for the discharges of interest. This is typically at a gaging station where the rating curve (depth–discharge relation) is known. Other possible starting points are at a weir, dam, or constriction where the flow becomes critical, or at the inflow to a lake or reservoir where the water-surface elevation is known. Where no known starting depth can be determined, you can begin the computations with an assumed depth at a point downstream (assuming subcritical flow) from the reach where profiles are needed. If you start far enough downstream and assume a depth not too different from the true depth, the computed profiles will converge to the correct profile by the time you get to the reach of interest. Bailey and Ray (1966) give equations for estimating the distance x^* required for convergence: For M1 profiles

$$x^* = (0.860 - 0.640\mathbb{F}^2)\frac{Y_n}{S_0}, \tag{9-36a}$$

and for M2 profiles

$$x^* = (0.568 - 0.788\mathbb{F}^2)\frac{Y_n}{S_0}, \tag{9-36b}$$

where $\mathbb{F}$ is the Froude number of the flow, Y_n is the normal depth, and S_0 is the bed slope. Equations 9-36 are based on the assumption that the assumed starting depth is greater than 75% and less than 125% of the actual depth at that point. The exercises that follow give some feeling for the values of x^* likely to be found in nature.

Exercises

9-1. Consider a channel with $S_0 = 0.00800$, $n = 0.030$, and $w = 15$ m. (a) Compute the normal depth Y_n for discharges ranging from 0.100 to 1000 $m^3\ s^{-1}$ and plot the results on three- by five-cycle log–log paper. (b) Compute the critical depth Y_c as a function of discharge and plot the results on the same graph. (c) Assuming uniform flow, at which discharge will the flow become critical?

9-2. Compute the water-surface profile for a channel carrying a discharge of 2000 $ft^3\ s^{-1}$ over a distance of 1000 ft upstream of where it enters a reservoir with a fixed elevation. The channel is assumed to be rectangular, with a width of 70 ft and a roughness n of 0.045. Other data needed for the computation are listed in the accompanying table. Plot the results on a graph (with vertical exaggeration) and show the normal and critical depths.

Distance upstream from initial point (ft)	Measured bed elevation† (ft)	Water-surface elevation† (ft)
0	843.14	853.35‡
100	843.25	
200	843.60	
400	844.05	
600	844.57	
1000	845.81	

†Elevations are expressed as distance above sea level.
‡Reservoir surface.

9-3. Compute the water-surface profile for a channel carrying a discharge of 4.00 $m^3 s^{-1}$ over a distance of 100 m upstream of the point where it becomes critical. The channel is assumed to be rectangular, with a width of 3.0 m and a roughness n of 0.035. Other data needed for the computation are given in the accompanying table. Plot the results on a graph (with vertical exaggeration) and show the normal and critical depths.

Distance upstream from initial point (m)	Measured bed elevation† (m)
0	22.04
5.0	22.08
10.0	22.13
20.0	22.19
50.0	22.49
100.0	22.90

†Elevations are expressed as distance above sea level.

9-4. Compute the water-surface profile for a channel carrying a discharge of 40.0 $m^3 s^{-1}$ over a distance of 1000 m upstream of where it enters a reservoir with a fixed elevation. The channel is assumed to be rectangular, with a width of 18.0 m and a roughness n of 0.035. Other data needed for the computation are given in the accompanying table. Plot the results on a graph (with vertical exaggeration) and show the normal and critical depths.

Distance upstream from initial point (m)	Measured bed elevation[†] (m)	Water-surface elevation[†] (m)
0	153.82	159.92
100	154.62	
200	155.12	
500	157.52	
1000	160.02	

[†]Elevations are expressed as distance above sea level.

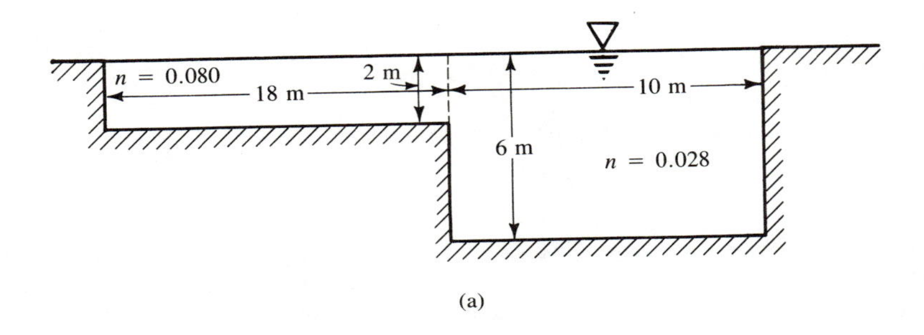

(a)

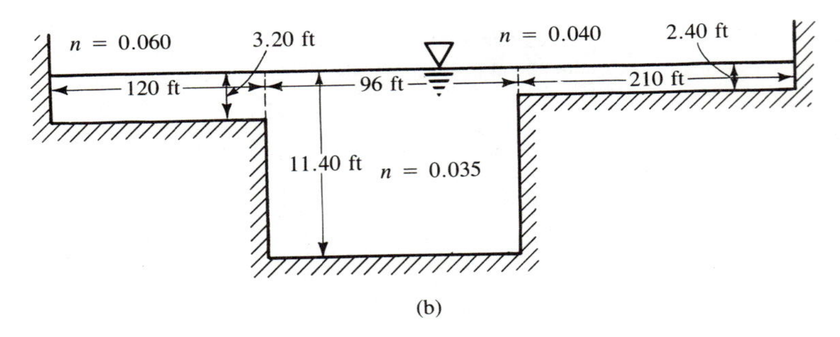

(b)

Figure 9.9
Composite cross sections for (a) Exercise 9-5 and (b) Exercise 9-6.

9-5. The discharge through the entire cross section shown in Figure 9.9a is 220 $m^3 s^{-1}$. (a) Compute the conveyances of each of the subsections. (b) Compute the energy slope through the section.

9-6. The discharge through the entire cross section shown in Figure 9.9b is 12,000 $ft^3 s^{-1}$.

(a) Compute the conveyances of each of the subsections. (b) Compute the energy slope through the section.

9-7. Decide whether the flows in the following table are of the M1 or M2 type. Then compute the distance x^* for water-surface profile calculations to converge to the true depth by starting from an assumed depth that is within 25% of the true depth.

Flow	Discharge ($m^3\ s^{-1}$)	Cross-sectional area (m^2)	Width (m)	Bed slope	Manning's n
a	66.2	121	74.1	0.00114	0.108
b	32.6	47.7	129	0.0014	0.033
c	89.7	124	123	0.00083	0.037
d	0.756	1.45	7.0	0.0091	0.055
e	607	949	140	0.000076	0.047
f	44.8	85.5	117	0.00083	0.038
g	68.3	61.8	53.4	0.000947	0.034
h	359	258	133	0.00083	0.037
i	102	117	144	0.00063	0.028
j	0.380	1.00	6.4	0.0091	0.055

CHAPTER **10**

Rapidly Varied Steady Flows

As noted in Table 5.3, flows in which the downstream rates of change of depth and velocity are relatively large are characterized as *rapidly varied*. In such flows the assumption of parallel streamlines (which implies a hydrostatic pressure distribution) that was invoked in deriving the energy equation (Equation 7-17) is significantly violated, and theoretical descriptions of flow profiles can only be approximate. As a consequence, rapidly varied flow is generally treated by considering various typical situations as isolated cases; applying the basic principles of conservation of mass, energy, and/or momentum as a starting point; and placing heavy reliance on empirical relations established in laboratory experiments.

Three cases of rapidly varied flows are of primary interest to surface-water hydrologists: (1) flows that experience abrupt changes in depth or width; (2) hydraulic jumps, which are standing waves that mark a rapid transition from supercritical to subcritical flow; and (3) flows over weirs. The latter are of special interest because weirs are often used to measure stream discharge; our objective will be to develop the equations that are the basis for such measurements. Ippen (1950) and particularly Chow (1959) cover a much larger range of rapidly varied flow conditions in the detail required for the engineer who must design structures associated with such flows.

CHANNEL TRANSITIONS

Approximate methods for determining water-surface profiles for steady flows through abrupt changes in depth and width are based on the principle of conservation of energy with the simplifying assumption that the total head H is the same throughout the transition. This assumption is often acceptable because the friction losses over a short distance will generally be very small in relation to the total head. However, if there is a zone of excessive turbulence (as in a hydraulic jump, discussed later) or eddies that cause the equivalent of shock losses (see the discussion of Equations 9-25 and 9-29), this assumption cannot be made.

In the developments in this section, the principle of conservation of energy is applied in the form of Equation 7-17 to enable us to compute the flow depth just downstream from an abrupt change in width or depth when the upstream depth is known. As noted earlier, the depths through the transition cannot be computed by these methods because the assumptions of parallel streamlines and small surface curvature used in developing Equation 7-17 are significantly violated. Thus we can only sketch an approximation of the water surface connecting the upstream and downstream depths. Furthermore, if the flow through a transition reaches the critical state (the Froude number is equal to one), those assumptions will also be violated downstream from the transition, so that we cannot calculate even the downstream depth by application of this approach. Aspects of this situation are covered later in the sections on hydraulic jumps and weirs.

Changes in Depth Consider first an abrupt drop in the elevation of a rectangular channel that carries a constant discharge, has a constant width, and has the very low slope that is typical of natural streams (Figure 10.1). Applying the energy equation (Equation 7-17) to cross section 1 just upstream of the transition and cross section 2 just downstream of it, we obtain

$$H = Z_1 + Y_1 + \frac{V_1^2}{2g} = Z_2 + Y_2 + \frac{V_2^2}{2g} \tag{10-1}$$

if we assume that H does not change significantly through the transition ($\cos\theta$ and α are assumed equal to one throughout this discussion). If the datum for elevation head is taken as the channel bottom on the upstream side, Equation 10-1 becomes

$$Y_1 + \frac{V_1^2}{2g} = Z_2 + Y_2 + \frac{V_2^2}{2g}, \tag{10-2}$$

where Z_2 is a negative number. If discharge Q, width w, and the upstream depth Y_1 are known, the upstream velocity can be found from the continuity relation:

$$V_1 = \frac{Q}{wY_1}. \tag{10-3}$$

Since Z_2 is determined by the geometry of the transition, only Y_2 and V_2 are unknown in Equation 10-2. However,

$$V_2 = \frac{Q}{wY_2}, \tag{10-4}$$

so Equation 10-2 actually has only one unknown, the downstream depth Y_2:

$$Y_2 + \frac{Q^2}{2gw^2Y_2^2} = Y_1 + \frac{V_1^2}{2g} - Z_2, \tag{10-5a}$$

or

$$Y_2 + \frac{Q^2}{2gw^2Y_2^2} = Y_1 + \frac{Q^2}{2gw^2Y_1^2} - Z_2. \tag{10-5b}$$

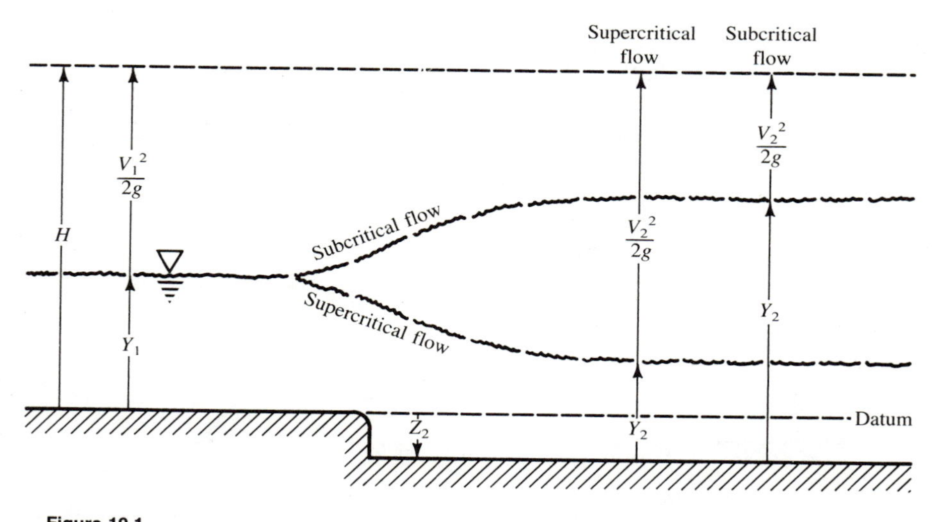

Figure 10.1
Definitions of terms for the analysis of water-surface profiles due to an abrupt decrease in the channel-bottom elevation Z_2. Datum is the upstream channel bottom. Total head H is constant throughout. Alternate values of the velocity head and pressure head depend on whether the flow is supercritical (left-hand values) or subcritical (right-hand values).

One way to solve Equation 10-5 is by trial and error. However, if we recognize that it can be written as

$$H_{s2} = H_{s1} - Z_2, \tag{10-6}$$

where H_{s1} and H_{s2} are the specific heads at the two cross sections (see Equation 7-22), a specific-head diagram for the flow can be utilized to advantage. In Figure 10.2a (cf. Figure 7.4), if H_{s1} is known and the upstream flow is subcritical, Y_1 can be found as shown. Using Equation 10-6 and the given value of Z_2, we can find H_{s2}. Since Z_2 is a negative number, $H_{s2} > H_{s1}$ and the downstream depth Y_2 can be found from the intersection of a vertical line from H_{s2} to the subcritical arm of the specific-head diagram for the discharge. Clearly, if the upstream flow is subcritical, the flow

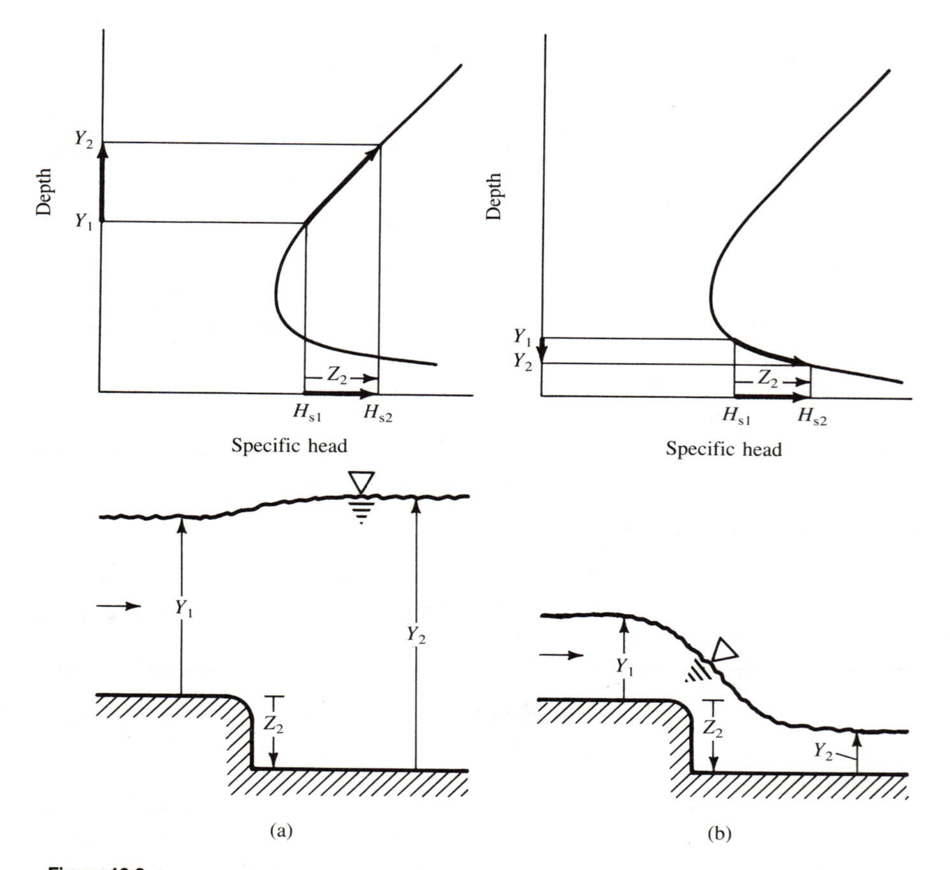

Figure 10.2
Energy relations (specific-head diagram) and depth changes due to an abrupt decrease in channel elevation: (a) subcritical flow; (b) supercritical flow.

downstream from an abrupt drop will also be subcritical, but at a higher depth and lower velocity.

If the upstream flow is supercritical, the lower arm of the specific-head diagram is used in the same way to estimate the downstream depth. Figure 10.2b shows that in this case the downstream depth will be smaller and the velocity higher.

Exactly the same procedures are applied in estimating changes in flow profiles due to abrupt increases in channel-bottom elevation. In this case, however, Z_2 is a positive number, so that $H_{s2} < H_{s1}$ (Equation 10-6). In computing the downstream depth, we therefore move to the left on the appropriate arm of the specific-head diagram (Figure 10.3). Thus if the upstream flow is subcritical, the downstream depth is less than the upstream depth (Figure 10.3a); if the upstream flow is supercritical, the downstream depth is greater than the upstream depth (Figure 10.3b).

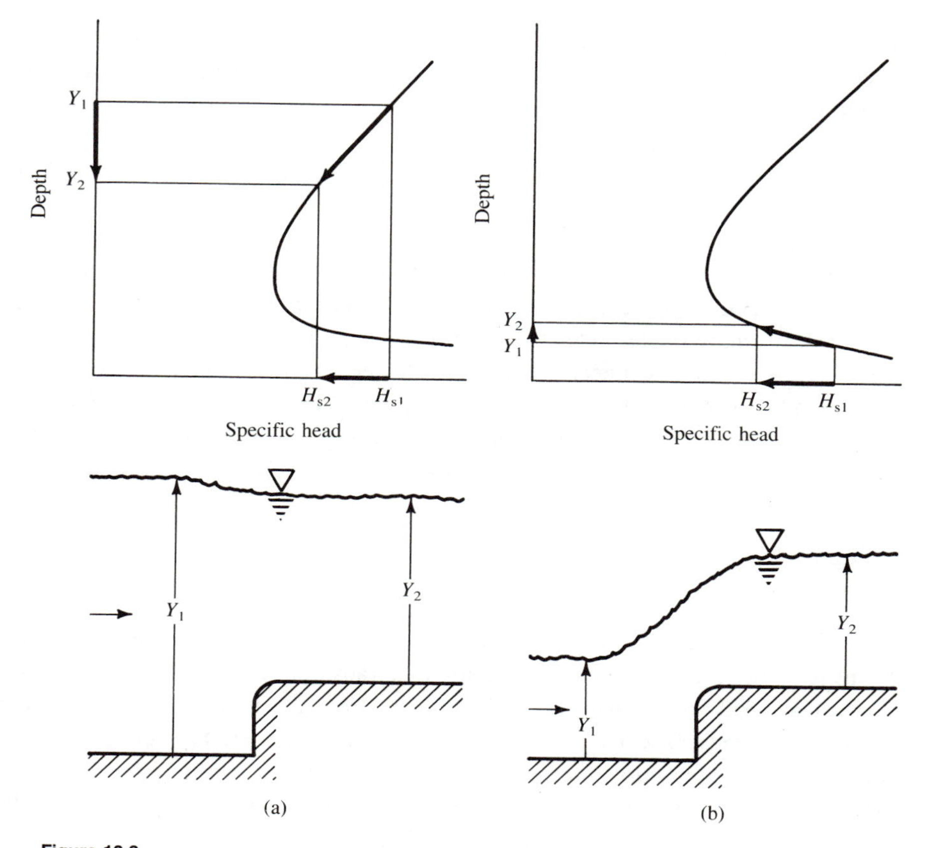

Figure 10.3
Energy relations (specific-head diagram) and depth changes due to an abrupt increase in channel-bottom elevation: (a) subcritical flow; (b) supercritical flow.

However, an important constraint arises in the case of an increase in the channel-bed elevation: The specific head for a given discharge cannot decrease below the minimum value that occurs when the flow becomes critical (see the discussion related to Equation 7-25). This minimum value H_{sc} is given by

$$H_{sc} = Y_c + \frac{V_c^2}{2g}, \tag{10-7}$$

where Y_c is the critical depth and V_c is the velocity when the flow is critical. Substituting Equation 7-24 into Equation 10-7 gives

$$H_{sc} = Y_c + \frac{Y_c}{2} = 1.5Y_c, \tag{10-8}$$

where Y_c can be found from Equation 9-3. Because of Equation 10-8, Equation 10-6 must be constrained as follows when it is to be applied to increases in channel-bottom elevation:

$$H_{s2} = H_{s1} - Z_2 \geq H_{sc} = 1.5Y_c. \tag{10-9}$$

In the situations we have been considering, all the values in Equation 10-9 are fixed by the given discharge and channel geometry. It is possible to select values for Q, w, and Y_1 such that the constraint in Equation 10-9 cannot be satisfied. Such flows are physically impossible under the conditions of nearly parallel streamlines and negligible energy loss assumed here, and only lower discharge or higher head values than initially selected will satisfy the assumptions. The existence of a situation violating the constraint can be determined from a rearrangement of Equation 10-9: Flow is possible at the selected conditions if

$$Z_2 \leq H_{s1} - H_{sc}; \tag{10-10}$$

otherwise it is not.

It should be emphasized that the foregoing approach will give the upstream and downstream depths measured close enough to a transition such that the assumption of constant head is reasonable, but far enough from the transition that extreme curvatures of the surface and streamlines are no longer present. The surface profile in the region of high curvature near the point of transition can only be sketched approximately. In addition, any zones of flow separation and eddying that occur will produce deviations from the theory discussed here.

Solutions to problems involving abrupt channel transitions are facilitated by the construction of a dimensionless specific-head diagram. We can construct such a diagram by dividing the terms of Equation 7-22 by Y_c:

$$\frac{H_s}{Y_c} = \frac{Y}{Y_c} + \frac{Q^2}{2gw^2Y^2Y_c}. \tag{10-11}$$

From Equation 7-23,

$$\begin{aligned}\frac{H_s}{Y_c} &= \frac{Y}{Y_c} + \frac{Y_c^3}{2Y^2Y_c} \\ &= \frac{Y}{Y_c} + \frac{1}{2}\left(\frac{Y_c}{Y}\right)^2.\end{aligned} \tag{10-12}$$

Figure 10.4 shows a plot of Equation 10-12.

The relation of the Froude number $\mathbb{F}$ to Y/Y_c can be determined from the definition of specific head (Equation 7-20):

$$\begin{aligned}\frac{V^2}{2g} &= H_s - Y; \\ V^2 &= 2g(H_s - Y).\end{aligned} \tag{10-13}$$

From the definition of $\mathbb{F}$,

$$\mathbb{F}^2 = \frac{V^2}{gY}, \tag{10-14}$$

and substituting Equation 10-13 into Equation 10-14 gives

$$\begin{aligned}\mathbb{F}^2 &= \frac{2(H_s - Y)}{Y} \\ &= 2\left(\frac{H_s}{Y} - 1\right).\end{aligned} \tag{10-15}$$

Writing H_s and Y in dimensionless form transforms Equation 10-15 to

$$\mathbb{F}^2 = 2\left(\frac{H_s/Y_c}{Y/Y_c} - 1\right). \tag{10-16}$$

Now the relation between H_s/Y_c and Y/Y_c (Equation 10-12) can be substituted into Equation 10-16, and after some algebraic manipulation we have

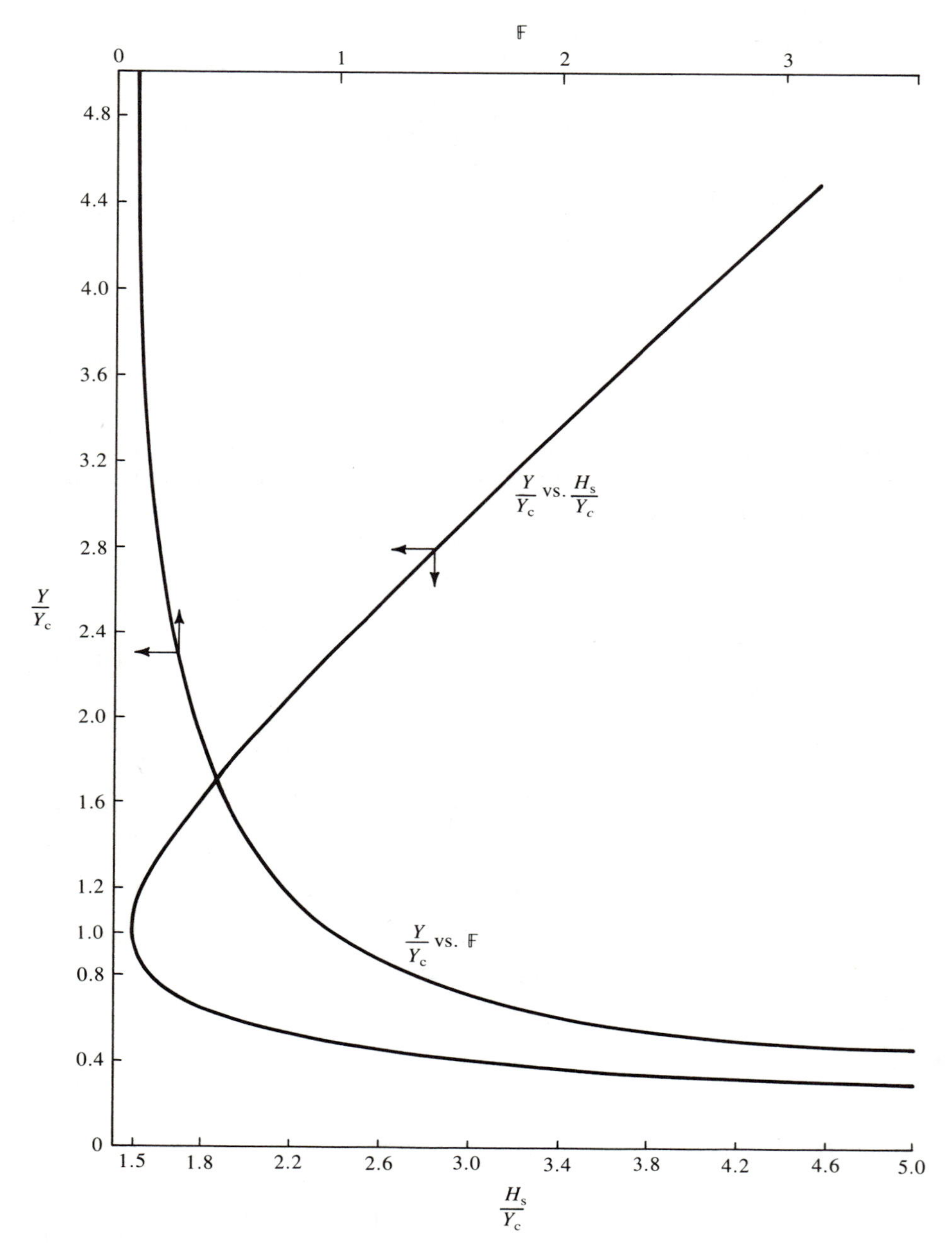

Figure 10.4
Dimensionless specific-head diagram (Y/Y_c vs. H_s/Y_c; Equation 10-12) and relation between Y/Y_c and the Froude number $\mathbb{F}$ (Equation 10-17).

$$\mathbb{F} = \left(\frac{Y_c}{Y}\right)^{3/2}. \tag{10-17}$$

Some examples will illustrate the use of the relations we have just established. Consider a rectangular channel with a width of 5.00 m and a low slope. The channel bottom drops suddenly by 0.80 m. The channel is carrying a discharge of 10.0 m^3 s^{-1} at an upstream depth of 1.60 m, and we want to compute the downstream depth. First it is useful to compute the critical depth, which from Equation 9-3 is

$$Y_c = \left(\frac{10.0}{5 \cdot (9.8)^{1/2}}\right)^{2/3} = 0.74 \text{ m}.$$

The upstream specific head can be found from Figure 10.4: With $Y_1/Y_c = 1.60/0.74 = 2.16$, $H_s/Y_c = 2.26$ and $H_{s1} = 1.67$ m. The channel elevation $Z_2 = -0.80$ m, so from Equation 10-6

$$H_{s2} = 1.67 - (-0.80) = 2.47 \text{ m},$$

and thus $H_{s2}/Y_c = 3.34$. From Figure 10.4 the corresponding $Y/Y_c = 3.29$; thus $Y_2 = 2.43$ m.

As a second example, consider a rectangular channel of low slope, 12.00 ft wide and carrying a discharge of 500 ft^3 s^{-1} at an upstream depth of 2.70 ft. There is an abrupt rise in the channel of 0.80 ft. Equation 9-3 gives $Y_c = 3.80$ ft, so the flow is supercritical and $Y/Y_c = 0.711$. Figure 10.4 gives $H_{s1}/Y_c = 1.76$ and thus $H_{s1} = 6.69$ ft. The minimum allowable value of specific head for this discharge is $(1.5)(3.80) = 5.70$ ft. From Equation 10-10, Z_2 must satisfy

$$Z_2 < 6.69 - 5.70 = 0.99 \text{ ft}.$$

Since $Z_2 = 0.80$ ft, the flow is physically possible and the calculation can proceed. Equation 10-6 gives

$$H_{s2} = 6.69 - 0.80 = 5.89 \text{ ft},$$

so that $H_{s2}/Y_c = 1.55$, and from Figure 10.4, we see that $Y_2/Y_c = 0.84$ and thus $Y_2 = 3.19$ ft.

Changes in Width The problem of computing downstream changes in depth (and velocity) due to abrupt changes in channel width can also be approached by making use of Figure 10.4, provided that we can neglect energy losses due to eddies. Typically, we would know the width, discharge, and depth at an upstream section

(w_1, Q, and Y_1, respectively) and the width at a downstream section (w_2), and would want to compute the downstream depth Y_2. To do this, we must assume that the total head H, as well as the discharge, is constant through the transition; and that the slope of the channel is very small, so that differences in specific head are effectively equal to differences in total head. The given information is sufficient to allow computation of the critical depth Y_{c1} for the upstream section from Equation 9-3. Figure 10.4 can then be used to find the specific head H_{s1} at the upstream section. Since Q is the same at both sections, Equation 9-3 again gives Y_{c2}. From the assumption that $H_{s2} = H_{s1}$, the ratio H_{s2}/Y_{c2} is known. Then Figure 10.4 can be used to find Y_2/Y_{c2} and hence Y_2. If the flow passes through the critical state, however, flow under the assumed conditions is not possible.

Again, an example will help clarify the procedure. Consider a flow of $Q = 400 \text{ ft}^3 \text{ s}^{-1}$ in a channel in which $w_1 = 100$ ft, $Y_1 = 2$ ft, and $V_1 = 2 \text{ ft s}^{-1}$. What will be the flow characteristics immediately downstream of an abrupt narrowing of the channel to a width $w_2 = 50$ ft? From Equation 9-3, the critical depth in the upstream section $Y_{c1} = 0.792$ ft, so the upstream flow is subcritical. The upstream specific head $H_{s1} = 2.06$ ft is assumed to remain constant through the transition. The downstream critical depth can also be computed via Equation 9-3: $Y_{c2} = 1.26$ ft; thus $H_{s2}/Y_{c2} = 1.64$. Since this ratio exceeds 1.5, the flow does not pass through the critical state and the computations can proceed. From Figure 10.4 we find that the ratio $Y_2/Y_{c2} = 1.36$ corresponds to the value $H_{s2}/Y_{c2} = 1.64$ for subcritical flow. Thus $Y_2 = (1.36)(1.26) = 1.71$ ft. The mean velocity downstream of the transition is therefore $V_2 = 4.68 \text{ ft s}^{-1}$. Note that the downstream flow is still subcritical.

We emphasize that the methods presented here apply only to situations in which the assumptions of parallel streamlines, constant discharge, constant head, and rectangular channels are not seriously violated. Chow (1959) presents an alternative approach to width transitions based on the principle of conservation of momentum; his method has some advantages over the use of energy relations when there is a significant loss of energy through the transition. Ippen (1950) and Chow (1959) discuss theoretical approaches and empirical studies of many types of depth and width transitions encountered in the designing of bridge piers, culverts, pilings, spillways, trash racks, various types of flow-control gates, and sharp bends. Davidian et al. (1962) present detailed laboratory studies of flows through width constrictions, with particular emphasis on relating estimates of discharge to water-surface profiles.

HYDRAULIC JUMPS

Natural supercritical flows are uncommon, but they do occur locally, or at high discharges, in stream reaches with relatively steep slopes. They are also observed fairly often in melt-water channels on glaciers, where the ice provides very low channel resistance and flow velocities are consequently high even with modest slopes (Figure 10.5.) The change from supercritical flow to subcritical steady flow is usually brought about by a decrease in channel slope, depth, or width, as discussed earlier in this chapter. Even if such changes are gradual, the point at which the flow reaches the critical state is marked by an abrupt increase in depth followed by a relatively short zone of very high turbulence and an irregular surface. Such a pronounced standing wave is called a *hydraulic jump* (Figure 10.6). Downstream from this zone the flow is reestablished as a uniform or gradually varied subcritical flow.

As with the depth and width transitions discussed earlier, the rapidly varied flow profile within the jump itself cannot be computed, but application of basic principles does enable us to calculate the relation between the depths immediately upstream and downstream from the jump. Experimental studies have revealed regularities in the basic forms and downstream extents of jumps, so that they can be well characterized by the combination of theoretical and empirical analysis.

The principle of conservation of energy cannot be usefully applied in the analysis of depths upstream and downstream of hydraulic jumps, as it was for depth and width transitions, because (1) the extreme turbulence causes significant energy losses, so we cannot assume a constant head; and (2) these losses are due to eddies (shock losses) rather than to boundary resistance, so we cannot account for them by a factor analogous to Manning's n. The principle of conservation of momentum (see

Figure 10.5
A channel eroded in ice in central Alaska. The very low channel resistance permits supercritical flow to develop. Note the irregular water surface.

Figure 10.6
Supercritical flow over a run-of-river dam, with an irregular hydraulic jump at the base of the dam.

Chapter 3) *can* be applied, however, and is formulated for a rectangular channel of low slope carrying a constant discharge by reference to Figure 10.7. Momentum is mass times velocity, so the rate at which momentum enters the jump at cross section 1, $\Delta M_1/\Delta t$, is

$$\frac{\Delta M_1}{\Delta t} = \frac{\Delta m\, V_1}{\Delta t}, \qquad (10\text{-}18)$$

where m is the mass entering the volume in time Δt and V_1 is the mean velocity at section 1. It is clear, however, that

$$\frac{\Delta m}{\Delta t} = \rho Q, \qquad (10\text{-}19)$$

where ρ is the mass density of water and Q is the discharge. Thus

$$\frac{\Delta M_1}{\Delta t} = \rho Q V_1 = \rho w Y_1 V_1^2, \qquad (10\text{-}20a)$$

where w is the width, which is assumed constant, and Y_1 is the depth. The same considerations lead to the expression for the rate of momentum leaving the jump at section 2:

$$\frac{\Delta M_2}{\Delta t} = \rho Q V_2 = \rho w Y_2 V_2^2, \tag{10-20b}$$

where the symbols are analogously defined.

It is of interest that Equations 10-20 also represent the change in turbulent or inertial force in the downstream direction. This is seen by comparison with Equation 5-28: The turbulent force per unit mass F_t/m could be written as

$$\frac{F_t}{m} = V\left(\frac{V}{L}\right) = \frac{V^2}{L}, \tag{10-21}$$

where V represents the average velocity in the jump, L is the downstream extent of the jump, and V/L is the longitudinal velocity gradient through the jump. Multiplying Equation 10-21 by the mass of water in the jump gives

$$F_t = \rho L w Y\left(\frac{V^2}{L}\right) = \rho w Y V^2, \tag{10-22}$$

where Y represents the average depth of water in the jump. The form of Equation 10-22 is identical to that of Equation 10-20a,b.

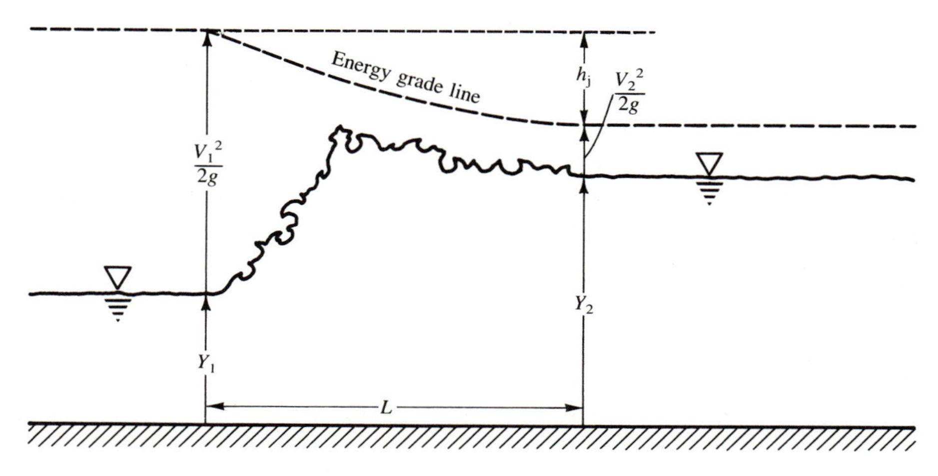

Figure 10.7
Definitions of terms for analyzing hydraulic jumps.

With a constant discharge, Equation 10-20a,b shows that the difference in the rates at which momentum enters and leaves the jump is due only to the difference in velocities at the entrance and exit sections:

$$\frac{\Delta M_1}{\Delta t} - \frac{\Delta M_2}{\Delta t} = \rho Q(V_1 - V_2). \qquad (10\text{-}23)$$

A change in velocity is clearly an acceleration, and from Newton's second law a net force must act on the mass of flowing water to produce such a change. Thus

$$\Sigma F = \rho Q(V_1 - V_2), \qquad (10\text{-}24)$$

where ΣF represents the net force.

From the discussion in Chapter 5, we recall that there are three forces that are generally significant in a turbulent open-channel flow: turbulent, gravitational, and pressure gradient. We have just seen that the change in turbulent force is represented by the right-hand side of Equation 10-24. This change must be balanced by changes in the other two forces. However, if the channel being considered has very low slopes, as we have assumed, the gravitational force (proportional to $g \sin \theta$, where θ is the slope) can be neglected. The pressure-gradient force is due to the difference in depth across the jump (see Equation 5-30), and if it is assumed that pressure increases linearly with depth (i.e., the pressure distribution is hydrostatic), we can write

$$P_1 = \gamma\left(\frac{Y_1}{2}\right) \qquad (10\text{-}25a)$$

and

$$P_2 = \gamma\left(\frac{Y_2}{2}\right), \qquad (10\text{-}25b)$$

where P_1 and P_2 are the average pressures on the upstream and downstream faces, respectively, of the jump. To find the total pressure force F_{p1} and F_{p2} on each of these faces, we must multiply the pressures by the area of each face:

$$F_{p1} = \gamma\left(\frac{Y_1}{2}\right)Y_1 w \qquad (10\text{-}26a)$$

and

$$F_{p2} = \gamma\left(\frac{Y_2}{2}\right)Y_2 w. \qquad (10\text{-}26b)$$

Thus the net force causing the change in momentum (or, put another way, the net force balancing the change in turbulent force) is equal to the difference between F_{p1} and F_{p2}. Thus

$$\Sigma F = F_{p2} - F_{p1} = \frac{\gamma w}{2}(Y_2^2 - Y_1^2)$$

and Equation 10-24 becomes

$$\frac{\gamma w}{2}(Y_2^2 - Y_1^2) = \rho Q(V_1 - V_2). \tag{10-27}$$

Equation 10-27 is the expression that determines the relation between the upstream and downstream depths, and it only remains to manipulate it into a more explicit and useful form. The form that is most convenient relates the ratio Y_2/Y_1 to the Froude number $\mathbb{F}_1$ of the upstream flow. This relation is arrived at by first dividing Equation 10-27 by w and Y_1^2 to obtain

$$\frac{Y_2^2}{Y_1^2} - 1 = \frac{2qV_1}{gY_1^2} - \frac{2qV_2}{gY_1^2}, \tag{10-28}$$

where

$$q \equiv \frac{Q}{w} = V_1Y_1 = V_2Y_2. \tag{10-29}$$

Using Equation 10-29, we can make some judicious substitutions in Equation 10-28 to arrive at

$$\left(\frac{Y_2}{Y_1}\right)^2 - 1 = \frac{2V_1^2}{gY_1} - \frac{2q^2}{gY_1^2Y_2}. \tag{10-30}$$

From the definition of the Froude number, and again using Equation 10-29, we obtain

$$\left[\left(\frac{Y_2}{Y_1}\right)^2 - 1\right]\left(\frac{Y_2}{Y_1}\right) = 2\mathbb{F}_1^2\left(\frac{Y_2}{Y_1}\right) - 2\mathbb{F}_1^2 = 2\mathbb{F}_1^2\left(\frac{Y_2}{Y_1} - 1\right). \tag{10-31}$$

Dividing both sides of Equation 10-31 by $(Y_2/Y_1 - 1)$ and rearranging, we have

$$\left(\frac{Y_2}{Y_1}\right)^2 + \frac{Y_2}{Y_1} - 2\mathbb{F}_1^2 = 0. \tag{10-32}$$

Equation 10-32 is a quadratic equation in Y_2/Y_1 with one positive root and one negative root. The negative root is of no physical significance, but the positive root is

$$\frac{Y_2}{Y_1} = \frac{(1 + 8\mathbb{F}_1^2)^{1/2} - 1}{2}, \tag{10-33}$$

which is valid for $\mathbb{F}_1 > 1$. This is the expression we have been seeking; it allows us to find the depth (and hence the velocity) at the downstream end of a hydraulic jump when the upstream flow characteristics are known. Figure 10.8 shows a plot of Equation 10-33.

Chow (1959) describes empirical studies showing that hydraulic jumps have characteristic forms that depend on the value of $\mathbb{F}_1$, as shown in Figures 10.9 and 10.10. In most natural streams Froude numbers rarely exceed 2, so only the undular and weak jumps are likely to be observed outside of man-made structures.

The lengths of hydraulic jumps, that is, the distance L between the cross sections where gradually varied flow exists, have also been investigated experimentally, and according to Chow (1959) the results are most conveniently expressed as a plot of L/Y_2 versus $\mathbb{F}_1$. This relation is also shown in Figure 10.8.

The final aspect of hydraulic jumps that we will explore here is the loss of energy h_j in the jump. The relationships established so far provide sufficient information to enable us to compute this loss for any case meeting the assumptions of the analysis, since knowledge of the upstream flow characteristics gives $\mathbb{F}_1$ and Equation 10-33 allows us to calculate the downstream depth and velocity. The values can then be substituted into the energy equation (Equation 7-17) to give

$$h_j = \frac{V_1^2}{2g} + Y_1 - \frac{V_2^2}{2g} - Y_2. \tag{10-34}$$

It is possible, however, to derive a more general dimensionless expression relating h_j/Y_1 to $\mathbb{F}_1$. To do so we start with Equation 10-34, make substitutions of the same relations we used in arriving at Equation 10-33, and use Equation 10-33 itself to arrive at

$$\frac{h_j}{Y_1} = \frac{\mathbb{F}_1^2}{2}\left\{1 - \left[\frac{(1 + 8\mathbb{F}_1^2)^{1/2}}{2} - \frac{1}{2}\right]^{-2}\right\} - \frac{(1 + 8\mathbb{F}_1^2)^{1/2}}{2} + \frac{3}{2}. \tag{10-35a}$$

Chow (1959) states that it is also possible to relate h_j/Y_1 to Y_2/Y_1 in a simpler expression:

$$\frac{h_j}{Y_1} = \frac{1}{4}\left(\frac{Y_2}{Y_1}\right)^2 - \frac{3}{4}\left(\frac{Y_2}{Y_1}\right) - \frac{1}{4}\left(\frac{Y_1}{Y_2}\right) + \frac{3}{4}. \tag{10-35b}$$

Equation 10-35a is plotted in Figure 10.11.

The preceding discussion thus allows us to compute the flow characteristics downstream from a hydraulic jump when the upstream characteristics are known, to

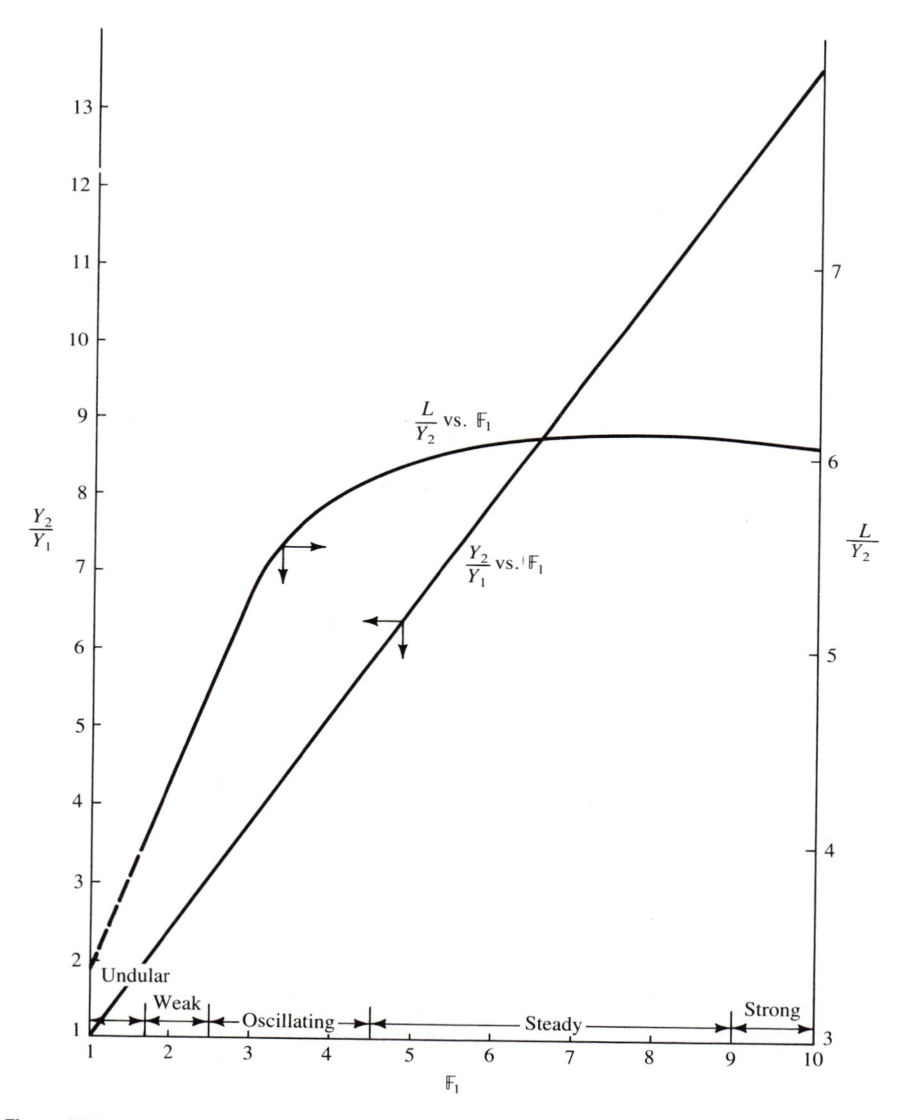

Figure 10.8
Relation of upstream depth Y_1 and downstream depth Y_2 of a hydraulic jump to the Froude number $\mathbb{F}_1$ of the upstream flow (Equation 10-33), and experimentally determined relations between the relative length of a jump L/Y_2 and $\mathbb{F}_1$. The type of jump is indicated at the bottom (see Figures 10.9 and 10.10).

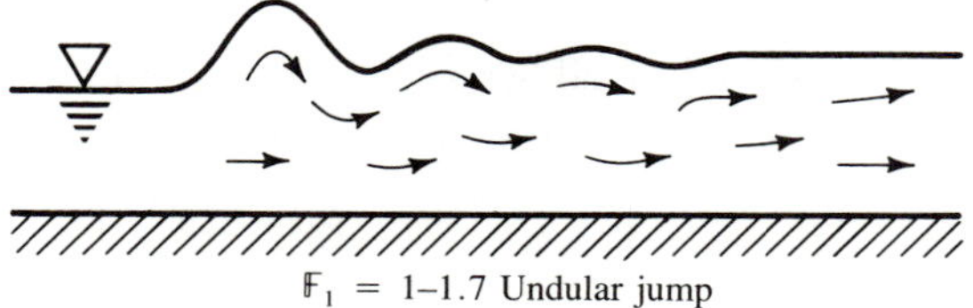

$\mathbb{F}_1$ = 1–1.7 Undular jump

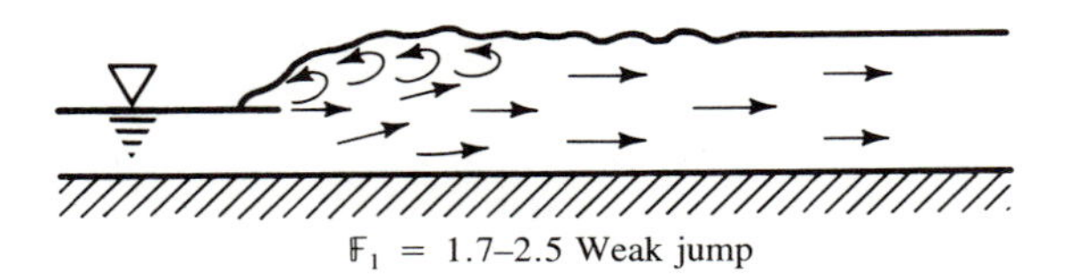

$\mathbb{F}_1$ = 1.7–2.5 Weak jump

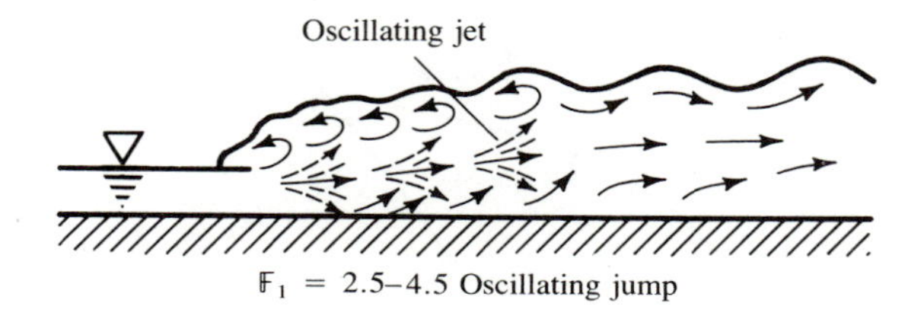

$\mathbb{F}_1$ = 2.5–4.5 Oscillating jump

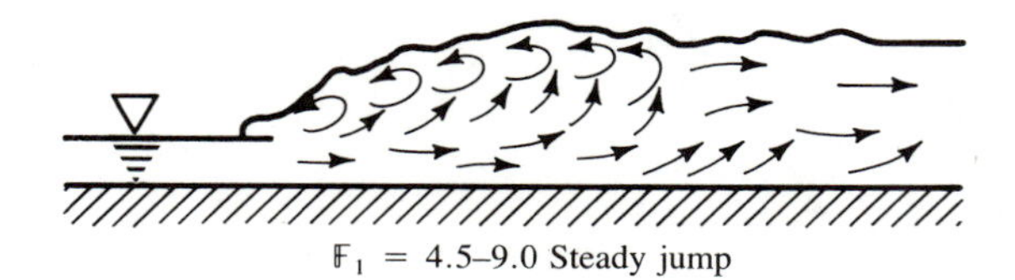

$\mathbb{F}_1$ = 4.5–9.0 Steady jump

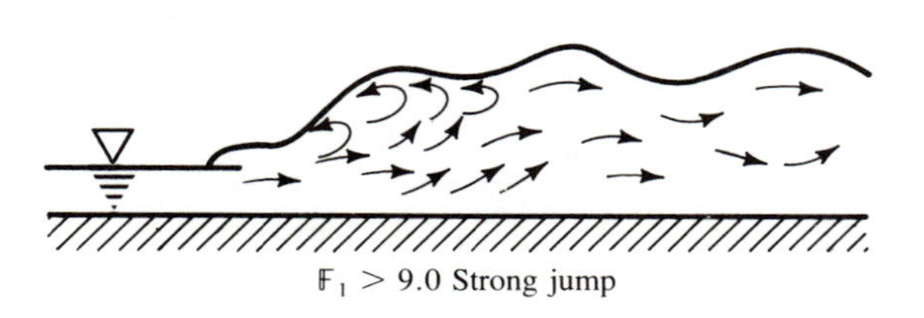

$\mathbb{F}_1 > 9.0$ Strong jump

Figure 10.9
Types of hydraulic jumps. From V. T. Chow, *Open Channel Hydraulics,* © 1959, p. 395. Reproduced by permission of McGraw-Hill Book Company, New York.

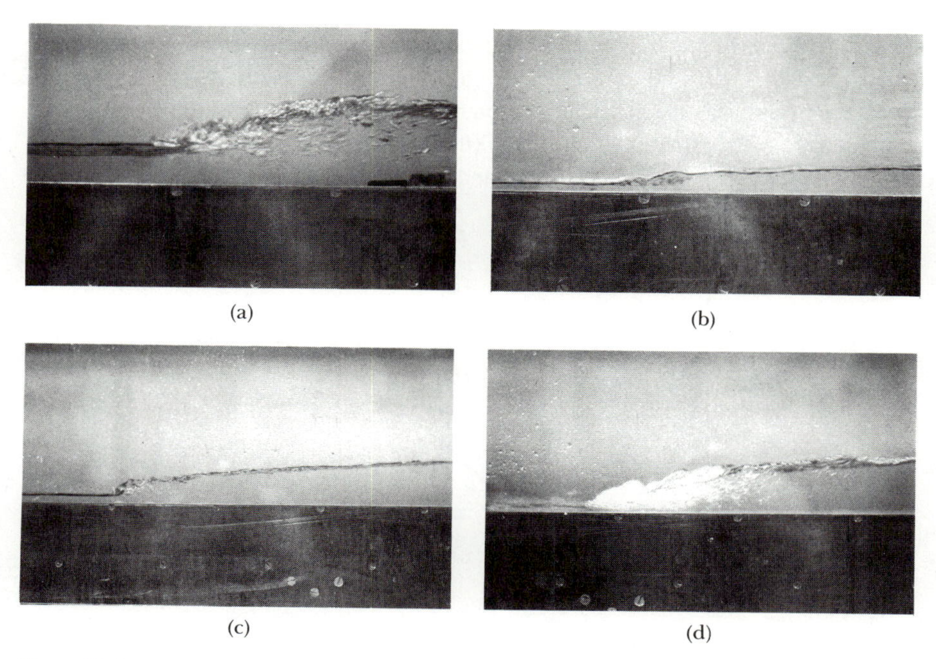

Figure 10.10
Hydraulic jumps in a laboratory flume: (a) weak; (b) oscillating; (c) steady; (d) strong.

estimate the length of the jump, and to compute the energy losses due to turbulence in the jump for approximately rectangular channels of low slope. For steep slopes we would also have to consider the gravitational forces due to the weight of the water in the jump in developing an equation analogous to Equation 10-27. However, taking these gravitational forces into account requires prior knowledge of the form of the jump, and is thus an intractable theoretical problem. Rouse (1961) and Chow (1959) present results of empirical studies of hydraulic jumps on steep slopes. Fortunately, it turns out that deviations from the results found herein assuming a horizontal channel are not important for the slopes usually encountered in natural streams.

Rouse (1961) and Chow (1959) also discuss many other aspects of hydraulic jumps that are of interest in engineering design, and Rouse (1961) presents an interesting discussion of the hydraulic jump as a gravity-wave phenomenon.

WEIRS

Weirs are damlike devices constructed across channels in order to control or measure flow rates. They are of particular interest to hydrologists because they are generally the most practical means for continuous measurement of discharge where accuracy is of special concern, such as on research watersheds. Weirs provide this accuracy by assuring a consistent relation between the elevation of the water surface a short distance upstream of the weir (the *weir head,* denoted by H_w in Figure 10.12), which can be readily measured and recorded, and the discharge passing over

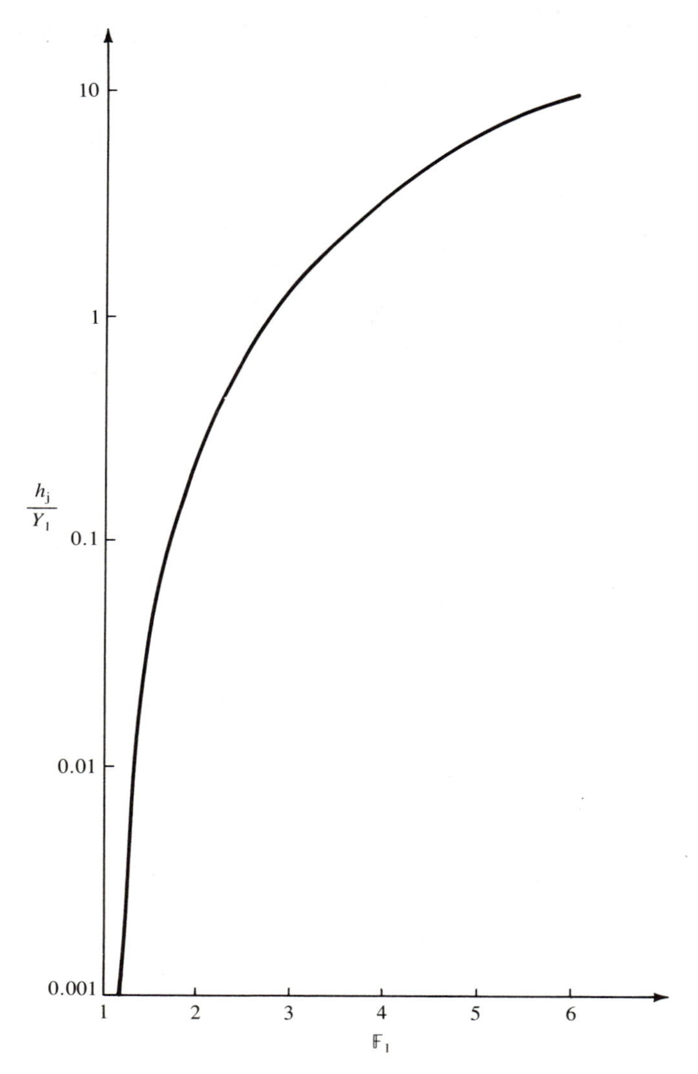

Figure 10.11
Relation between relative head loss h_j/Y_1 through a hydraulic jump and Froude number $\mathbb{F}_1$ of the upstream flow (Equation 10-35a).

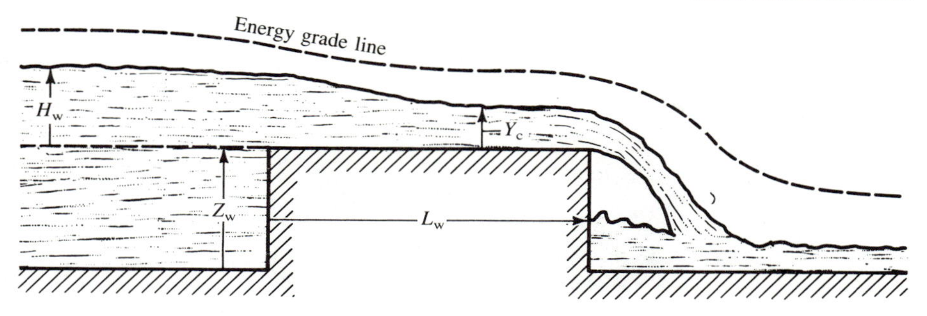

Figure 10.12
Definitions of terms for analysis of a broad-crested weir. The location of the critical depth Y_c varies with discharge and the nature of the weir surface.

the weir, which is virtually impossible to measure directly. This consistent relation is attainable because weirs are designed to force the flow to pass from the subcritical to the supercritical state; at the point where the flow is critical there is a fixed relation between discharge and flow depth that is independent of channel slope and roughness (Equation 9-3). However, the geometry of the weir does affect the relation between weir head and discharge, as will be discussed later.

Weirs are described in terms of (1) their relative "thickness" (i.e., the ratio H_w/L_w in Figure 10.12), and (2) the shape of the opening through which the water passes over the weir crest. If H_w/L_w is less than about 1.5–2, the weir is *broad crested* (Figure 10.13a–c). At higher values of H_w/L_w the flow springs free from the upstream edge, and the weir is classified as *sharp crested* (Figure 10.14). Sharp-crested weirs, usually with triangular, trapezoidal, or rectangular openings, are the types usually installed for the specific purpose of discharge measurement. Here, however, we begin with an analysis of rectangular broad-crested weirs, since they provide further insight into abrupt channel transitions and lead naturally to an analysis of sharp-crested weirs. In addition, it is sometimes desirable to be able to estimate the flow rate over an obstacle that functions as a broad-crested weir, as for example in estimating the discharge of a past flood by using high-water marks upstream of a highway embankment or similar obstruction.

Broad-Crested Weirs In the first section of this chapter, we saw that when a subcritical flow encounters an abrupt rise in the channel bottom, its depth decreases and its velocity increases (Figure 10.3a). If the height Z_2 of the rise exceeds the constraint given in Equation 10-10, the flow will pass through the critical state and become supercritical. This transition is marked by a rapid decrease in depth and a highly curved surface near the upstream end of the rise (Figure 10.13). If the upstream end of the obstruction is not rounded, a zone of flow separation and high turbulence may be generated along the weir crest. Because of these features, there is a significant loss of energy through the transition.

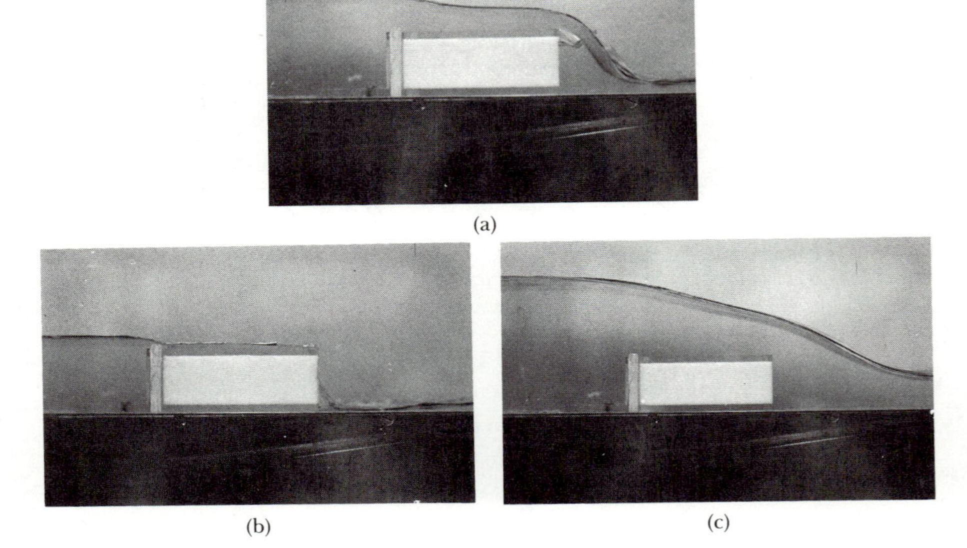

(a)

(b)

(c)

Figure 10.13
Flows over broad-crested weirs in a laboratory flume: (a) normal; (b) long; (c) short.

Figure 10.14
Flow over a rectangular sharp-crested weir in a laboratory flume.

Figure 10.13 shows the typical water-surface profile at the upstream end of a broad-crested weir. Based on studies summarized by Tracy (1957), if $0.4 > H_w/L_w > 0.08$, the surface remains subhorizontal over the entire weir crest (Figure 10.13a). When H_w/L_w is less than about 0.08 (Figure 10.13b), the weir is described as *long*. Under these conditions, the frictional forces exerted by the weir crest may be sufficient to decelerate the flow to the point where the critical state is again reached, in which case a hydraulic jump forms. When H_w/L_w exceeds about 0.4 the surface is entirely curvilinear, and the weir is *short* (Figure 10.13c). Finally, if the ratio

H_w/L_w reaches values of about 1.5–2, the flow at the upstream end becomes detached and the weir is effectively sharp crested (Figure 10.14).

If we look again at Figure 10.12 and assume a rectangular channel of constant width w, we see that Equation 9-3 gives the exact relation between the discharge Q and critical depth Y_c for steady flows with nearly parallel streamlines and low slopes:

$$Q = wg^{1/2}Y_c^{3/2}. \tag{10-36}$$

The difficulty in exploiting this relation in order to determine the discharge over a broad-crested weir is that (1) the flow near the critical point is highly curvilinear; (2) we do not know exactly where the flow reaches the critical depth; and (3) the location of the critical depth changes as discharge changes. Experiments have shown, however, that Y_c is proportional to H_w, and H_w can be readily measured upstream of the weir influence. Thus

$$Y_c = \beta_b H_w, \tag{10-37}$$

where β_b is the constant of proportionality. Substituting Equation 10-37 into Equation 10-36 yields

$$Q = wg^{1/2}\beta_b^{3/2}H_w^{3/2} \tag{10-38a}$$

or

$$q = g^{1/2}\beta_b^{3/2}H_w^{3/2}. \tag{10-38b}$$

It is convenient to define a dimensionless weir coefficient C_b as

$$C_b \equiv \beta_b^{3/2}; \tag{10-39}$$

with this definition, Equation 10-38b becomes

$$q = C_b g^{1/2}H_w^{3/2}. \tag{10-40}$$

Equation 10-40 is the standard form of the discharge equation for a rectangular broad-crested weir. For normal and long weirs C_b tends to have a relatively constant value of about 0.465. As shown in Figure 10.15, however, C_b increases as H_w/L_w increases beyond about 0.4. The degree of rounding of the upstream edge of the weir and the approach velocity also affect the value of C_b, but to a lesser extent (Tracy, 1957). If it is possible to calibrate a particular weir, doing so is of course the best method for determining C_b; otherwise we can refer to Figure 10.15 or the experimental results summarized by Tracy (1957).

In many texts a discharge coefficient C_d is defined as

$$C_d \equiv C_b g^{1/2} = \beta_b^{3/2} g^{1/2}, \tag{10-41}$$

so that the broad-crested weir equation (Equation 10-40) becomes

$$q = C_d H_w^{3/2}. \tag{10-42}$$

Clearly, C_d has dimensions of $[L^{1/2}T^{-1}]$, and its numerical value will vary depending on the system of units. You should keep this relation between numerical value and units firmly in mind, because many texts and handbooks give the impression that C_d is dimensionless. For example, American texts commonly state that C_d varies from 2.67 to 3.05 (e.g., Chow, 1959), although this is only true when H_w is in feet and q is in square feet per second.

Sharp-Crested Weirs As noted earlier, a weir is sharp crested if the lower flow boundary springs free from the solid boundary at the upstream end. Sharp-crested

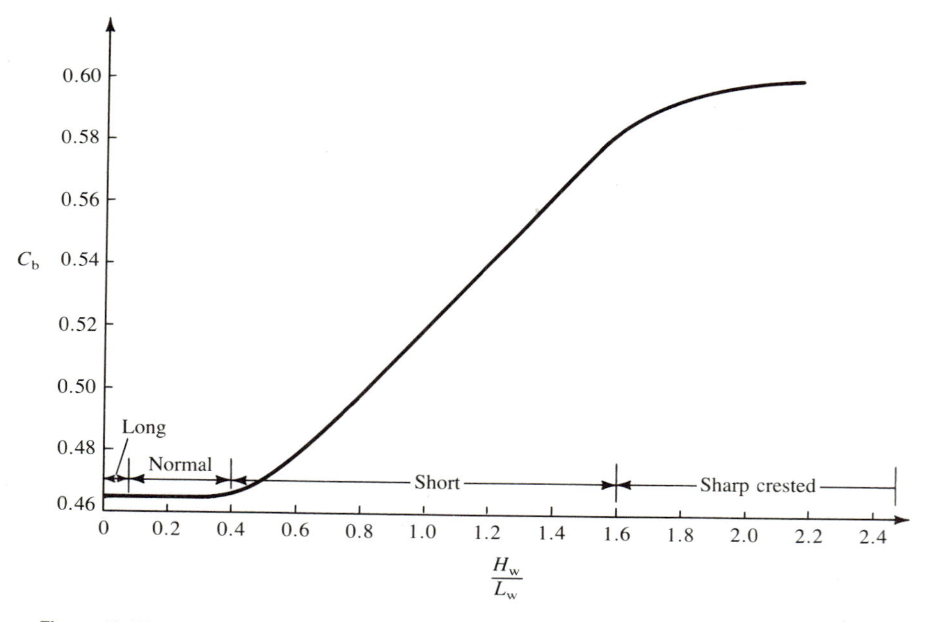

Figure 10.15
Weir coefficient C_b for rectangular broad-crested weirs as a function of relative weir height H_w/L_w. Data from Tracy (1957).

weirs occur when H_w/L_w is greater than about 1.5. For discharge measurements, weirs are usually constructed to have a crest that is literally sharp (i.e., a knife-edge), so that $H_w/L_w \rightarrow \infty$ and the flow will spring free even for very small values of H_w (Figure 10.14).

Theoretically, the point where the water springs free from a sharp-crested weir (the *brink*) should be the section of minimum specific energy (Chow, 1959), and hence the locus of the critical depth. However, experiments by Rouse (1936) showed that the critical depth Y_c as computed by Equation 9-3 occurs at a distance of about $3Y_c$–$4Y_c$ upstream from the brink, and the depth at the brink has a value equal to about $0.71Y_c$. The true section of minimum specific energy is in fact at the brink, but because Equation 9-3 applies strictly only to gradually varied flow, it does not give the correct value for the critical depth in the region of rapidly varied flow near the brink.

It is clear from the earlier discussion and Figure 10.15 that rectangular broad-crested weirs are transitional to rectangular sharp-crested weirs, and presumably can be described by an equation identical to Equation 10-40. This is correct, but in order to treat sharp-crested weirs of other than rectangular shape it is useful to approach the formulation of the weir equation by taking advantage of the fact that the flow passes through the critical state at the brink. When flow is critical, specific head is at a minimum value and the velocity head is equal to twice the pressure head, as given by Equation 7-24:

$$Y_b = \frac{\alpha V_b^2}{g}, \tag{10-43}$$

where the subscript b indicates the value at the brink and the energy coefficient α is included to account explicitly for the deviation from parallel streamlines. Solving Equation 10-43 for V_b gives

$$V_b = \left(\frac{1}{\alpha}\right)^{1/2} g^{1/2} Y_b^{1/2}. \tag{10-44}$$

Multiplication of Equation 10-44 by the brink depth gives the discharge per unit width q:

$$q = \left(\frac{1}{\alpha}\right)^{1/2} g^{1/2} Y_b^{3/2}. \tag{10-45}$$

From the preceding discussions it is again reasonable to state that

$$Y_b = \beta_s H_w, \tag{10-46}$$

where β_s is a constant of proportionality appropriate for sharp-crested weirs. Substituting Equation 10-46 into Equation 10-45 gives

$$q = \left(\frac{1}{\alpha}\right)^{1/2} g^{1/2}\beta_s^{3/2}H_w^{3/2}. \tag{10-47}$$

Once again, the numerical constants can be combined into a single sharp-crested weir coefficient C_s, whereupon Equation 10-47 becomes

$$q = C_s g^{1/2} H_w^{3/2}, \tag{10-48}$$

which is of the same form as Equation 10-40, as expected.

Clearly, C_s accounts chiefly for the influences of weir geometry. While Tracy (1957) reported that the ratio H_w/Z_w of the weir head to height of the weir above the channel bottom has little effect on the coefficient for broad-crested weirs, it has a significant effect on the sharp-crested weir coefficient. For a rectangular sharp-crested weir that extends across the full width of the channel, experiments have shown that this relation can be well predicted by the empirical equation

$$C_s = 0.57 + 0.071\left(\frac{H_w}{Z_w}\right) \tag{10-49}$$

for values of H_w/Z_w up to about 10 (Chow, 1959; Daily and Harleman, 1966). Kindsvater and Carter (1959) conducted a series of experiments on rectangular sharp-crested weirs and determined that the relative width w_b/w of the weir opening (Figure 10.16) also has an important effect on C_s. Figure 10.16 shows their results, and indicates the combined effects of H_w/Z_w and w_b/w on C_s for values of H_w/Z_w up to 2.4 (Equation 10-48 is also plotted in this figure). Note that the presence of *contractions* (Figure 10.16) causes C_s to decrease, and that for highly contracted openings C_s decreases as H_w/Z_w increases. Again, it is always advisable to calibrate any weir that is to be used for precise discharge measurements.

Sharp-crested weirs with triangular openings (Figure 10.17) are commonly used to provide greater sensitivity for discharge measurements than do rectangular weirs. In the diagram in Figure 10.18 the cross-sectional area of flow A_b through a triangular opening is related to the water depth Y_b and the vertex angle θ_b as

$$A_b = Y_b^2 \tan\left(\frac{\theta_b}{2}\right). \tag{10-50}$$

Multiplying the area by the velocity as given by Equation 10-44 yields the discharge Q:

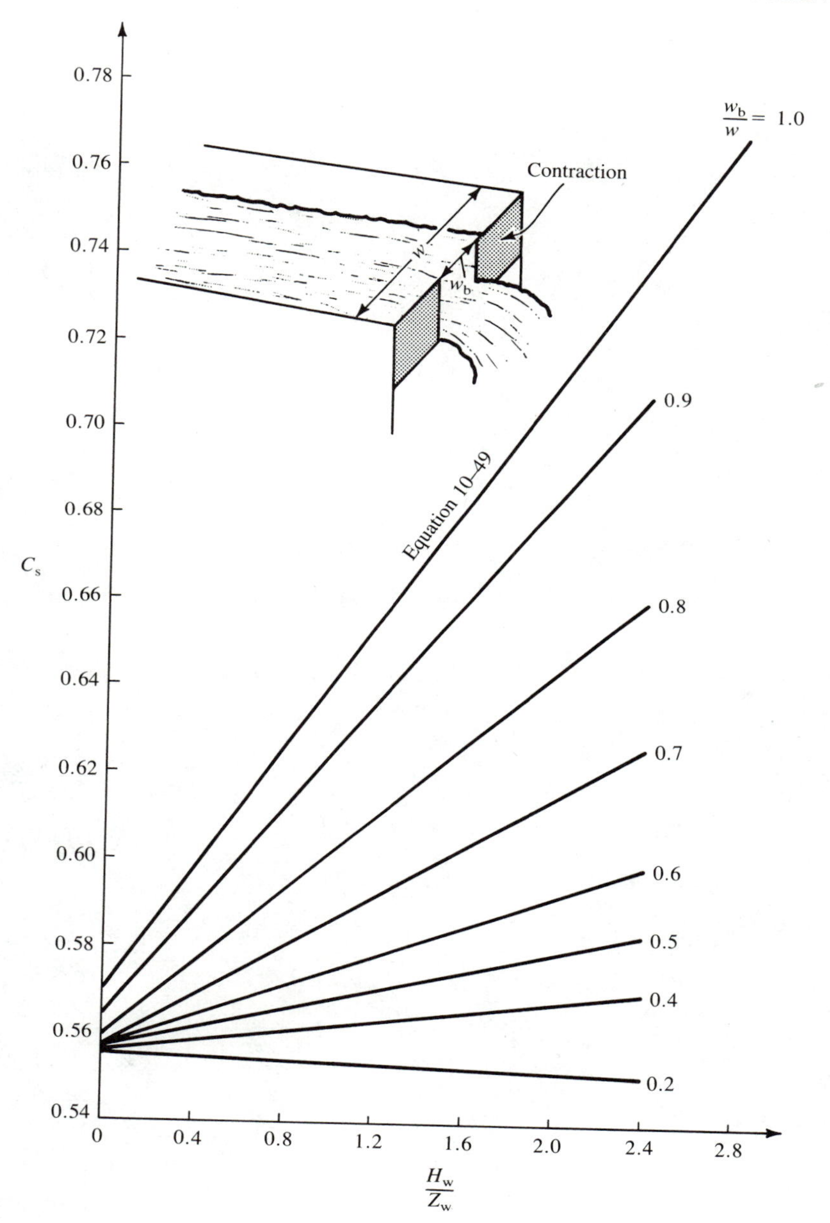

Figure 10.16
Variation of weir coefficient C_s for rectangular sharp-crested weirs as a function of relative weir head H_w/Z_w and degree of contraction w_b/w. Data from Kindsvater and Carter (1959).

(a)

(b)

Figure 10.17
Stream-gaging weirs in research watersheds: (a) 90° V-notch sharp-crested weir in central Alaska; (b) concrete 120° V-notch weir (which is effectively sharp crested for most flows) in Vermont.

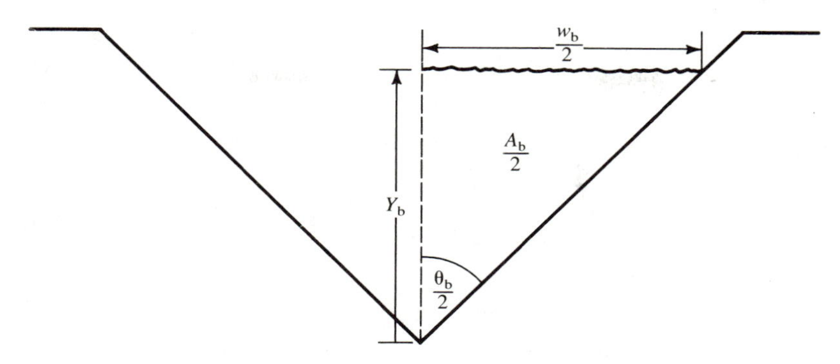

Figure 10.18
Derivation of the relation between flow depth Y_b and cross-sectional area A_b in a triangular weir notch (Equation 10-50). Since $\tan(\theta_b/2) = w_b/2Y_b$, $w_b = 2Y_b \tan(\theta_b/2)$. The area $A_b/2 = \frac{1}{2}Y_b(w_b/2)$, so $A_b = \frac{1}{2}w_bY_b$. Combining these relations yields $A_b = Y_b^2 \tan(\theta_b/2)$.

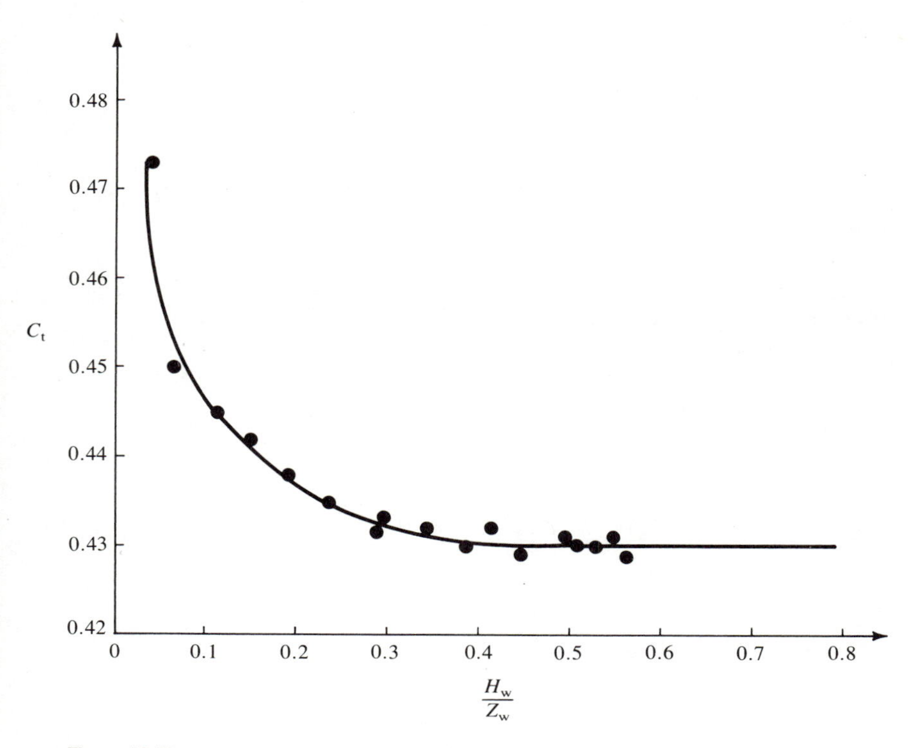

Figure 10.19
Weir coefficient C_t as a function of relative weir head H_w/Z_w as determined by laboratory calibration of the 90° V-notch weir shown in Figure 10.17a.

$$Q = \left(\frac{1}{\alpha}\right)^{1/2} g^{1/2} \tan\left(\frac{\theta_b}{2}\right) Y_b^{5/2}; \tag{10-51}$$

Alternatively, again utilizing Equation 10-46 and a weir coefficient, we have

$$Q = C_t g^{1/2} \tan\left(\frac{\theta_b}{2}\right) H_w^{5/2}, \tag{10-52}$$

where C_t is a coefficient for triangular sharp-crested weirs.

No comprehensive studies of how C_t varies with various geometric configurations have been published. For $\theta_b = 90°$, however, it is generally reported that $C_t \simeq 0.43$ no matter what the value of H_w/Z_w. This value appears to be a very good approximation above some minimum value of H_w/Z_w, which must be determined for each weir. When H_w/Z_w is small, minor irregularities of the notch affect the value of C_t. Such effects are illustrated in Figure 10.19, which shows the results of calibration studies of a 90° V-notch weir that was used in a research watershed. Note that C_t decreases as H_w/Z_w increases, as is the case for highly contracted rectangular sharp-crested weirs (Figure 10.16); but the decrease in this case is highly nonlinear.

Exercises

10-1. The accompanying table gives the depths (Y_1), velocities (V_1), and widths (w) just upstream of abrupt increases ($Z_2 > 0$) or decreases ($Z_2 < 0$) in channel-bottom elevation. Compute the Froude number of the upstream flow and the depths, velocities, and Froude numbers of the flows just downstream of the transition. Width is constant and slope and energy losses are negligible.

Flow	Y_1 (m)	V_1 (m s^{-1})	w (m)	Z_2 (m)
a	3.57	2.32	89.0	+0.80
b	0.41	0.64	104	+0.10
c	1.80	6.80	10.1	+0.35
d	0.90	4.50	8.20	+0.10
e	0.85	1.10	13.1	+0.20
f	3.65	1.62	10.0	+1.20
g	4.60	0.50	180	−1.60
h	0.80	3.40	15.0	−1.10

i	0.64	0.95	200	−0.80
j	0.40	0.84	34.0	−0.50
k	0.45	3.20	21.0	−0.50
l	1.00	0.70	76.0	−0.40

10-2. The accompanying table gives the depths (Y_1) and velocities (V_1) just upstream of abrupt changes in width from w_1 upstream to w_2 downstream. Compute the Froude number of the upstream flow and the depths, velocities, and Froude numbers of the flows just downstream of the transition. Slopes and energy losses are negligible.

Flow	Y_1 (m)	V_1 (m s^{-1})	w_1 (m)	w_2 (m)
a	0.39	1.20	4.20	3.79
b	0.16	1.00	2.30	2.63
c	0.22	6.50	6.80	3.68
d	0.93	3.63	4.00	6.52
e	0.22	2.35	5.00	13.90
f	0.061	0.52	2.00	1.87
g	0.23	6.60	12.00	10.60
h	0.13	0.85	3.40	4.04

10-3. (a) Show the steps that lead from Equation 10-16 to Equation 10-17. (b) Compute the Froude numbers of the flows in Exercise 9-7 by using the standard formula and Equation 10-17.

10-4. Show that Equation 10-33 is a solution of Equation 10-32. (*Hint:* Use the general formula for the solution of quadratic equations.)

10-5. The accompanying table gives the depths (Y_1) and velocities (V_1) of flows just upstream of hydraulic jumps in rectangular channels with negligible slopes and constant widths. (a) Estimate the lengths of the jumps and compute the depths and velocities just downstream of them. (b) Compute the energy loss in each jump and compare this value with the loss that would occur in the same length of channel with no jump, assuming a Manning's n of 0.020. (Use Equation 9-16 for the latter.)

Flow	Y_1 (m)	V_1 (m s^{-1})
a	1.08	6.50
b	1.23	5.20
c	0.95	5.50
d	0.76	6.00
e	0.33	3.95
f	1.22	4.50

10-6. Select two of the cases in Exercise 10-5 and show that the energy loss is the same when computed by Equations 10-35a and 10-35b.

10-7. The accompanying table gives the weir heads (H_w), lengths (L_w), and widths (w) of flows over rectangular broad-crested weirs. (a) Compute the discharge for each case, using Figure 10.15. (b) For flow a, compute and plot the *rating curve* (i.e., the relation between the weir head and the discharge) for weir heads ranging from 0 to 8.00 m.

Flow	w (m)	L_w (m)	H_w (m)
a	12.0	5.50	2.40
b	2.40	1.55	0.40
c	34.5	6.00	0.50
d	0.70	0.10	0.12
e	12.6	4.80	6.05
f	6.20	1.85	2.20

10-8. Compute and plot the rating curve for a rectangular sharp-crested weir with a width of 2.00 m and a height (Z_w) of 0.80 m for $0 \leq H_w \leq 1.00$ m.

10-9. Recompute the rating curve for the weir in Exercise 10-8, but assume that the width of the opening is reduced to 1.00 m by contractions of 0.50 m on each side. Plot this curve on the same graph as in Exercise 10-8.

10-10. Compute the rating curve for a 90° V-notch sharp-crested weir over the same range of H_w used in Exercises 10-8 and 10-9. Assume that C_t varies as shown in Figure 10.19. Plot this curve on the same graph used for the previous two exercises.

10-11. Explain why you would choose the weir described in Exercise 10-10 over those described in Exercises 10-8 and 10-9 for precise measurement of discharge in a small research watershed. Do the rectangular weirs have any advantage relative to the triangular weir?

CHAPTER 11

Gradually Varied Unsteady Flows

The solution of many hydrologic problems requires the development of models that simulate a streamflow hydrograph, that is, the time variations of discharge at a stream cross section. One important class of such problems involves predicting the effects on streamflows of changes in watershed land use (such as deforestation, various agricultural practices, or urbanization) and thus involves modeling flow paths on the watershed generally as well as in the stream-channel network. A second important class, called *streamflow routing* or *channel routing,* is concerned with predicting the hydrograph at the downstream end of a channel reach when you are given a hydrograph at the upstream end. Problems of this type arise when we want to estimate flood flows downstream of a gaging station where flows have been measured, or when we would like to predict the effects on streamflows of such activities as filling the floodplain, constructing flood-protection levees, or modifying the channel configuration.

Figure 11.1 is a conceptualization of the flow paths on a watershed during a rainstorm; it shows that streamflow at a cross section is an integrated response to spatial and temporal variations in rainfall rates and surface and subsurface flow rates upstream of the section. In watershed modeling, the characteristics of flow paths, such as length, slope, and flow resistance, are specified to represent the conditions of interest. A common feature of all these flow paths, however, is that they are the cumulative results of lateral inflows, as illustrated in Figure 11.2. This representation is particularly apt for overland flows (in which the lateral inflows are rain) and channel flows (in which the lateral inflows are overland flows and/or subsurface flows). Our objective in this chapter is to develop equations for spatially and temporally varying flows subject to lateral inflows.

In watershed modeling, we might ideally want to simulate streamflows by linking together the basic equations of motion so as to represent the pathways distributed over the entire watershed and channel network. Perhaps the closest approach to this ideal was the work of Freeze (1974), who modeled a rather small hillslope and adjacent stream channel on a very large computer, and was thereby able to demonstrate some important aspects of streamflow generation. Such an approach, however, is not practical for most problems because of the immense computer storage capacity and computation time it requires. Instead, hydrologists must usually simplify their models by ignoring flow paths that are relatively unimportant and by combining the important ones into a small number of representative paths. Examples

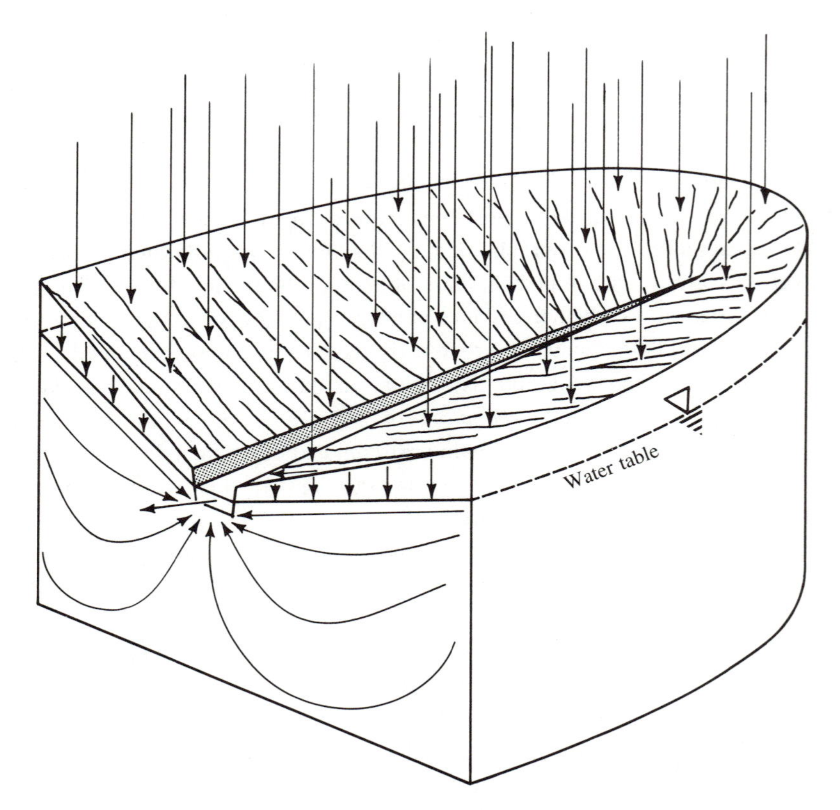

Figure 11.1
Schematic diagram of the flow paths in a typical small watershed.

of the development and application of such relatively simplified models can be found in Lane and Woolhiser (1977), Rovey et al. (1977), and Ross et al. (1979). The general concept of flows subject to lateral inflows, as illustrated in Figure 11.2, is retained in these models.

Most streamflow routing methods are simplified versions of the basic physical equations developed in this chapter. They often include empirical constants that are determined for the stream reach of interest by estimation (analogous to the estimation of Manning's n, as described in Chapter 8) or by calibration studies of the reach. Empirical routing methods are of great practical importance, but because they are thoroughly discussed in many texts (e.g., Linsley et al., 1949; Chow, 1959; Lawler, 1964; Gray and Wigham, 1970; Viessman et al., 1977), we will not present them here.

Before proceeding to the development of the basic equations, we explore some important related aspects of unsteady flows: the relations between changes in discharge and changes in velocity, width, and depth; gravity waves; and flood waves.

HYDRAULIC GEOMETRY

Hydraulic geometry is the term coined by Leopold and Maddock (1953) to refer to the ways in which changes in discharge are apportioned among changes in the components of discharge: width, mean depth, and mean velocity. At a given cross section, discharge changes with time, and the corresponding changes in the components define the *at-a-station hydraulic geometry*. At any given instant, discharge in a river system also changes spatially, usually increasing (because of increasing

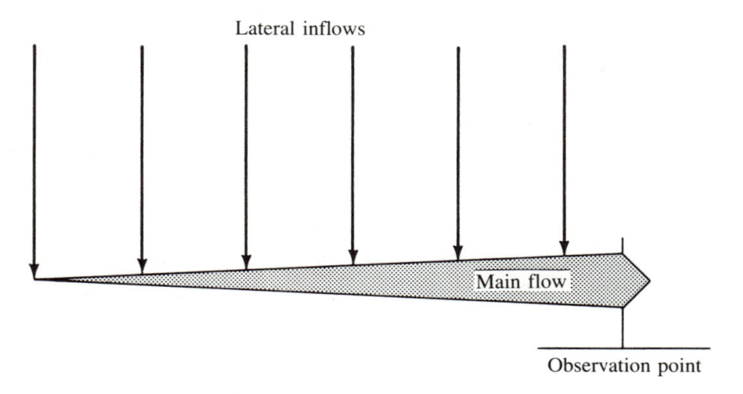

Figure 11.2
Essential features of a typical hydrologic flow receiving lateral inflows. Time variation of the flow at the observation point depends on the time variation of the inflows and the characteristics of the main flow path.

drainage area) as we move downstream. The corresponding spatial changes in width, mean depth, and mean velocity define the *downstream hydraulic geometry.*

Discharge Q is defined as

$$Q \equiv wYV, \tag{11-1}$$

where w is the width of the water surface; Y is the hydraulic, or mean, depth; and V is the mean velocity. Thus it is mathematically true that

$$w = aQ^b, \tag{11-2}$$

$$Y = cQ^f, \tag{11-3}$$

and

$$V = kQ^m, \tag{11-4}$$

where a, c, and k are constants that give the value of the width, depth, and velocity, respectively, for a discharge of one unit; and b, f, and m are exponents whose values are proportional to the rates at which the width, depth, and velocity, respectively, change as discharge changes. It can be readily shown for Equations 11-1–11-4 that

$$ack = 1 \tag{11-5}$$

and

$$b + f + m = 1. \tag{11-6}$$

It is clear that Equations 11-2–11-4 are not dimensionally homogeneous. Recalling Rules 14a and 14b of Chapter 3, we see that the values of a, c, and k depend on the system of units, whereas the values of b, f, and m do not. Thus the relative values of these exponents provide basic information about how changes in discharge are apportioned among changes in width, depth, and velocity. In fact, it can be shown that these exponents are equal to the ratios of the percentage changes in the components to the percentage change in discharge. It should also be noted that equations of the form of Equations 11-2–11-4 plot as straight lines on logarithmic (i.e., log–log) graph paper, with the slopes of the lines equal to the values of the exponents (see Exercise 11-1).

To explore at-a-station hydraulic geometry, consider a cross section of a wide rectangular channel at which flows are always uniform and roughness does not change. In this case we can substitute the hydraulic depth for the hydraulic radius and rearrange the Manning equation as written for discharge (Exercise 6-9) to

$$Y = \left(\frac{n}{u_M S_0^{1/2} w}\right)^{3/5} Q^{3/5}. \tag{11-7}$$

The velocity form of the Manning equation is

$$V = \left(\frac{u_M S_0^{1/2}}{n}\right) Y^{2/3}, \tag{11-8}$$

and substituting Equation 11-7 into 11-8 yields

$$V = \left(\frac{u_M^{3/5} S_0^{3/10}}{n^{3/5} w^{2/5}}\right) Q^{2/5}. \tag{11-9}$$

Thus for these assumed conditions $f = 0.6$ (Equation 11-7), $m = 0.4$ (Equation 11-9), and (of course) $b = 0$, since the width was assumed to be constant. Note that Equation 11-6 is satisfied.

Natural channels are not rectangular, so their widths as well as their depths and velocities change with discharge. Thus we would expect the value of b to be greater than zero and (keeping in mind Equation 11-6) the values of f and m to be somewhat less than those just computed. As Table 11.1 shows, these expectations are borne out by data from natural channels in many widely scattered locations. Changes in discharge at a station tend to be allocated about evenly between depth and velocity, with a much smaller fraction of the change allocated to width.

Table 11.1
At-a-Station and Downstream Hydraulic Geometries of Natural Streams
Data from Leopold et al. (1964) and Dunne and Leopold (1978).

	At a station			Downstream		
Location	b	f	m	b	f	m
Average, midwestern United States	0.26	0.40	0.34	0.5	0.4	0.1
Brandywine Creek, Pennsylvania	0.04	0.41	0.55	0.42	0.45	0.05
Ephemeral streams, western United States	0.29	0.36	0.34	0.5	0.3	0.2
Appalachian streams	—	—	—	0.55	0.36	0.09
Average, 158 locations in the United States	0.12	0.45	0.43	—	—	—
Average, ten locations on the Rhine River, Western Europe	0.13	0.41	0.43	—	—	—
Green River, Wyoming	0.16	0.38	0.44	0.55	0.35	0.10

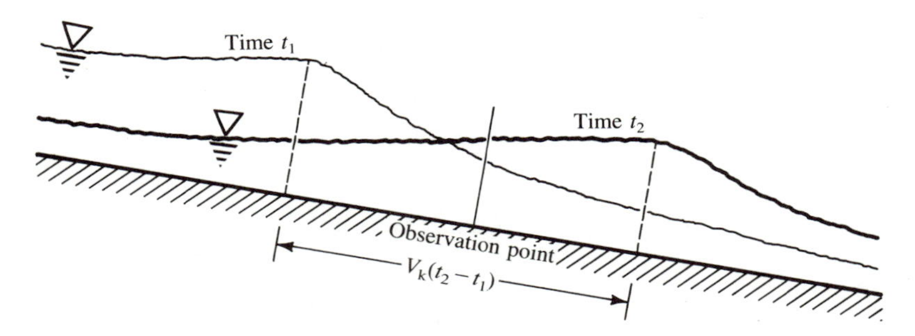

Figure 11.3
Movement of a flood wave with a velocity V_k. Note that at the observation point the water-surface slope is steeper when the discharge is increasing (time t_1) than when it is decreasing (time t_2).

It should also be noted that, in general, neither slope nor roughness remains constant as discharge varies at a particular cross section in a natural stream. Changes in these factors usually show some consistent relation to discharge. In Chapter 8 it was noted that roughness usually decreases as discharge increases. Furthermore, as a flood wave passes through a section, the water-surface slope is larger before the peak than after (Figure 11.3). Thus it is usually true that for a given value of mean depth, the discharge is higher when discharge is increasing than when it is decreasing. These factors cause a certain amount of scatter in plots of width, depth, or velocity against discharge.

The capacities of most natural channels tend to be equal to discharges that are exceeded with an average frequency of once every 1.5–3 years. Thus the values of width, depth, and velocity at this capacity, or *bank-full discharge,* are used to define the downstream hydraulic geometry. Downstream increases in bank-full discharge are typically reflected in a relatively rapid increase in width, a somewhat lesser increase in depth, and very little increase in velocity (Table 11.1). The relative values of the exponents in the width and depth relations are determined in large part by the erodibility of the channel bed and banks, and the modest change in velocity occurs partly because stream slopes tend to decrease downstream.

Further discussions of at-a-station and downstream hydraulic geometry can be found in Leopold and Maddock (1953), Leopold et al. (1964), and Dunne and Leopold (1978), and extensive hydraulic data for natural streams are given in Leopold and Wolman (1957).

WAVES IN OPEN CHANNELS

There are two broad classes of waves that occur in open channels: (1) *dynamic waves,* which arise as a result of the forces that act in fluids with free surfaces; and

(2) *kinematic waves,* which arise from continuity considerations alone and in which the net forces are zero.

Dynamic waves may be generated by elastic forces, surface-tension forces, and gravity forces. Elastic waves are due to the compressibility of the fluid, and aside from noting that sound waves belong to this class, we need not consider them further here. Capillary waves due to surface tension are important in small-scale flows, but are completely dominated by gravity and kinematic waves in phenomena of hydrologic importance. Thus we concentrate here on only these two types. An excellent comprehensive discussion of all types of dynamic waves is given by Rouse (1961).

Gravity Waves It was stated in Chapter 5 that the celerity C_g (i.e., the velocity of a wave relative to the velocity of the fluid) of a gravity wave in shallow water is given by

$$C_g = (gY)^{1/2}, \tag{11-10}$$

where g is the acceleration due to gravity (see Equation 5-49). Following Chow (1959), we can determine the value of C_g in more detail by considering a solitary wave caused by the sudden horizontal movement of a vertical plate in a horizontal rectangular channel in which the water is stationary (Figure 11.4a). To a stationary observer the wave motion is an unsteady flow, but the problem can be transformed to a case of steady flow by considering that the observer moves along with and at the same speed as the wave. Thus, relative to the observer, the water is moving to the left with a velocity C_g (Figure 11.4b). The one-dimensional energy equation for steady flow can now be applied between a section in the undisturbed flow and a section at the wave crest, neglecting friction losses:

$$Y + \frac{C_g^2}{2g} = Y + h_w + \frac{V_w^2}{2g}, \tag{11-11}$$

where $V_w^2/2g$ is the velocity head at the wave crest and h_w is the wave height. (The channel bottom is the datum and α is assumed equal to one.) The discharge (relative to the moving observer) is the same at both sections, as in previous applications of the energy equation. Writing this equivalence in terms of discharge per unit width, we have

$$YC_g = (Y + h_w)V_w \tag{11-12}$$

and solving for V_w gives

$$V_w = \frac{C_g Y}{Y + h_w}. \tag{11-13}$$

Substituting Equation 11-13 into Equation 11-11 yields

$$Y + \frac{C_g^2}{2g} = Y + h_w + \frac{[C_g Y/(Y + h_w)]^2}{2g},$$

and thus

$$C_g = \left[\frac{2g(Y + h_w)^2}{2Y + h_w}\right]^{1/2}. \tag{11-14}$$

If h_w is small relative to Y, Equation 11-14 reduces to Equation 11-10. Thus, strictly speaking, the equations given previously for the celerity of a gravity wave (Equations 5-49 and 11-10) are true only for waves of very small amplitude. For

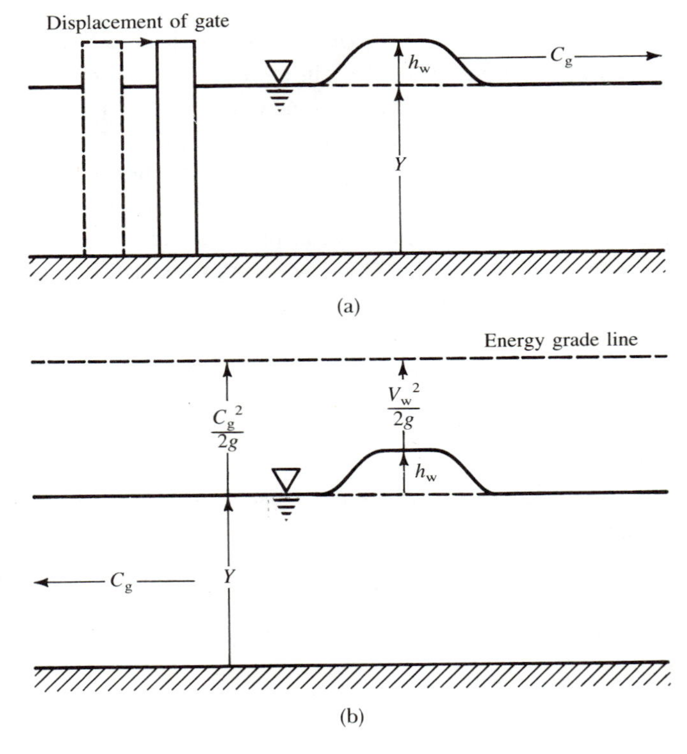

Figure 11.4
Generation of a solitary gravity wave by the sudden displacement of a gate in a stationary body of water. C_g is the wave celerity and h_w the wave height; see Equations 11-11–11-14. (a) Unsteady motion as it appears to a stationary observer. (b) The same motion seen as steady by an observer moving with the wave. After Chow (1959).

larger values of h_w, the numerator in Equation 11-14 increases more rapidly than the denominator, so larger gravity waves have larger celerities (see Exercise 11-5).

As noted in Chapter 10, hydraulic jumps are standing gravity waves traveling upstream relative to the flow. Because their amplitudes are usually large relative to the depth, their celerities are most appropriately calculated by means of Equation 11-14. For a jump to be stationary, its celerity must be just equal to the downstream flow velocity.

It is of interest that the general equation for the celerity of gravity waves was derived in 1845 by a British geophysicist, Sir George A. Airy, as

$$C_g = \left[\frac{g\lambda}{2\pi}\tanh\left(\frac{2\pi Y}{\lambda}\right)\right]^{1/2}, \tag{11-15}$$

where λ is the wavelength, $\pi = 3.14159\ldots$, and tanh( ) represents the *hyperbolic tangent* function. This function is defined as

$$\tanh(\chi) \equiv \frac{e^{\chi} - e^{-\chi}}{e^{\chi} + e^{-\chi}}, \tag{11-16}$$

where $e = 2.718\ldots$, the base of natural logarithms, and in this case $\chi \equiv 2\pi Y/\lambda$. It is clear from Equation 11-16 that $\tanh(2\pi Y/\lambda)$ approaches 1 when $2\pi Y/\lambda$ is large. Thus in deep water, where $Y \gg \lambda$, Equation 11-15 reduces to

$$C_g = \left(\frac{g\lambda}{2\pi}\right)^{1/2}, \tag{11-17}$$

which shows that the celerity of ocean waves depends only on their wavelength and not on the depth. In shallow water Y/λ is small, and it can be shown from Equation 11-16 that as $2\pi Y/\lambda \rightharpoonup 0$, $\tanh(2\pi Y/\lambda) \rightharpoonup 2\pi Y/\lambda$. In this case Equation 11-15 becomes

$$C_g = \left[\frac{g\lambda}{2\pi}\left(\frac{2\pi Y}{\lambda}\right)\right]^{1/2} = (gY)^{1/2}, \tag{11-18}$$

which is the familiar expression for gravity-wave celerity in rivers (Equation 11-10).

Kinematic Waves As indicated earlier, the properties of kinematic waves can be derived from continuity, or conservation-of-mass, considerations along with an equation of motion. The gravitational forces are assumed to be balanced by the frictional resistance due to inertial and viscous forces, so the net force is zero and is not included in the formulation. It is the absence of force considerations that distinguishes kinematic waves from dynamic waves.

The basic principles for formulating a continuity equation were discussed in Chapter 3. For the present case, the volume of concern is the arbitrary segment of the channel shown in Figure 11.5, which has a small downstream extent equal to dx. The mass of water M_{in} entering this volume in a short time increment dt is

$$M_{in} = \rho VA\,dt + \rho q\,dx\,dt, \tag{11-19}$$

where ρ is the mass density of water, V is the mean velocity at the upstream face, A is the cross-sectional area at the upstream face, and q is the rate of lateral inflow per unit length of channel.

Both the velocity and the area change with distance, and their rates of change are written as $\partial V/\partial x$ and $\partial A/\partial x$, respectively[1]. Thus the velocity and area at the downstream face are $V + (\partial V/\partial x)\,dx$ and $A + (\partial A/\partial x)\,dx$, respectively; and the mass of water M_{out} leaving the volume element during dt is

$$M_{out} = \left(V + \frac{\partial V}{\partial x}dx\right)\left(A + \frac{\partial A}{\partial x}dx\right)dt. \tag{11-20}$$

The change-in-storage term is accounted for as the change in the volume of the element during dt. Since the volume is approximately equal to $A\,dx$ and dx is fixed,

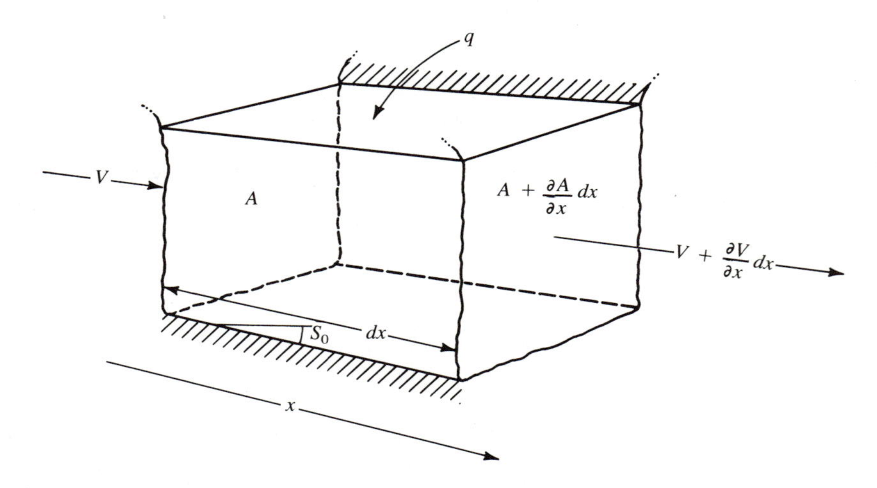

Figure 11.5
Definitions of terms used in deriving the continuity equation for unsteady gradually varied flow (Equation 11-30).

[1]See the footnote on partial derivative notation on page 97.

the time rate of change of volume is $(\partial A/\partial t)\,dx$. Thus the change in mass ΔM stored in the element during dt is

$$\Delta M = \rho \frac{\partial A}{\partial t}\,dx\,dt. \tag{11-21}$$

The continuity equation is now written by substituting the right-hand sides of Equations 11-19, 11-20, and 11-21 into Equation 3-14:

$$\rho VA\,dt + \rho q\,dx\,dt - \rho\left(V + \frac{\partial V}{\partial x}dx\right)\left(A + \frac{\partial A}{\partial x}dx\right)dt = \rho\,dx\frac{\partial A}{\partial t}\,dt. \tag{11-22}$$

Equation 11-22 can be simplified by eliminating ρ and dt from each term and expanding the bracketed terms to obtain

$$VA + q\,dx - VA - V\frac{\partial A}{\partial x}dx - A\frac{\partial V}{\partial x}dx - \frac{\partial V}{\partial x}\frac{\partial A}{\partial x}(dx)^2 = \frac{\partial A}{\partial t}dx. \tag{11-23}$$

This equation is further simplified by noting that the VA terms cancel and by reasoning that since dx is small, $(dx)^2$ is very small and can therefore be considered effectively equal to zero. Thus Equation 11-23 becomes

$$q\,dx - V\frac{\partial A}{\partial x}dx - A\frac{\partial V}{\partial x}dx = \frac{\partial A}{\partial t}dx,$$

$$q - V\frac{\partial A}{\partial x} - A\frac{\partial V}{\partial x} = \frac{\partial A}{\partial t}. \tag{11-24}$$

Equation 11-24 is one form of the continuity equation. However, it can be put into a more useful form by recalling that, from the definition of a derivative of the product of two dependent variables,

$$V\frac{\partial A}{\partial x} + A\frac{\partial V}{\partial x} \equiv \frac{\partial(VA)}{\partial x}. \tag{11-25}$$

Since velocity times area equals discharge (Q), Equation 11-25 is equivalent to

$$V\frac{\partial A}{\partial x} + A\frac{\partial V}{\partial x} = \frac{\partial Q}{\partial x}, \tag{11-26}$$

and substituting Equation 11-26 into Equation 11-24 gives

$$q - \frac{\partial Q}{\partial x} = \frac{\partial A}{\partial t}. \tag{11-27}$$

The final step is to note that

$$A = wY, \tag{11-28}$$

where w is the water-surface width and Y is the mean depth. Based on the previous discussion of hydraulic geometry, it is reasonable to state that width changes little with time at a given cross section, and if it is considered essentially constant,

$$\frac{\partial A}{\partial t} = w\frac{\partial Y}{\partial t}. \tag{11-29}$$

Substituting Equation 11-29 into Equation 11-27 gives a useful version of the continuity equation:

$$q - \frac{\partial Q}{\partial x} = w\frac{\partial Y}{\partial t}. \tag{11-30}$$

In words, this equation states that the time rate of change of depth at a given cross section is equal to the difference between the rate of lateral inflow and the downstream rate of change of discharge, both divided by width.

Now consider a flood wave moving along a channel with a velocity V_k (Figure 11.3). To simplify the development, assume there is no lateral inflow. An observer moving along with the wave crest will observe no change in discharge with time or distance, and so can write

$$dQ = 0. \tag{11-31}$$

From the definition of dQ,

$$dQ \equiv \frac{\partial Q}{\partial x}dx + \frac{\partial Q}{\partial t}dt = 0. \tag{11-32}$$

Equation 11-32 can be rearranged to give

$$\frac{dx}{dt} = -\frac{\partial Q/\partial t}{\partial Q/\partial x}. \tag{11-33}$$

Since dx/dt is the velocity of the observer and the flood wave, it is also true that

$$V_k = -\frac{\partial Q/\partial t}{\partial Q/\partial x}. \tag{11-34}$$

From the properties of derivatives we know that

$$\frac{\partial Q}{\partial t} = \frac{\partial Q}{\partial Y}\frac{\partial Y}{\partial t}. \tag{11-35}$$

If $q = 0$, Equation 11-30 can be solved for $\partial Y/\partial t$:

$$\frac{\partial Y}{\partial t} = -\frac{1}{w}\frac{\partial Q}{\partial x}. \tag{11-36}$$

Now Equation 11-36 can be put into Equation 11-35 to give

$$\frac{\partial Q}{\partial t} = -\frac{1}{w}\frac{\partial Q}{\partial x}\frac{\partial Q}{\partial Y}. \tag{11-37}$$

Finally, when the numerator of Equation 11-34 is replaced with Equation 11-37, the velocity of the flood wave is found to be

$$V_k = -\frac{-\frac{1}{w}\frac{\partial Q}{\partial x}\frac{\partial Q}{\partial Y}}{\frac{\partial Q}{\partial x}}$$

$$= \frac{1}{w}\frac{\partial Q}{\partial Y}. \tag{11-38}$$

Thus the velocity of a kinematic wave is proportional to the rate of change of discharge with depth. If the flow is turbulent, this rate can be evaluated via the Manning equation:

$$Q = \frac{u_M w S_0^{1/2} Y^{5/3}}{n}, \tag{11-39}$$

assuming a wide rectangular channel. From Equation 11-39,

$$\frac{\partial Q}{\partial Y} = \frac{5}{3}\left(\frac{u_M w S_0^{1/2} Y^{2/3}}{n}\right) \tag{11-40}$$

if the slope and roughness are constant. Putting this equation into Equation 11-38 gives the final expression for the velocity of a kinematic wave:

$$V_k = \frac{5}{3}\left(\frac{u_M S_0^{1/2} Y^{2/3}}{n}\right). \tag{11-41}$$

Since the mean flow velocity V is given by Equation 11-8, we can make the interesting observation that

$$V_k = \tfrac{5}{3}V; \tag{11-42}$$

that is, the velocity of a kinematic flood wave is 1.67 times the mean velocity of the water itself. Note that this conclusion was reached by invoking only two relations, the continuity equation and an equation of motion, which in this case was the Manning equation. No forces or accelerations were involved in the derivation.

The preceding deıivation assumed a wide rectangular channel, and the ratio V_k/V varies for different geometries. However, it is always greater than one unless significant overbank flow occurs. Gray and Wigham (1970) showed how the overbank effects can be estimated. Figure 11.6 is a cross section of a channel with two overbank zones. Treating the three subsections as rectangular, we can write

$$V_k = \frac{1.67}{w}(w_1V_1 + w_2V_2 + w_3V_3), \tag{11-43}$$

where w_i is the width of the ith subsection, V_i is the mean flow velocity of each subsection, and $w = w_1 + w_2 + w_3$. Now suppose that the mean velocities of the two overbank zones are essentially zero. Then $V_1 = V_3 = 0$ and

$$V_k = \left(\frac{1.67w_2}{w}\right)V_2. \tag{11-44}$$

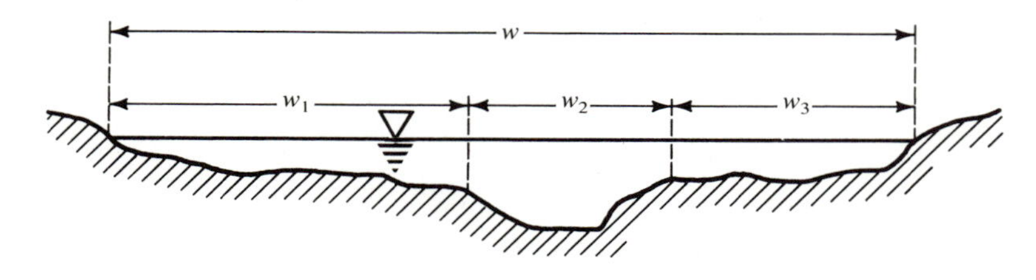

Figure 11.6
Definitions of terms used in deriving the effects of overbank flows on flood-wave velocity (Equation 11-43). After Gray and Wigham (1970).

If $w_2/w = 0.6$, then $V_k = V_2$ and the flow velocity and wave velocity are equal. When $w_2/w < 0.6$, the flood wave travels at a slower speed than the water velocity in the main channel. Thus, by providing areas for storage of water as a flood wave passes downstream, overbank zones reduce the velocity of the flood wave.

Although kinematic waves had been studied by a few Europeans and observed in the Mississippi River by 1900, two English mathematicians, M. J. Lighthill and G. B. Whitham, gave the first comprehensive treatment of the subject in 1955. They stated that such waves are a general occurrence that arises in any flow in which there is a functional relationship among (1) the flow rate (discharge); (2) the amount of flowing substance in a segment of the flow (cross-sectional area or depth); and (3) the location in the flow (distance along the channel). Such a relationship is expressed in the equation of motion (e.g., the Manning equation) and the continuity equation (Equation 11-30). Interestingly, Lighthill and Whitham (1955) showed that an analogous relationship exists for automobile traffic, and devoted the second part of their seminal paper to a discussion of traffic flow. In traffic, the flow rate is inversely, rather than directly, proportional to the amount of flowing substance, as your own experience will no doubt verify. Because of this, kinematic waves in traffic travel "upstream" rather than downstream as in rivers.

Chow (1959) discusses kinematic waves under the heading of "monoclinal rising waves."

Relations between Gravity Waves and Kinematic Waves Gravity waves travel in all directions in the liquid, and the wave celerity given in Equation 11-10 or Equation 11-14 is the speed of the wave *relative to the liquid*. Thus the downstream velocity V_g of a gravity wave is

$$V_g = V + C_g. \tag{11-45}$$

A kinematic wave travels only in the downstream direction, with a velocity V_k. From Equations 11-42 and 11-45, the downstream velocities of the two types of waves are equal when

$$C_g = \tfrac{2}{3}V. \tag{11-46}$$

Using Equation 11-10 and the definition of the Froude number, we see that it is true that when

$$\mathbb{F} = \tfrac{3}{2}, \qquad V_g = V_k; \tag{11-47a}$$

when

$$\mathbb{F} > \tfrac{3}{2}, \qquad V_g < V_k; \tag{11-47b}$$

and when

$$\mathbb{F} < \tfrac{3}{2}, \qquad V_g > V_k. \qquad (11\text{-}47c)$$

Thus under conditions normally found in stream channels, flood waves travel at slower speeds than do gravity waves. However, Lighthill and Whitham (1955) showed that gravity waves attenuate rapidly because of friction and disappear quickly, whereas kinematic waves usually dissipate slowly and hence dominate in flows when $\mathbb{F} < \frac{3}{2}$. At $\mathbb{F} \simeq \frac{3}{2}$, other studies have shown that flows become unstable; steady, or even gradually varied, flows are not possible, and *roll waves* form spontaneously. Roll waves have been observed during floods in steep mountain streams (Chow, 1959) and can sometimes be seen on relatively steep road surfaces during rainstorms. In general, however, the kinematic flood wave is the only hydrologically important wave phenomenon.

Table 11.2 summarizes the most important features of gravity waves and kinematic waves.

Table 11.2
Characteristics of Gravity Waves and Kinematic Waves in Natural Channels

	Gravity waves	Kinematic waves
Cause	Any disturbance that causes a local change in surface elevation	Local accumulation of water, such as is due to an influx of runoff
Direction of travel	All directions	Downstream
Downstream velocity	$V + (gY)^{1/2}$	$\frac{5}{3}V$
Dissipation	Nonstationary waves dissipate rapidly as a result of frictional resistance; standing waves like hydraulic jumps persist as long as flow conditions persist.	Generally dissipate gradually; overbank flow increases the rate of dissipation.

DEVELOPMENT OF THE UNSTEADY GRADUALLY VARIED FLOW EQUATIONS

Equation 11-30 is the continuity equation for gradually varied unsteady flow in wide channels. It is written in terms of two dependent variables (Q and Y), two independent variables (x and t), and two parameters that are assigned values appropriate to a particular time and stream reach (q and w). Because of the two dependent variables (the unknowns), another relation between Q and Y is required in order to obtain solutions. This second equation expresses the conservation of momentum (see Chapter 3). The momentum equation can be developed in two ways: (1) by an approach similar to that used in Equations 10-18–10-27, but accounting for changes with time; and (2) by an equivalent approach beginning with energy considerations (Strelkoff, 1969). Following Chow (1959), we will use the second of these two approaches because it can be related most readily to material covered previously.

Figure 11.7 is an energy diagram that is identical to the one constructed in the initial development of mechanical energy concepts in Chapter 7 (Figure 7.3), except for a change in notation. Considering a small channel segment of length dx and writing the changes in the energy terms between the upstream and downstream ends of the reach in differential form yields

$$\frac{V^2}{2g} + Y + Z = \left[\frac{V^2}{2g} + d\left(\frac{V^2}{2g}\right)\right] + (Y + dY) + (Z + dZ) + dH, \quad (11\text{-}48)$$

where dH is the total head loss through the reach and α and $\cos\theta = 1$ (cf. Equation 7-17). Note that $d(V^2/2g)$ and dY may have either positive or negative values (or may be zero for uniform flow), whereas dZ will almost always be negative for natural channels and dH must always be positive.

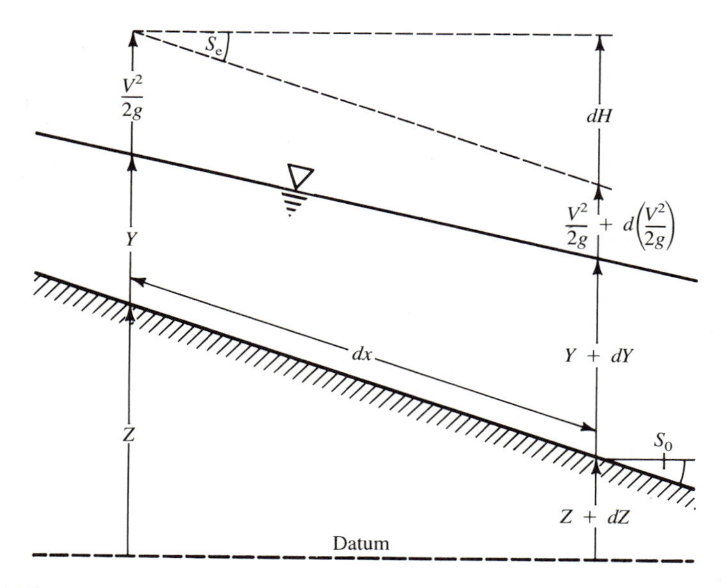

Figure 11.7
Definitions of terms used in deriving the momentum equation for unsteady gradually varied flow (Equation 11-58). After Chow (1959).

Equation 11-48 reduces immediately to

$$dH = -\left[d\left(\frac{V^2}{2g}\right) + dY + dZ\right]$$
$$= -\left[\frac{1}{2g}d(V^2) + dY + dZ\right];$$

and since

$$d(V^2) = 2V\,dV,$$

we have

$$dH = -\left(\frac{V}{g}dV + dY + dZ\right). \qquad (11\text{-}49)$$

At this point we can divide Equation 11-49 by dx to obtain the downstream rate of change of total head:

$$\frac{dH}{dx} = -\left(\frac{V}{g}\frac{dV}{dx} + \frac{dY}{dx} + \frac{dZ}{dx}\right). \qquad (11\text{-}50)$$

As noted, Equation 11-48 and hence Equation 11-50 are merely restatements of the energy equation for gradually varied steady flow as developed in Chapter 7 (Equation 7-17). In such flows the change in total head has three components, due to downstream changes in (1) velocity; (2) depth; and (3) elevation of the channel bottom.

For unsteady flows, there is an additional component due to changes of velocity with time. Recall from Chapter 6 (Equation 6-5) that, in general, acceleration a has two components:

$$a = \underbrace{V\frac{\partial V}{\partial x}}_{\text{convective acceleration}} + \underbrace{\frac{\partial V}{\partial t}}_{\text{local acceleration}}.$$

Clearly, the first term on the right-hand side of Equation 11-50 represents the head loss due to convective acceleration. The expression for head loss due to local acceleration is developed by invoking Newton's second law:

$$F_\ell = \rho B\frac{\partial V}{\partial t}, \qquad (11\text{-}51)$$

where ρ is the mass density of the fluid and F_ℓ is the force exerted on a volume B of fluid undergoing the local acceleration. The work done, or energy expended, in accelerating this volume is the force times the downstream distance dx, so

$$dE_\ell = \rho B \frac{\partial V}{\partial t} dx, \quad (11\text{-}52)$$

where dE_ℓ represents the energy lost as a result of the local acceleration. This energy loss is transformed to a head loss dH_ℓ by dividing by the weight of the water:

$$dH_\ell = \frac{\rho B}{\gamma B} \frac{\partial V}{\partial t} dx$$

$$= \frac{1}{g} \frac{\partial V}{\partial t} dx, \quad (11\text{-}53)$$

where γ is the weight density. Finally, of course, the downstream rate of head loss due to local acceleration is

$$\frac{dH_\ell}{dx} = \frac{1}{g} \frac{\partial V}{\partial t}. \quad (11\text{-}54)$$

Introducing this component of head loss into Equation 11-50 and switching to partial derivative notation throughout gives the equation for the downstream rate of head loss in unsteady gradually varied flows:

$$\frac{\partial H}{\partial x} = -\left(\frac{V}{g}\frac{\partial V}{\partial x} + \frac{\partial Y}{\partial x} + \frac{\partial Z}{\partial x} + \frac{1}{g}\frac{\partial V}{\partial t}\right). \quad (11\text{-}55)$$

Now note the following identities:

$$\frac{\partial H}{\partial x} \equiv S_e \quad (11\text{-}56)$$

and

$$\frac{\partial Z}{\partial x} \equiv -S_0, \quad (11\text{-}57)$$

where S_e and S_0 are the energy and channel slopes, respectively. With these terms, Equation 11-55 becomes

$$S_e = S_0 - \frac{V}{g}\frac{\partial V}{\partial x} - \frac{\partial Y}{\partial x} - \frac{1}{g}\frac{\partial V}{\partial t} \tag{11-58a}$$

or

$$S_0 - S_e = \frac{V}{g}\frac{\partial V}{\partial x} + \frac{\partial Y}{\partial x} + \frac{1}{g}\frac{\partial V}{\partial t}. \tag{11-58b}$$

Although Equation 11-58 was derived from energy considerations, we would have arrived at the same relationship if we had developed expressions for the conservation of momentum in an unsteady flow. Thus Equation 11-58 is usually referred to as the *momentum equation.* It shows that the difference between the channel slope and the energy slope is due to inertial forces arising from convective and local acceleration and to pressure forces resulting from downstream changes in depth. When these accelerations and depth changes are negligible (or if they fortuitously cancel each other), $S_0 = S_e$ and the flow is uniform.

Strictly speaking, there is another term that should be included in Equation 11-58: the flow resistance due to the upstream and downstream components of the velocity of the lateral inflows. In hydrologic situations, these inflows could be due to rain falling into an overland flow, infiltration out of or seepage into an overland flow, entrance of overland flow or groundwater into a channel, or seepage or overbank flow out of a channel. In all these cases we can usually assume that the inflow velocities are small (except perhaps in the case of rain into overland flow) and that the inflow or outflow is at right angles to the main flow. Thus this component generally makes a negligible contribution to the one-dimensional energy equations and can safely be neglected.

Equations 11-30 and 11-58 together—the continuity and momentum equations—are the system of partial differential equations that describes one-dimensional, gradually varied, unsteady open-channel flows. These equations were first developed and published by Jean-Claude Barré de Saint-Venant (1797–1886) in France in 1848, and are known as the *Saint-Venant equations.* Strelkoff (1969) gives a detailed derivation and reexamination of those equations.

APPLICATION OF THE SAINT-VENANT EQUATIONS

Figure 11.8 illustrates the essential features of the application of this system of equations, which has two dependent variables (V or Q and Y) and two independent variables (x and t). Partial differential equations of this type can only be solved numerically. To do this, we must specify the values of the dependent variables at $t = 0$ (initial conditions) and two boundary conditions. These boundary conditions would typically be (1) the value of the dependent variables at the upstream end of the channel, and (2) specification of either normal or critical flow conditions at the downstream end (Freeze, 1974). The equations are then written in algebraic form, as difference equations rather than partial differential equations, for use with one of several numerical schemes adapted for a computer. Further discussion of these

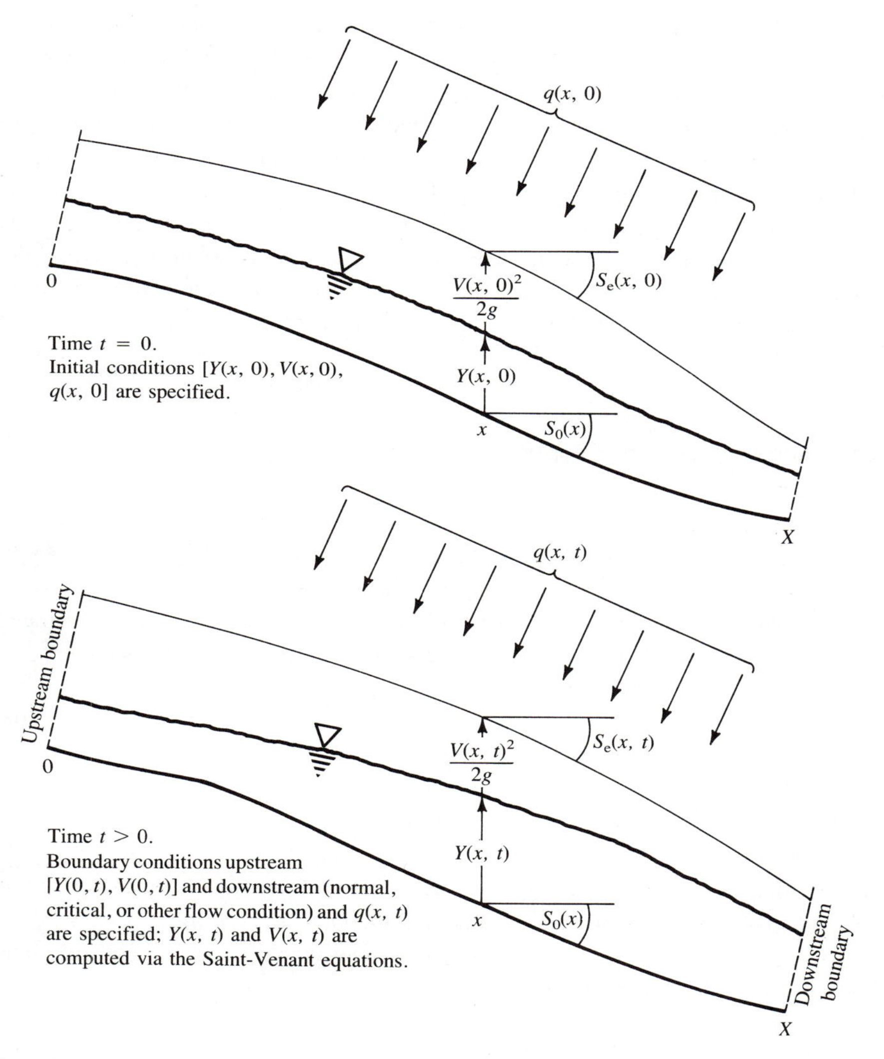

Figure 11.8

Definitions of terms for application of the Saint-Venant equations for continuity (Equation 11-30) and momentum (Equation 11-58b).

numerical methods is beyond the scope of this text, but descriptions of such methods can be found in Brakensiek (1966), Ragan (1966), Strelkoff (1970), Rovey et al. (1977), Ross et al. (1979), Weinmann and Laurenson (1979), and Smith (1980).

Laboratory experiments by Ragan (1966) provide an excellent illustration of the ability of the Saint-Venant equations to model, or route, channel flows with lateral inflows. These experiments were done in a 20-cm-wide, 22-m-long tiltable flume in which water was continually supplied at the upstream end and additional water could be supplied from a series of lateral-inflow pipes distributed along the channel, representing the runoff contributions from a watershed. Figure 11.9 shows the comparison of actual hydrographs at the downstream end of the flume with the hydrographs computed by numerical solution of the Saint-Venant equations for four different spatial patterns and two different temporal patterns of lateral inflows.

Results of using the Saint-Venant equations to model streamflow in a field situation are shown in Figure 11.10. The actual hydrograph was measured at the outlet of a 9.2-ha watershed in Wisconsin. In this case, the Saint-Venant equations were applied twice, first to simulate the overland flow generated by the particular storm, and then to model the channel flow by using the simulated overland flow as the lateral inflow.

SIMPLIFICATION OF THE UNSTEADY MOMENTUM EQUATION

As noted earlier, the sum of the terms on the right-hand side of Equation 11-58b equals the difference between the channel slope and the energy slope. These terms represent the convective acceleration forces, the hydrostatic pressure forces, and the local acceleration forces. In steady uniform flows there are no changes with time or distance, so each of these terms equals zero, and Equation 11-58b becomes

$$S_0 - S_e = 0 \qquad \text{(11-59a)}$$

or

$$S_0 = S_e, \qquad \text{(11-59b)}$$

consistent with our observation in Chapter 6 that uniform flow is characterized by equality of the slopes of the energy line and the channel bottom.

In steady gradually varied flows there are changes with distance but not with time. Then Equation 11-58b becomes

$$\frac{V}{g}\frac{dV}{dx} + \frac{dY}{dx} = S_0 - S_e, \qquad \text{(11-60)}$$

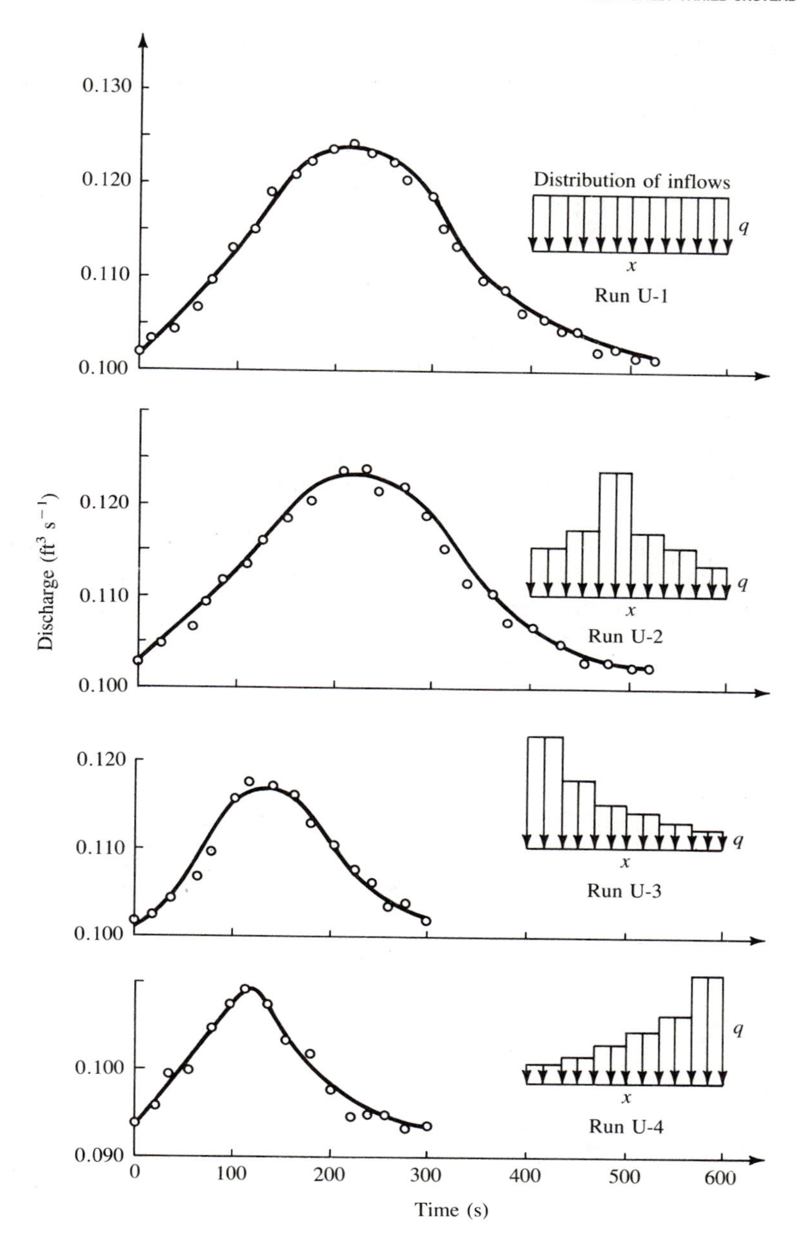

Figure 11.9
Comparison of measured hydrographs (open circles) and hydrographs simulated by numerical solution of the Saint-Venant equations (solid lines) for different spatial patterns of lateral inflows (insets). From Ragan (1966).

indicating that the difference between the channel slope and the energy slope equals the sum of the convective acceleration and the acceleration due to pressure forces. From the definition of specific head H_s (Equation 7-20) it is apparent that

$$\frac{dH_s}{dx} = \frac{V}{g}\frac{dV}{dx} + \frac{dY}{dx}, \tag{11-61}$$

so it can also be stated that

$$\frac{dH_s}{dx} = S_0 - S_e, \tag{11-62}$$

which is identical to Equation 9-7.

Thus it is clear that Equation 11-58b is a general expression relating channel slope and energy slope; steady uniform and gradually varied flows are simply special cases. Further insight into these relations can be developed by following the approach of Weinmann and Laurenson (1979), which involves transforming the momentum equation into the form of a rating curve, that is, an expression relating discharge and depth. To do this, the Manning equation is first written as

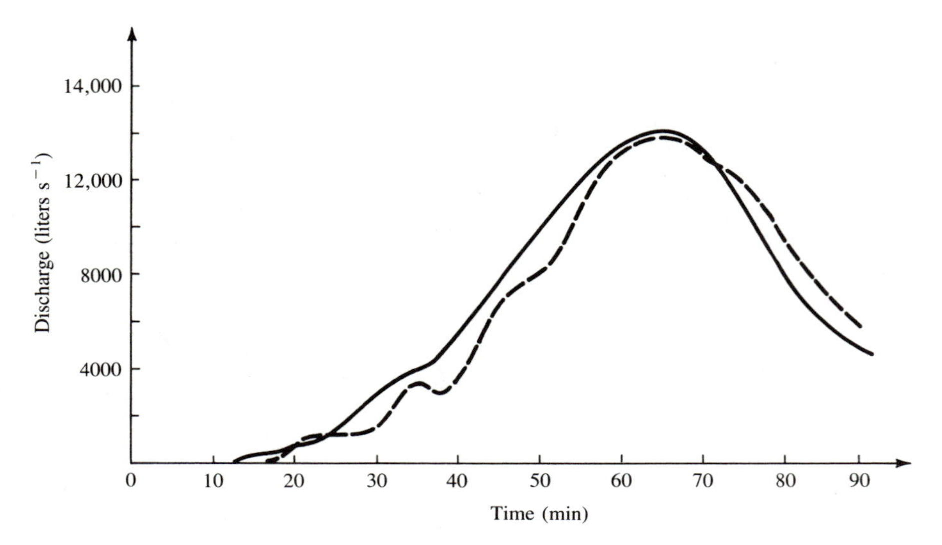

Figure 11.10
Comparison of measured hydrograph (solid line) and hydrograph simulated by numerical solution of the Saint-Venant equations (dashed line) for a storm on a 9.2-ha watershed in Wisconsin. After Morgali (1963).

$$Q = \frac{u_M w Y^{5/3} S_e^{1/2}}{n}. \tag{11-63}$$

Solving Equation 11-58a for S_e gives

$$S_e = S_0 - \frac{\partial Y}{\partial x} - \frac{V}{g}\frac{\partial V}{\partial x} - \frac{1}{g}\frac{\partial V}{\partial t}, \tag{11-64}$$

and upon putting Equation 11-64 into Equation 11-63, we have

$$Q = \frac{u_M w Y^{5/3}}{n}\left(S_0 - \frac{\partial Y}{\partial x} - \frac{V}{g}\frac{\partial V}{\partial x} - \frac{1}{g}\frac{\partial V}{\partial t}\right)^{1/2}. \tag{11-65}$$

For uniform flow the Manning equation applies strictly, in the form

$$Q = \frac{u_M w Y^{5/3} S_0^{1/2}}{n}, \tag{11-66}$$

so that again all terms in the parentheses in Equation 11-65 are zero except the first. In this case, discharge is a single-valued function of depth only, since n, w, and S_0 are considered fixed. If the other terms have nonzero values, the discharge depends on other factors in addition to depth, and Equation 11-65 describes a looped rating curve. Figure 11.11 shows schematically the relation between the looped rating curve and the single-valued uniform-flow curve that approximates it. The smaller the acceleration and pressure terms, the smaller the width of the loop and the better an assumption of uniform flow will approximate the actual conditions. The uniform-flow approximation is the basis for the so-called kinematic-wave approximation to the momentum equation, which is widely used in hydrologic simulation and which is discussed in detail in a later subsection.

The relation between the looped rating curve, Equation 11-65, and the previous discussion of scatter in at-a-station hydraulic geometry relations should now be apparent.

All streamflow routing models use the continuity equation in one of the forms derived earlier (Equation 3-14, 3-16, 11-27, or 11-30). As implied earlier, however, there may be cases where it is acceptable to simplify the momentum equation by eliminating the terms with negligible magnitudes. Weinmann and Laurenson (1979) provide a unified discussion of these simplifications, and show that it is possible to classify them on the basis of the number of terms included in Equation 11-65:

$$Q = \underbrace{\underbrace{\underbrace{\frac{u_M w Y^{5/3}}{n}\left(S_0\right.}_{\text{kinematic-wave models}} - \frac{\partial Y}{\partial x}}_{\text{diffusion-analogy models}} - \frac{V}{g}\frac{\partial V}{\partial x} - \left.\frac{1}{g}\frac{\partial V}{\partial t}\right)^{1/2}}_{\text{complete dynamic models}}. \tag{11-67}$$

Some aspects of applying the complete dynamic equations of Saint-Venant were discussed earlier; now we briefly examine the diffusion-analogy and kinematic-wave simplifications.

Diffusion Analogy Defining the normal discharge Q_n as

$$Q_n \equiv \frac{u_M w Y^{5/3} S_0^{1/2}}{n} \tag{11-68}$$

and neglecting the acceleration terms in Equation 11-67 yields

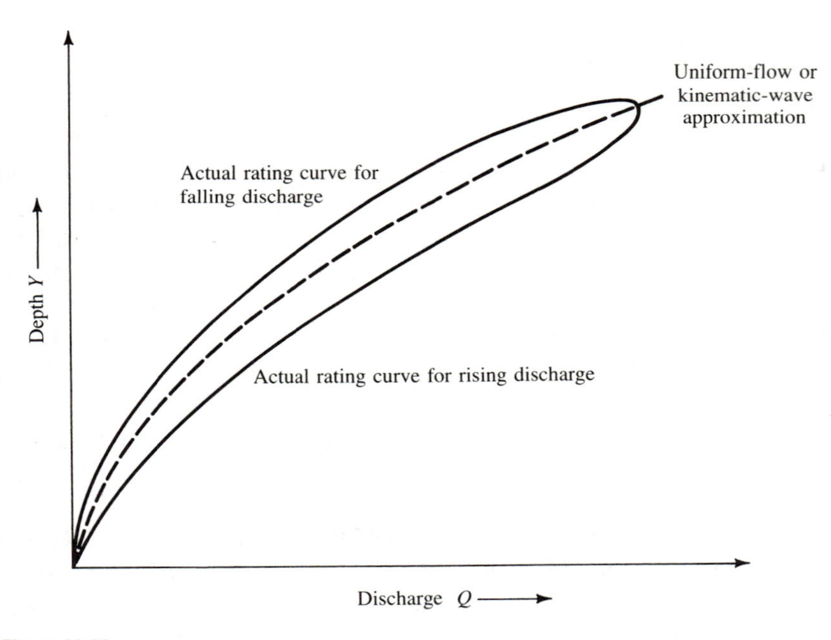

Figure 11.11
Relation between actual looped rating curve and kinematic-wave (i.e., uniform-flow) approximation. After Weinmann and Laurenson (1979).

$$Q = Q_n\left(1 - \frac{1}{S_0}\frac{\partial Y}{\partial x}\right)^{1/2}. \tag{11-69}$$

Substitution of Equation 11-69 into the continuity equation (Equation 11-30) and a lengthy series of algebraic manipulations (see Appendix A) eventually leads to

$$\frac{\partial Y}{\partial t} + \frac{1}{w}\frac{\partial Q_n}{\partial x} - \frac{Q_n}{2wS_0}\frac{\partial^2 Y}{\partial x^2} = \frac{q}{w}. \tag{11-70}$$

This equation can also be written as

$$\frac{\partial Y}{\partial t} + \frac{1}{w}\frac{\partial Q_n}{\partial Y}\frac{\partial Y}{\partial x} - \frac{Q_n}{2wS_0}\frac{\partial^2 Y}{\partial x^2} = \frac{q}{w}; \tag{11-71}$$

or, with Equations 11-40 and 11-41, as

$$\frac{\partial Y}{\partial t} + V_k\frac{\partial Y}{\partial x} - \frac{Q_n}{2wS_0}\frac{\partial^2 Y}{\partial x^2} = \frac{q}{w}. \tag{11-72}$$

Equation 11-72 is called a *convective-diffusion equation* because by defining

$$D \equiv \frac{Q_n}{2wS_0} \tag{11-73}$$

it can be written as

$$\frac{\partial Y}{\partial t} + V_k\frac{\partial Y}{\partial x} - D\frac{\partial^2 Y}{\partial x^2} = \frac{q}{w}, \tag{11-74}$$

which has the form of an equation that arises in many physical situations, including the movement of dissolved substances in water and the flow of heat. The first term on the left represents the time rate of change of depth (or, in other situations, of concentration or temperature) at a point. For steady-state conditions, this term equals zero. The second term represents the ''advective'' term, which appears when there is a flow velocity that bodily carries the substance (water, dissolved materials, or heat) along with it.

In general, the third term in Equation 11-74 represents the diffusion of the substance in the medium, and its dimensions are indeed $[L^2T^{-1}]$, which is true of all types of diffusivities (see the discussion of Equations 3-21–3-27). For floods, D expresses the rate of attenuation or dissipation of the flood wave. This can be shown by noting that $\partial Y/\partial x = 0$ at the crest of a flood wave. Taking $q = 0$ for the moment, we find that Equation 11-72 reduces to

$$\frac{\partial Y}{\partial t} = \frac{Q_n}{2wS_0}\frac{\partial^2 Y}{\partial x^2} \tag{11-75}$$

at the crest. Since the crest is obviously a maximum point in the relation between Y and x, it is true from calculus that $\partial^2 Y/\partial x^2 < 0$ at that point. Thus the right-hand side of Equation 11-75 is negative, and represents the rate at which the height of the flood wave decreases with time. Since the wave is moving downstream with time at a velocity V_k, it also expresses the downstream rate of decrease.

The diffusion term will be negligible if $Q_n/2wS_0$ and/or $\partial^2 Y/\partial x^2$ are small. In general, the first condition will be true if

$$Q_n \ll 2wS_0 \tag{11-76a}$$

or, by substitution of Equation 11-68, if

$$u_M Y^{5/3} \ll 2nS_0^{1/2}. \tag{11-76b}$$

Substitution of some values representing typical natural streams will show that condition 11-76 seldom, if ever, holds for such flows. It may, however, hold for overland flows on steep slopes. As shown in Appendix B, the second condition will be true if either the amplitude of the flood wave is small or the wavelength is large.

It is also worth noting that even if the diffusion term is negligible, but $q \neq 0$, Equation 11-74 shows that

$$\frac{\partial Y}{\partial t} = \frac{q}{w} \tag{11-77}$$

at the crest. Thus if $q < 0$, which occurs if there is an outflow of water due to seepage into the channel bed or to overbank flows onto the floodplain, the flood wave will also dissipate as it moves downstream. When $q > 0$, the height of the flood wave will of course increase downstream.

Equation 11-74 is often applied for flood routing in channels. It is much less complex to solve numerically than are the full Saint-Venant equations, and gives acceptable results for a wide range of channel flows in which the acceleration terms are negligible. Methods for numerical solution of the diffusion-analogy model are discussed by Lighthill and Whitham (1955) and in articles cited by Weinmann and Laurenson (1979).

Kinematic-Wave Approximation If all the force terms in Equation 11-67 are negligible, the momentum equation reduces to

$$Q = \frac{u_M w Y^{5/3} S_0^{1/2}}{n} = Q_n. \tag{11-78}$$

As indicated earlier, Q is a single-valued function of Y in this case, and if a coefficient K is defined as

$$K \equiv \frac{u_M S_0^{1/2}}{n}, \tag{11-79}$$

then

$$Q = KwY^{5/3}. \tag{11-80}$$

Recall from the discussion of at-a-station hydraulic geometry that the exponent $\frac{5}{3}$ is appropriate when roughness, slope, and width do not change with discharge. For cases when these restrictions do not hold, Equation 11-80 can be generalized as

$$Q = KwY^{1/f}, \tag{11-81}$$

where f is the empirical exponent in the at-a-station hydraulic geometry relation (Equation 11-3). By noting that

$$\frac{\partial Q}{\partial x} = \frac{\partial Q}{\partial Y}\frac{\partial Y}{\partial x}, \tag{11-82}$$

we can write the continuity equation (Equation 11-30) as

$$\frac{\partial Y}{\partial t} + \frac{1}{w}\frac{\partial Q}{\partial Y}\frac{\partial Y}{\partial x} = \frac{q}{w}. \tag{11-83}$$

From Equation 11-81,

$$\frac{\partial Q}{\partial Y} = \left(\frac{1}{f}\right)KwY^{1/f-1}, \tag{11-84}$$

and putting Equation 11-84 into Equation 11-83 yields

$$\frac{\partial Y}{\partial t} + \left(\frac{1}{f}\right)KY^{1/f-1}\frac{\partial Y}{\partial x} = \frac{q}{w}. \tag{11-85}$$

From Equations 11-38 and 11-84, however, it is clear that

$$V_{\mathrm{k}} = \left(\frac{1}{f}\right) K Y^{1/f-1}, \tag{11-86}$$

so Equation 11-85 can also be written as

$$\frac{\partial Y}{\partial t} + V_{\mathrm{k}} \frac{\partial Y}{\partial x} = \frac{q}{w}. \tag{11-87}$$

Equation 11-87 is called the *kinematic-wave approximation* to the Saint-Venant equations because all the force terms in the flow are assumed to be negligible, and only the continuity equation (Equation 11-30 or 11-83) and an equation of motion (Equation 11-86) are involved in its derivation. It is identical to Equation 11-74 if $D = 0$, and thus describes a flood wave that does not spontaneously dissipate as it moves downstream. From our discussion of the diffusion-analogy approximation in the preceding section it appears that diffusive effects can be neglected for overland flows and for flood waves in channels when the waves are of low amplitude relative to their wavelength. Since the kinematic-wave approximation is based on uniform flows, it will not give satisfactory results where there are significant deviations from such conditions, as for example when backwater (M1 profiles discussed in Chapter 9) occurs upstream of a lake or tributary junction.

Woolhiser and Liggett (1967) found from numerical modeling studies that Equation 11-87 is a good approximation to the Saint-Venant equations when

$$\frac{gLS_0}{V^2} > 10, \tag{11-88}$$

where L is the length of channel reach or overland-flow path being modeled and V is the mean flow velocity (see also Eagleson, 1970).

The early explorations of the kinematic-wave approximation by Wooding (1965) and Brakensiek (1966, 1967) are instructive for further understanding its applications to hydrologic problems. Subsequently, the relative simplicity of the kinematic-wave approximation has been exploited in many hydrologic studies, including those of Woolhiser and Liggett (1967), Freeze (1974), Rovey et al. (1977), Lane and Woolhiser (1977), Beven et al. (1979), Beven (1979), Smith (1980), and Morris and Woolhiser (1980).

SUMMARY

Although hydrologic phenomena generally show pronounced variations with time, we saw in previous chapters that important aspects of them can be usefully modeled by ignoring temporal variation. However, estimation of watershed and stream response to rain or snowmelt inherently involves modeling changes of discharge and its components (width, mean depth, and mean velocity) with time.

Empirically, the allocation of temporal changes in discharge among these components is reflected in the equations describing at-a-station hydraulic geometry (Equations 11-2–11-4). The Manning equation can be invoked to suggest the allocation of discharge variations to mean depth and mean velocity when constant width is assumed. Actual streams tend to behave in rough accord with this analysis, with depth and velocity approximately equally absorbing 80–85% of any change, and width absorbing the remainder.

Wave motion is an important class of unsteady flows. Two wave phenomena are of special interest to the hydrologist: gravity waves and kinematic waves. Gravity waves are dynamic phenomena that travel in all directions with a celerity that depends on the flow depth (Equations 11-10, 11-14, and 11-15). Except for stationary hydraulic jumps, however, gravity waves dissipate rapidly and can generally be ignored in modeling stream response. Kinematic waves, however, are essential aspects of that response. They arise when lateral inflows, as from overland flow into a channel, cause accumulations of water. In contrast to gravity waves, kinematic waves can be modeled by ignoring dynamic relations and considering only the conservation of mass (Equation 11-30) and an equation of motion like the Manning equation. Kinematic waves move only in the downstream direction, with a velocity V_k about five-thirds times the mean flow velocity (Equation 11-42). They attenuate to the extent that pressure-gradient forces act (which is proportional to the amplitude-to-wavelength ratio) or in response to net outflows from the channel.

The fundamental equations describing unsteady gradually varied flow were originally formulated from the basic principles of conservation of mass (Equation 11-30) and momentum (Equation 11-58a or 11-58b) by J. C. B. de Saint-Venant in 1848. In their complete form, these equations account for inertial forces due to local and convective acceleration and pressure-gradient forces due to downstream changes in depth.

When the inertial forces are negligible, the Saint-Venant equations can be reduced to a single equation with exactly the same form as the diffusion equation that describes many types of flow phenomena (Equation 11-74). One term of this equation is the product of the kinematic-wave velocity V_k and the downstream rate of change of depth $\partial Y/\partial x$. The velocity V_k can be related directly to the depth by means of the at-a-station hydraulic geometry (Equation 11-86). Another term in the diffusion-analogy equation is the product of the "diffusivity" of the flood wave and $\partial^2 Y/\partial x^2$. This term represents the rate of attenuation of the wave crest due to pressure-gradient forces, which act to retard the flow upstream of the crest (where $\partial Y/\partial x > 0$) and to accelerate the flow downstream of the crest (where $\partial Y/\partial x < 0$). The effect of lateral inflows q on the downstream increase or decrease in the height of the crest is also readily accounted for in the diffusion-analogy equation: When

$q > 0$, the crest tends to increase downstream, whereas the reverse is true when $q < 0$.

When both the inertial- and pressure-gradient force terms in the Saint-Venant equations are ignored, those equations reduce to a single equation with all the terms of the diffusion-analogy equation except the diffusion term itself (Equation 11-87). As noted earlier this kinematic-wave equation can also be derived by considering only the conservation of mass and an equation of motion. Flood waves modeled with this simplification suffer no downstream attenuation due to pressure forces, but may increase or decrease in amplitude depending on the sign of the lateral inflow term q. Use of the kinematic-wave approximation is equivalent to assuming a single-valued relation between depth and discharge at each cross section of the flow, whereas the diffusion-analogy and full Saint-Venant equations model a looped, or hysteretic, relation.

Numerical solution by means of a digital computer is required in order to solve the partial differential equations of unsteady gradually varied flow, even if simplified to the diffusion-analogy or kinematic-wave approximations. Many studies in the literature report the successful use of various methods in modeling runoff events.

Exercises

11-1. The following are some measured values of width (w), mean depth (Y), and mean velocity (V) from a gaging station on a small research watershed. Plot w, Y, and V versus the discharge on log–log paper and estimate the values of a, c, k, b, f, and m (Equations 11-2–11-4). Check to see that Equations 11-5 and 11-6 are satisfied. Compare your values with the data in Table 11.1.

w (ft)	Y (ft)	V (ft s^{-1})
2.9	0.50	0.37
4.1	0.83	0.63
3.0	0.67	0.47
2.6	0.38	0.18
2.6	0.33	0.13
2.6	0.26	0.07
3.0	0.73	0.46
2.7	0.31	0.10
2.6	0.23	0.03
2.8	0.40	0.17
3.6	0.80	0.37
2.6	0.24	0.03

11-2. Consider a trapezoidal channel with a bottom width of 1.7 m and side slopes of 2:1. The tops of the banks are at an elevation of 1.2 m above the bottom. Assuming the channel slope is 0.0018 and the roughness (Manning's n) is 0.042, use the Manning equation to compute the hydraulic geometry (i.e., a, c, k, b, f, and m in Equations 11-2–11-4) for the channel for all discharges up to bank full.

11-3. Do the downstream hydraulic geometry data in Table 11.1 indicate that width-to-depth ratios for channels generally increase or decrease as we proceed downstream to larger channels?

11-4. Show the derivation of Equation 11-14 from Equation 11-13.

11-5. (a) Using Equation 11-14 as a starting point, show that

$$\frac{C_g}{(gY)^{1/2}} = \left[\frac{2\left(1 + \frac{h_w}{Y}\right)^2}{2 + \frac{h_w}{Y}}\right]^{1/2}.$$

(b) Use this relationship to plot $C_g/(gY)^{1/2}$ as a function of h_w/Y for $0 \leq h_w/Y \leq 2$.

11-6. (a) Derive the velocity of a kinematic wave for turbulent flow governed by the Chézy equation. [*Hint:* Write the Chézy equation (Equation 6-48) for discharge, find $\partial Q/\partial Y$, and then use Equation 11-38.] (b) Express the relation between V_k and V when the Chézy equation applies, and compare it with Equation 11-42.

11-7. Repeat Exercise 11-6 for laminar flow. Use Equation 6-17 as a starting point.

11-8. Consider a river–floodplain cross section like the one shown in Figure 11.6, but with the following dimensions: $w_1 = 100$ m, $w_2 = 70$ m, $w_3 = 120$ m. Assume that the mean flow velocity V for the 100-yr flood is 2.4 m s^{-1} in the central channel and 0 in the two floodplain segments. Make a graph showing how the flood-wave velocity V_k will change as w_1 and w_3 are gradually reduced by floodplain filling. Make calculations at simultaneous steps of 10% reduction in width on each side until both segments are completely filled.

11-9. (a) Show that the ratio of the downstream velocity of a kinematic wave V_k to that of a gravity wave V_g can be written as the following function of the Froude number $\mathbb{F}$:

$$\frac{V_k}{V_g} = \frac{1.67\mathbb{F}}{\mathbb{F} + 1}.$$

(*Hint:* Use Equations 11-10, 11-42, and 11-45.) (b) Compute and plot the foregoing relation for $0 \leq \mathbb{F} \leq 2$.

11-10. (a) Show that Equation 11-73 can be written as

$$D = \frac{u_M Y^{5/3}}{2nS_0^{1/2}}.$$

(b) The following table presents some data for flood crests in a range of natural flows. Compute D for each flow.

Flow	V (m s^{-1})	Y (m)	S_0	n	$\partial^2 Y/\partial x^2$ (m^{-1})
			Channel Flows		
a	0.73	0.52	0.0014	0.033	2.7×10^{-9}
b	0.52	0.28	0.0011	0.028	5.3×10^{-9}
c	0.49	0.20	0.0023	0.033	1.4×10^{-7}
d	1.00	1.40	0.00083	0.037	3.8×10^{-9}
e	0.26	0.55	0.00080	0.074	5.1×10^{-7}
f	0.23	0.46	0.010	0.26	2.4×10^{-6}
g	0.67	6.80	0.000076	0.047	1.1×10^{-9}
h	0.05	0.14	0.0038	0.20	9.5×10^{-8}
i	0.19	0.25	0.0038	0.10	6.6×10^{-9}
			Overland Flows		
j	0.011	0.01	0.18	0.42	2.3×10^{-5}
k	0.0045	0.005	0.12	0.40	1.0×10^{-4}
l	0.0056	0.008	0.050	0.37	1.0×10^{-4}
m	0.0036	0.006	0.084	0.30	4.0×10^{-4}

11-11. The values of $\partial^2 Y/\partial x^2$ in the table in Exercise 11-10 were estimated from observations of hydrographs. (a) Use those values and the values of D obtained in Exercise 11-10 to compute the time rate of change of depth for the flood crests. (b) Compute the flood-wave velocity from Equation 11-42 and use it to compute the downstream rates of decrease of the flood peaks listed in Exercise 11-10. (c) Using Equation 11-88, compute the value of reach length L for which the kinematic-wave approximation is satisfactory in each of the flows listed in Exercise 11-10.

CHAPTER 12

Principles of Flow Through Porous Media

Most of the precipitation that falls on the earth's land surfaces infiltrates into the ground. Subsequently it may percolate downward through soil and/or rock material to recharge groundwater and ultimately flow to a lake or stream, or it may move back toward the surface and be vaporized by plant transpiration or direct evaporation from the soil. In any case, movement of water through porous earth material is a critically important link in the hydrologic cycle, and understanding the laws that govern its motion is essential for the hydrologist.

There are many excellent books on the hydraulic, hydrologic, and geologic aspects of groundwater (or, more generally, on flow through porous media). Although we cannot, of course, cover in a single chapter all the aspects of porous media flow that are important in hydrology, we can provide an intuitive but quantitatively sound understanding of the physics of hydrologically important porous media flows and illuminate the essential similarities between such flows and the surface-water flows discussed in previous chapters.

To do this, we first define the terms used to describe porous media and the major classes of flows that occur in them. In order to develop the basic physics of these flows, we first consider the case when all pores are completely filled with water under hydrostatic pressure. The forces causing and resisting motion are identified, and the flow in a single idealized pore is analyzed and used as the basis for understanding the equation of motion that governs flows in the highly irregular assemblages of pores that characterize natural media. These basic principles are then generalized to apply to flows in which the pores are only partially filled with water and the water is under tension. Finally, the equation of continuity for porous media flow is developed, and the ways in which this equation is combined with the equation of motion to model natural subsurface flows are indicated.

CHARACTERISTICS OF POROUS MEDIA

A porous medium is a solid material containing a system of interconnecting void spaces, or pores. The porous media encountered by hydrologists are aggregations of granular earth materials (clays, silts, sands, and gravels) and rocks containing interconnected fractures, joints, or solution channels. A fundamental property of these materials is their *porosity* n, defined as

$$n \equiv \frac{V_v}{V_T}, \tag{12-1}$$

where V_v is the volume of void spaces within a volume V_T of the material. An alternative index of the relative amount of pore space in a volume of porous medium is the *void ratio* e:

$$e \equiv \frac{V_v}{V_s}, \tag{12-2}$$

where V_s is the volume of solid material within a volume of the material. Clearly, both n and e are dimensionless. Since $V_T = V_v + V_s$, e and n are related as

$$e = \frac{n}{1 - n} \tag{12-3}$$

and

$$n = \frac{e}{1 + e}. \tag{12-4}$$

Table 12.1 gives typical ranges of porosities and void ratios for earth materials.

Flow in porous media occurs, of course, through the interconnected pores. In developing the principles of flow, it will be important to distinguish between situations in which all the pores are filled with water (*saturated conditions*) and those in which some of the pores contain air as well as water (*unsaturated conditions*). Thus it is useful to define two additional parameters, the *volumetric water content* θ,

Table 12.1
Ranges of Porosities (n) and Void Ratios (e) of Earth Materials
From Davis (1969)

Material	n	e
Granular materials		
Gravel	0.25–0.40	0.33–0.67
Sand	0.25–0.50	0.33–1
Silt	0.35–0.50	0.54–1
Clay	0.40–0.70	0.67–2.3
Sandstone	0.05–0.30	0.05–0.43
Limestone and dolomite	0–0.20	0–0.25
Shale	0–0.10	0–0.11
Rocks with joints, fractures, or solution channels		
Fractured basalt	0.05–0.50	0.05–1
Fractured crystalline rock	0–0.10	0–0.11
Dense crystalline rock	0–0.05	0–0.05
Karst	0.05–0.50	0.05–1

$$\theta \equiv \frac{V_w}{V_T}, \qquad (12\text{-}5)$$

and the *degree of saturation* s,

$$s \equiv \frac{V_w}{V_v}, \qquad (12\text{-}6)$$

where V_w is the volume of water contained within a volume of material V_T. Clearly, for saturated conditions $s = 1$ and $\theta = n$; for unsaturated conditions $s < 1$ and $\theta < n$.

CLASSIFICATION OF FLOWS

The principal classification of porous media flows is based on the pressure condition in the flowing water. A water surface exposed to the earth's atmosphere is at atmospheric pressure, which varies slightly about a standard absolute value of 101,000 N m^{-2} (14.7 lb $in.^{-2}$). When such a surface occurs in a porous medium, it is called a *water table*. By convention, atmospheric pressure is usually taken as the datum for pressure measurement, and is assigned a value of zero. Pressures measured with respect to this datum are referred to as *gage pressures,* so that, in general,

$$P_{gage} = P_{abs} - P_a, \qquad (12\text{-}7)$$

where P_{gage} is gage pressure, P_{abs} is absolute pressure, and P_a is atmospheric pressure.

The pressures referred to throughout this chapter are gage pressures. In order to avoid some terminological difficulties, we will refer to the major class of porous media flows in which the gage pressures are greater than zero as *groundwater flows,* and those in which gage pressures are less than zero (i.e., in which the water is under tension) as *soil-water flows.* In many other texts these classes are called *saturated flows* and *unsaturated flows,* respectively.

Groundwater Flows Consider the situation shown in Figure 12.1, in which a porous medium is saturated beneath a horizontal surface at atmospheric pressure. The principles of hydrostatics developed in Chapter 5 are not changed by the presence of the medium if all the pores are interconnected (i.e., if a path exists between any two locations in the pore spaces). Thus the hydrostatic pressure P at

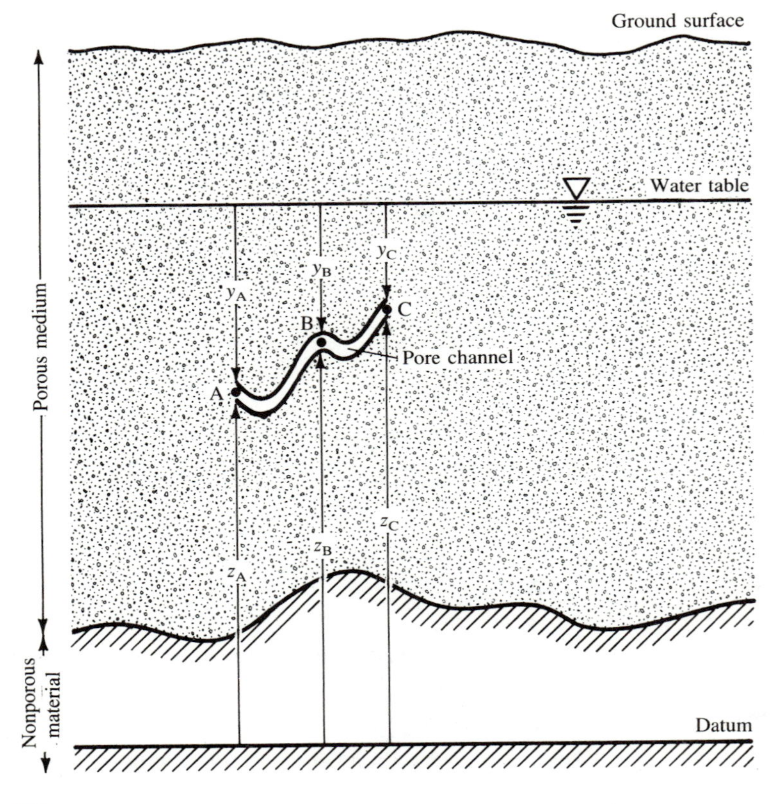

Figure 12.1
A portion of an unconfined groundwater body with a horizontal water table. Potential energy relations are shown in one segment of the interconnected pores. As in Figure 5.5, $z_A + y_A = z_B + y_B = z_C + y_C$, and the potential energy is identical in all the pore spaces below the water table.

any point in the pores is the same in all directions and increases linearly with depth according to

$$P = \gamma y, \tag{12-8}$$

where γ is the weight density of water and y is the vertical distance from the surface downward to the point. Equation 12-8 is identical to Equation 5-2, with $P_a = 0$.

Now consider the situation in Figure 12.2. Once again, the presence of a porous medium in the "conduit" between the two bodies of water does not affect the pressure relations, and the situation is equivalent to that shown in Figure 5.6a. The water in the conduit is under pressure, and the magnitude of the pressure is given by Equation 12-8 if y has its general definition as the vertical distance beneath the potential, or piezometric, head line.

Figures 12.1 and 12.2 depict the two subclasses of groundwater flows. When such flows are bounded by a surface at atmospheric pressure, they are called *free-surface, water-table,* or *unconfined* flows. When both the upper and lower boundaries are effectively nonporous, the flow is described as *confined* or *artesian.*[1]

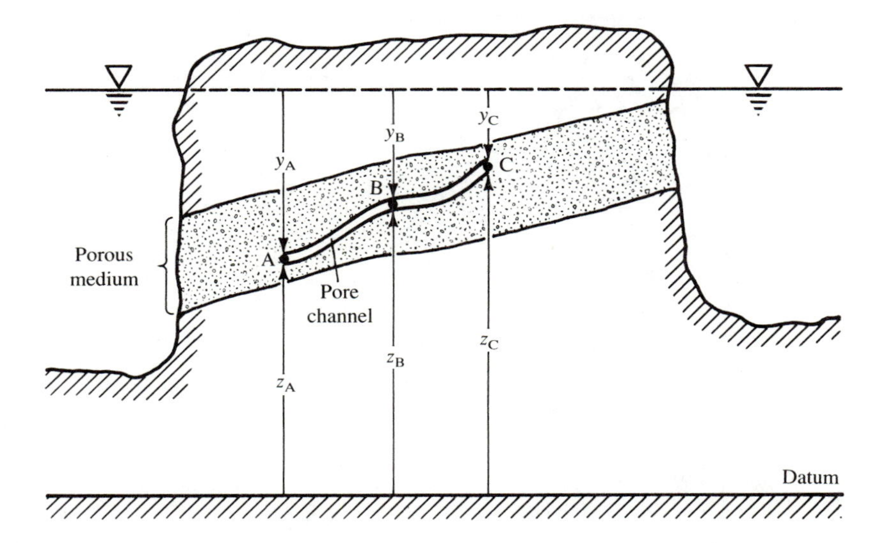

Figure 12.2
Confined groundwater shown as occurring in a pervious stratum between two lakes with the same surface elevation. Potential energy relations are shown in one segment of the interconnected pores; the dashed line is the piezometric surface. As in Figure 5.6, $z_A + y_A = z_B + y_B = z_C + y_C$, and the potential energy is identical throughout the medium.

[1]This term comes from the district of Artois in northern France, where the first wells tapping confined groundwater bodies were drilled and investigated around 1750 (Todd, 1959).

Soil-Water Flows Recall from our discussion of capillarity in Chapter 4 that if a bundle of glass tubes of small radii were placed vertically in a container of water, surface-tension forces would cause the water to rise in each tube to a height h inversely proportional to the tube's radius R (Equation 4-11):

$$h = \frac{2\sigma \cos \theta_c}{\gamma R}, \qquad (12\text{-}9)$$

where σ is the surface tension and θ_c is the characteristic contact angle between the meniscus and the solid wall. Since the radii of pores in earth materials range from about 5×10^{-4} mm to about 10 mm, and since θ_c for most minerals is near 0° (Table 4.3), Equation 12-9 indicates that there will often be a zone of significant thickness above the water table where all the pores are full of water that is suspended from menisci. Because of the variability of pore sizes in natural granular materials, this zone, called the *capillary fringe* or *tension-saturated zone*, has an irregular upper boundary (Figure 12.3).

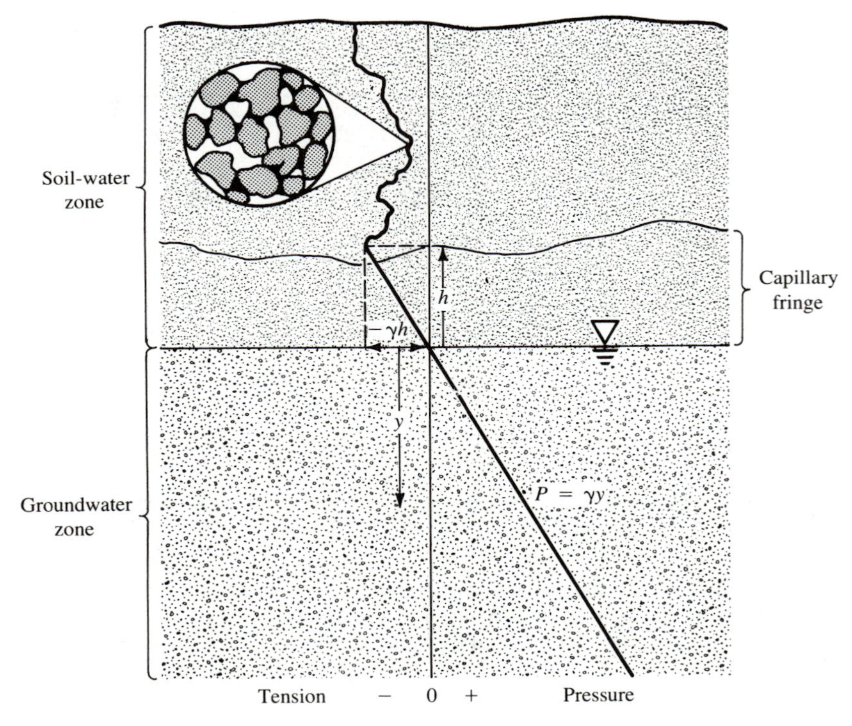

Figure 12.3
Pressure relations in a porous medium containing a water table. Insert shows detail of water held by surface tension in the unsaturated zone, where pressure is negative and depends on the pore size and water content.

Recalling Equation 4-14, we see that the pressure within the capillary fringe is given by Equation 12-8. In this case, however, we measure *upward* from the atmospheric pressure surface to the point of interest, so y is a negative number. Thus the pressure in the capillary fringe is also negative, reflecting the fact that the water is under tension. The tension at the top of the capillary fringe is called the *air-entry pressure* or *air-entry tension* P_{ae} and is computed from Equation 12-8 as

$$P_{ae} = -\gamma h, \tag{12-10}$$

where h is taken as the positive height of capillary rise.

Figure 12.3 shows the pressure conditions in the vicinity of a water table. Above the capillary fringe the medium is unsaturated and water is held by surface tension to the solid grains. From Equation 4-15, the tension, P_u, in the water adjacent to the menisci in these unsaturated pores is

$$P_u = -\frac{2\sigma \cos \theta_c}{r_m}, \tag{12-11}$$

where r_m is the radius of curvature of the meniscus. Clearly, for a given soil, the radii of the menisci are larger and hence the tension is lower (i.e., P_u is a smaller negative number) when the water content is higher (Figure 12.4a,b). Also, for a given water content, r_m is smaller and the tension is higher for finer-grained materials (Figure 12.4c,d). Because an awareness of the relations between tension and water content is essential in order to understand and model flow in the unsaturated zone, these relations are discussed in more detail later in this chapter.

As indicated earlier, movement of water in zones of porous media in which the water is under tension is commonly called unsaturated flow; since the capillary fringe is saturated, however, we refer to this second major class of porous media flows as soil-water flow.

FORCES IN POROUS MEDIA FLOW

In unconfined groundwater flows the roles of the various forces are identical to those in the open-channel flows discussed in Chapter 5, although (as will be shown later) their relative importance changes. Motion is induced by gravity when the water table is sloping, and by the pressure-gradient forces if the thickness of the zone of flow decreases in the direction of flow. The motion is resisted by friction transmitted by viscous and (rarely) turbulent forces, and if the flow thickness increases in the "downstream" direction, by the pressure-gradient forces as well. Surface tension acts at right angles to the flow at the surface, and aside from maintaining the capillary fringe, it plays no role in the bulk flow. As will be shown later, Coriolis and centrifugal forces can always be neglected because of the small velocities of groundwater flows.

In confined groundwater flows either the pressure-gradient or gravity forces, or both may induce motion. One of these forces may act to resist motion, depending on the configuration of the flow. The other forces act as in unconfined flows, except that because there are no free surfaces, surface-tension forces are zero.

In soil-water flows, surface tension assumes the role of a motivating force along with gravity. As will be shown in more detail later, a gradient of tension induces

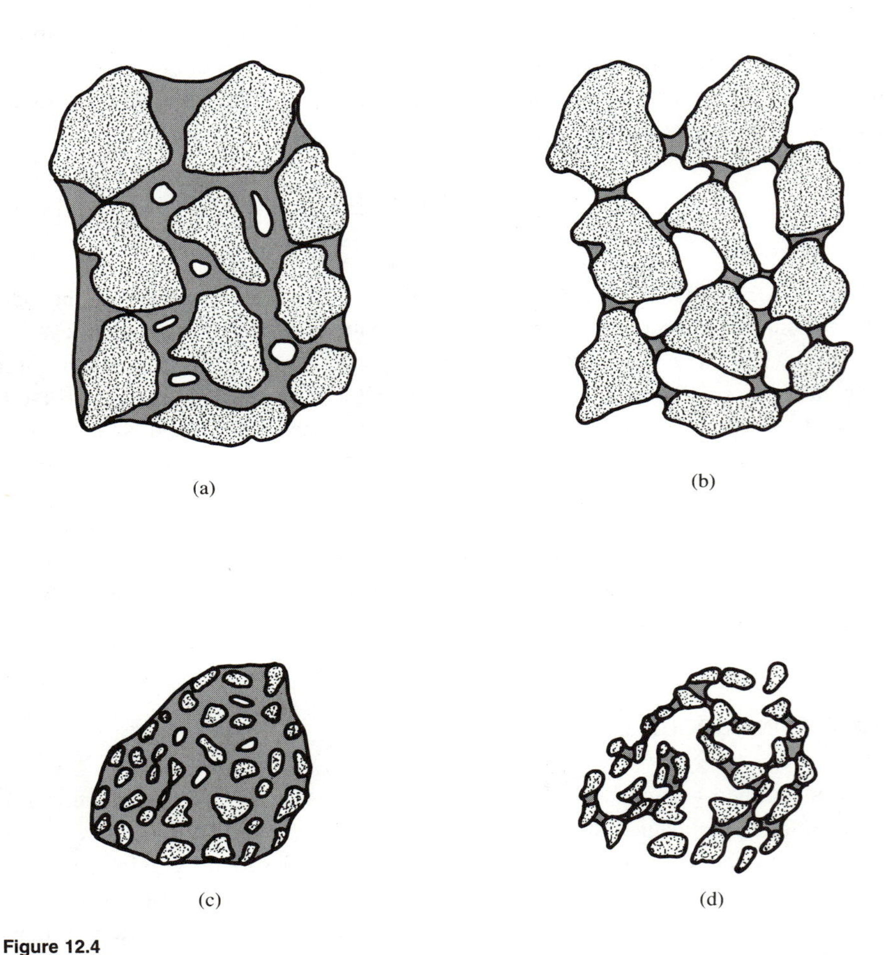

Figure 12.4
Dependence of radii of menisci on grain size—compare (a) with (c) and (b) with (d)—and on water content—compare (a) with (b) and (c) with (d)—in an unsaturated porous medium.

flow and, in fact, is physically identical to a pressure gradient. Again, motion is resisted by viscous and turbulent forces, and Coriolis and centrifugal forces are vanishingly small.

A more quantitative feeling for the relative importance of forces in porous media flow can be obtained by computing the forces per unit mass, for which purpose we use formulas analogous to those listed in Table 5.1. Since the flows of concern here take place through small pores, the appropriate length dimension in these formulas becomes the typical pore diameter rather than the bulk flow thickness. Table 12.2 shows the relative magnitudes of forces when Y in the equations of Table 5.1 is replaced by d_m, the representative pore diameter (essentially equal to the representative grain diameter) of a range of porous media, and values of velocity q_s appropriate to such flows are used. Table 12.2 shows some marked contrasts to Table 5.2, especially in the large relative magnitude of the surface-tension forces that act in soil-water flows. Viscosity provides by far the dominant resisting force in all but the largest porous media flows, so we can anticipate that we will be concerned almost exclusively with laminar flows in the remainder of this chapter.

POTENTIAL ENERGY IN POROUS MEDIA FLOW

Porous media flow, like all flow, occurs only in response to a gradient of potential energy. At any point, the flow direction is parallel to the potential energy gradient and the flow rate is directly proportional to the magnitude of the gradient and inversely proportional to the magnitude of the resisting forces. [It is interesting that the fundamental aspects of potential energy in porous media flow were first elucidated only relatively recently by the American geophysicist M. King Hubbert (1940).]

The general expression for potential energy at a point on a streamline in a porous medium is fundamentally identical to that for open-channel flow (Equation 7-6), but for reasons that will soon be apparent, we have changed the notation slightly:

$$H_p = z + \psi, \qquad (12\text{-}12)$$

where H_p is potential, or piezometric, head at the point; z is the distance to the point measured vertically upward from an arbitrary horizontal datum (the elevation head); and ψ is the pressure head at the point. As discussed next, the nature of ψ depends on whether the point is in the saturated zone below the top of the capillary fringe or in the unsaturated zone.

Saturated Zone In the saturated zone

$$\psi = y, \qquad (12\text{-}13a)$$

Table 12.2

Forces per Unit Mass in Porous Media Flows of Various Sizes

Fluid properties at 10°C are assumed.

	Typical flow characteristics		Forces per unit mass ($m\ s^{-2}$)				
Medium	Particle diameter (mm)	Specific discharge ($m\ s^{-1}$)	Gravitational	Viscous	Turbulent	Surface tension	Coriolis[†]
Clay	0.002	5.0×10^{-12}	9.8	1.6×10^{-6}	1.3×10^{-17}	1.9×10^{7}	4.7×10^{-16}
Silt	0.01	5.0×10^{-9}	9.8	6.6×10^{-5}	2.5×10^{-12}	7.4×10^{5}	4.7×10^{-13}
Sand	0.5	5.0×10^{-6}	9.8	2.6×10^{-5}	5.0×10^{-8}	3.0×10^{2}	4.7×10^{-10}
Gravel	2.0	2.5×10^{-3}	9.8	8.2×10^{-4}	3.1×10^{-3}	1.9×10^{1}	2.5×10^{-7}

[†] At latitude 40°.

where y is the distance measured vertically downward to the point from the surface of zero gage pressure. Thus $y > 0$ below the water table and throughout a confined groundwater flow, and for these cases we compute the total potential energy on a streamline exactly as for flows in channels. Consistent with the previous discussion of pressures, $y < 0$ in the capillary fringe, where the water is under tension.

The total potential head at a point in a groundwater flow is measured by means of a *piezometer,* which is simply an open-ended pipe with impermeable walls. When such a device is inserted into the flow, the static level above the datum of the water in the pipe is the value of H_p at the lower end of the pipe. Referring to Figure 12.5, we note that the water level in a piezometer inserted into the portion of the flow where there is a downward velocity component (a zone of net groundwater recharge) will be below the water table at that point, and that the opposite will be true where the streamlines have an upward component (a zone of net groundwater discharge). A *well,* which differs from a piezometer in that it has permeable walls that admit water over all or part of its length, measures the average potential head for the portion of the flow it penetrates.

In Chapters 8–11 open-channel flows were considered to be one-dimensional. Knowledge of variations of velocity with depth and width were not necessary to solve most hydrologic problems, so we could conceptually reduce the flow to a single representative streamline at the channel bed and could use the kinetic energy coefficient α to account for the variation of velocity within the flow. A one-dimensional representation is not adequate for most groundwater flows, but many can be usefully visualized as two-dimensional flows like the one shown in Figure 12.5. In these cases we can ''map'' the potential head throughout the flow by using a series of lines, each connecting points at which the potential head has a particular value. These lines of equal potential head are called *equipotential lines.* Following the principles of equations of motion, we draw streamlines orthogonally to the equipotential lines to produce a *flow net.* (Clearly, the streamlines in a porous medium flow are abstract representations of general flow directions rather than traces of the highly tortuous paths of actual water parcels negotiating an irregular network of pores.)

The construction of flow nets is a powerful tool for the analysis of groundwater flows. The principles for constructing these nets are well described in several texts, including those of Cedergren (1967) and Freeze and Cherry (1979), and will not be developed here. We should note, however, that the basis for the construction of flow nets is the solution of the complete equations of groundwater flow under steady-state conditions. These are second-order partial differential equations, and are derived at the end of this chapter. The solutions are expressions for the spatial variability of the total potential head, and in general they require that the equipotential lines be curved. This in turn gives rise to regions in which the streamlines converge and to other regions where they diverge. As in open-channel flow, a hydrostatic pressure distribution exists only where the streamlines are parallel. Thus in the flow shown in Figure 12.5, pressures are less than hydrostatic where the streamlines have a downward component (as at A) and greater than hydrostatic where they have an upward component (as at B). As noted earlier, a zone of net groundwater recharge exists where the streamlines have a downward component, and a zone of net groundwater discharge where they have an upward component.

Unsaturated Zone Above the capillary fringe, the pressure (tension) is not directly dependent on the elevation above the water table, but for a given soil depends primarily on the local water content, as indicated by Equation 12-11 and

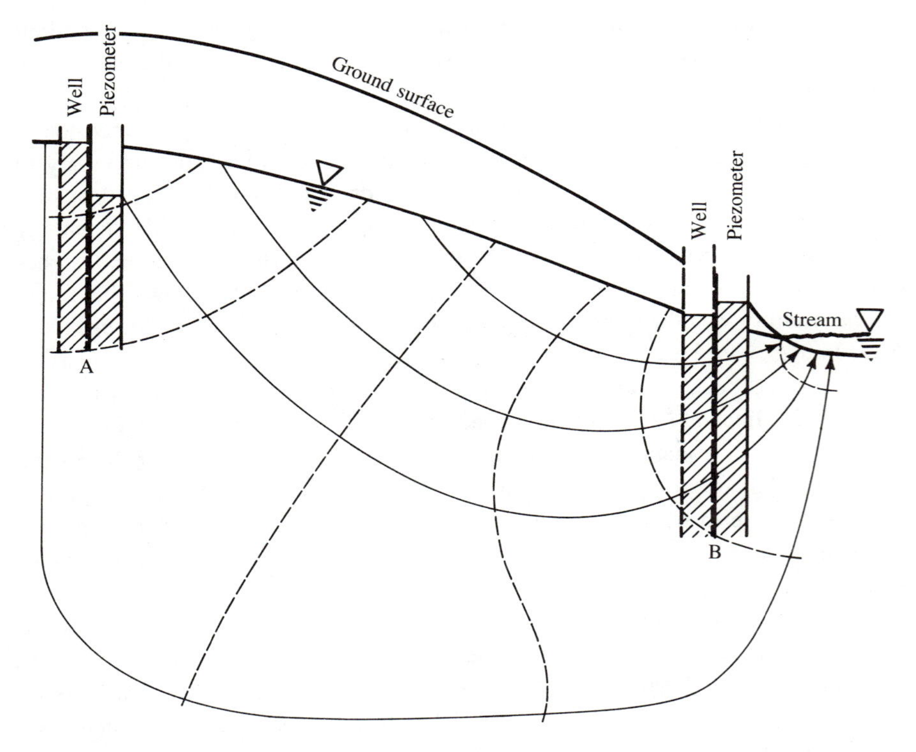

Figure 12.5
Example of a flow net in an unconfined groundwater flow to a stream. Dashed lines are equipotential lines; solid lines are streamlines. Piezometer at A measures a potential head lower than that where it penetrates the water table (shown in well A) in a region with a downward flow component. Piezometer at B measures a potential head higher than that where it penetrates the water table (shown in well B) in a region with an upward flow component.

Figure 12.3. Recalling that pressure head is defined as pressure divided by weight density (Equations 5-12 and 5-13), we find that Equation 12-11 leads to

$$\psi = -\frac{2\sigma}{\gamma r_m} \tag{12-13b}$$

for the unsaturated zone. Note that if we assume that $\cos \theta_c = 1$, then Equation 12-13b is identical to the expression for the height of capillary rise in a tube or pore of radius r_m (Equation 12-9).

Equation 12-13b states that the pressure head at a point in the unsaturated zone depends on the radius of the meniscus at that point. Although this equation is theoretically sound, its practical use is limited because there is no convenient way to measure r_m. Instead, we make use of the fact that r_m is related to the water content θ as indicated earlier, and establish an empirical relationship between ψ and θ for each soil of interest.

Figure 12.6 shows examples of such relationships, usually called *moisture tension curves* or *soil-water characteristics,* for particular coarse- and fine-grained soils. Note that the relationships are highly nonlinear, and that for each soil there is a range of tension head over which the water content is the saturated value (i.e., $\theta = n$). The maximum value of the tension head at saturation corresponds to the air-entry pressure P_{ae}, which was defined in Equation 12-10. Using that equation, we see that

$$\psi_{ae} = \frac{-P_{ae}}{\gamma} = h. \tag{12-14}$$

Thus the *air-entry tension head* ψ_{ae} for a soil is equal to the height of capillary rise for that soil. Since pore radii are much smaller in clays than in sands (Table 12.2), h and ψ_{ae} are considerably larger for the finer-grained soils. Note also that for any value of θ, $|\psi|$ is much larger in clays than in sands. This means that if a sand and a clay with the same water content are in contact, there will be a strong pressure gradient from the sand to the clay, and hence water will flow from the sand to the clay until the values of ψ become equal in both.

Actually, ψ is not a single-valued function of θ for most soils, as was indicated in Figure 12.6. Instead, the relationship is usually characterized by *hysteresis:* The tension corresponding to a given water content is higher when the soil is drying than when it is wetting, as shown in Figure 12.7. The reason for this can be seen by reference to Figure 12.8. The minimum radius in the right-hand "neck" is r_1, and that in the left-hand neck is a smaller value r_2. The maximum radius between the necks is r_3. If at an initial equilibrium state the large pore is filled, with a meniscus at the neck on the right (Figure 12.8a), the pressure head across the meniscus is given by Equation 12-13b as

$$\psi_1 = -\frac{2\sigma}{\gamma r_1}. \tag{12-15}$$

If the pressure head of the pore water is decreased algebraically (tension increased) to a value ψ_2 ($\psi_2 < \psi_1$), the water will move in the direction of decreasing potential energy until it encounters a pore of radius r_2, where

$$r_2 = -\frac{2\sigma}{\gamma \psi_2} \tag{12-16}$$

(Figure 12.8b). At this point equilibrium is established again, with the large central pore empty. If the pressure head is now increased back to its original value of ψ_1 and

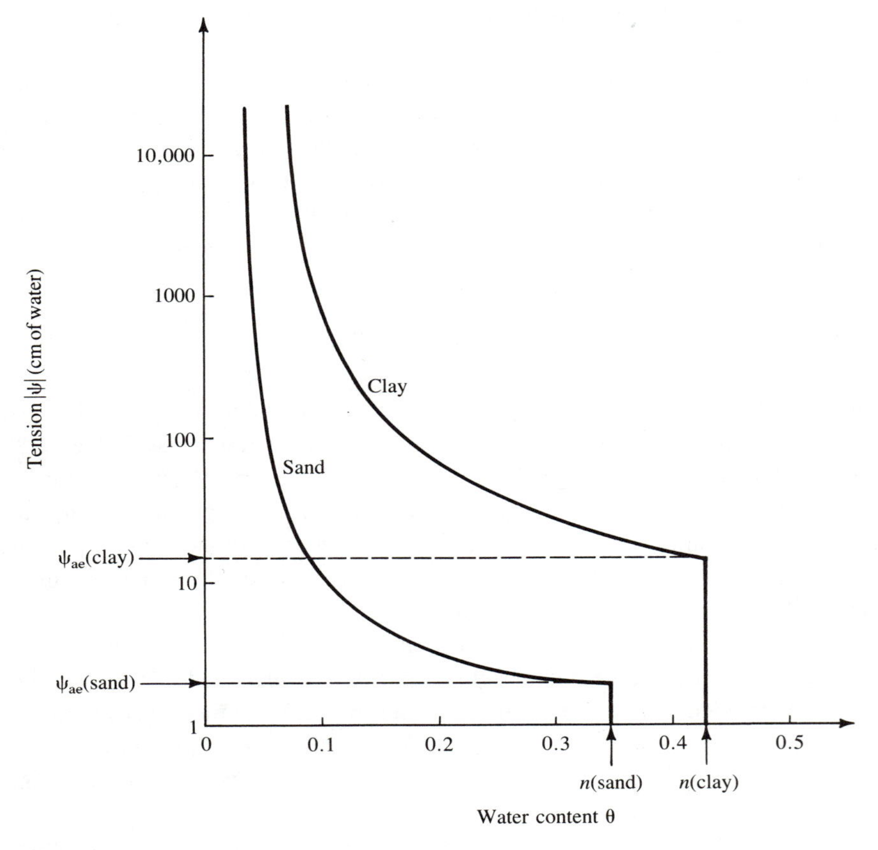

Figure 12.6
Schematic soil-water–tension curves for a sand and a clay, showing that the finer-grained soil has a higher tension at a given water content; ψ_{ae} is the air-entry tension head and n is the porosity.

wetting begins, the meniscus will move to the right until it encounters the radius r_1, where the equilibrium given by Equation 12-15 again holds. This time, however, the large pore is not filled (Figure 12.8c). Thus the water content at pressure head ψ_1 is greater when the soil is drying than when it is wetting, or (as stated earlier) the head for a given water content is lower when the soil is drying than when it is wetting.

Clearly, moisture tension curves are of vital importance for hydrologic studies in which detailed modeling of the unsaturated zone is required. These curves are usually obtained by experiments on soil samples in special laboratory devices called *pressure plates* (see Hillel, 1980). Attempts have been made to represent moisture tension curves mathematically for use in computer models; an example is the equation suggested by Brooks and Corey (1966):

$$\frac{\psi}{\psi_{ae}} = \left(\frac{n - \theta_r}{\theta - \theta_r}\right)^{1/\lambda}, \tag{12-17}$$

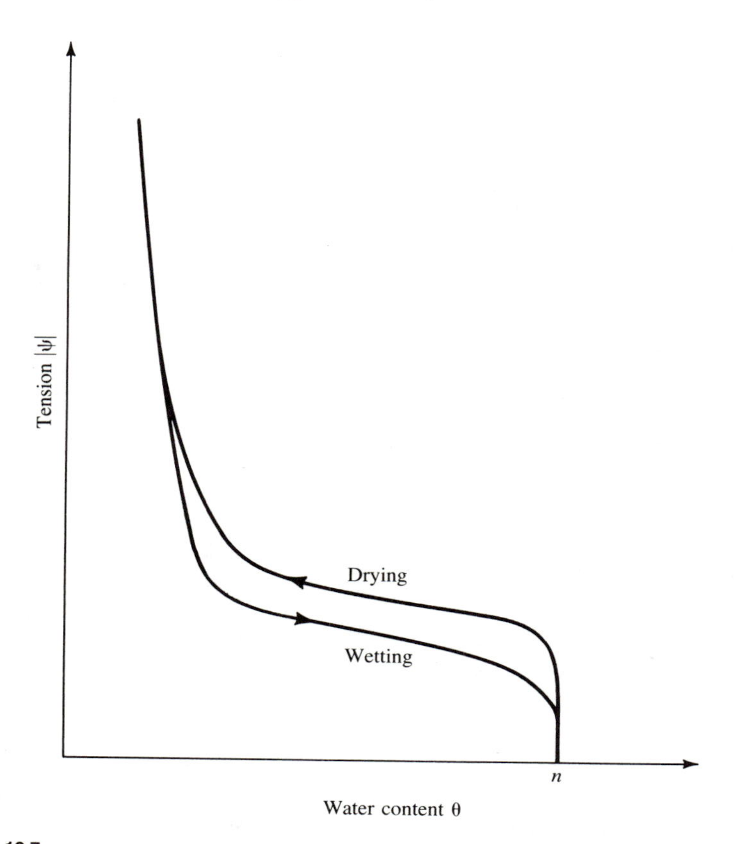

Figure 12.7
Schematic soil-water–tension curve illustrating hysteresis. For a given water content, tension is higher when the soil is drying than when it is wetting.

where θ_r is the water content at which the water films in the soil become discontinuous (approximately the point where the slope of the soil-water characteristic curve becomes vertical), n is porosity, and λ is an empirical factor that depends on the grain-size distribution. Other approaches to this problem are cited by Hillel (1980).

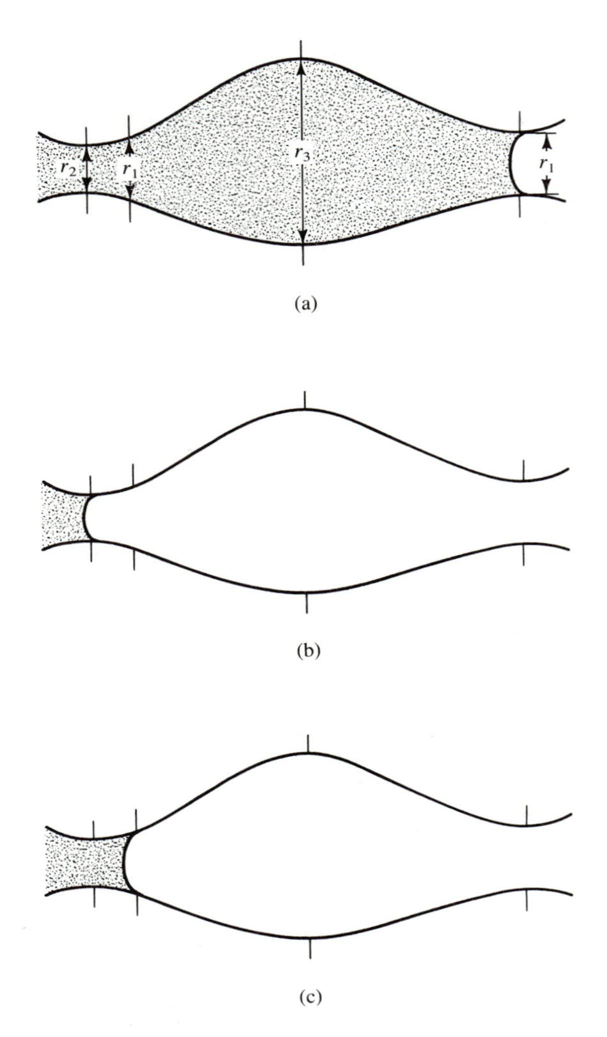

Figure 12.8
Schematic diagram of a soil pore showing the reason for hysteresis in moisture–tension relations. In (a) the pore is full of water at a tension corresponding to r_1. When the tension is increased, the pore drains until a new equilibrium is reached, corresponding to r_2 (b). When the tension is subsequently decreased to its original value, equilibrium again occurs at r_1, but with the large pore empty (c).

In the field, pressure head in the unsaturated zone is measured by means of *tensiometers*. These instruments consist of an airtight vertical tube filled with water, with a porous membrane (usually a ceramic cup) at the bottom and a vacuum gage near the top (Figure 12.9). The tube is inserted into the soil with the membrane at the point of measurement. Initially the water in the tube is under hydrostatic pressure, so the potential gradient causes a flow of water out of the tensiometer through the porous cup. The water level in the tube consequently falls, creating a vacuum that is registered by the gage. Flow out of the instrument continues until the strength of the suction created by the vacuum equals the tension in the soil water, at which point the gage is read. The total head at the point of measurement is then computed

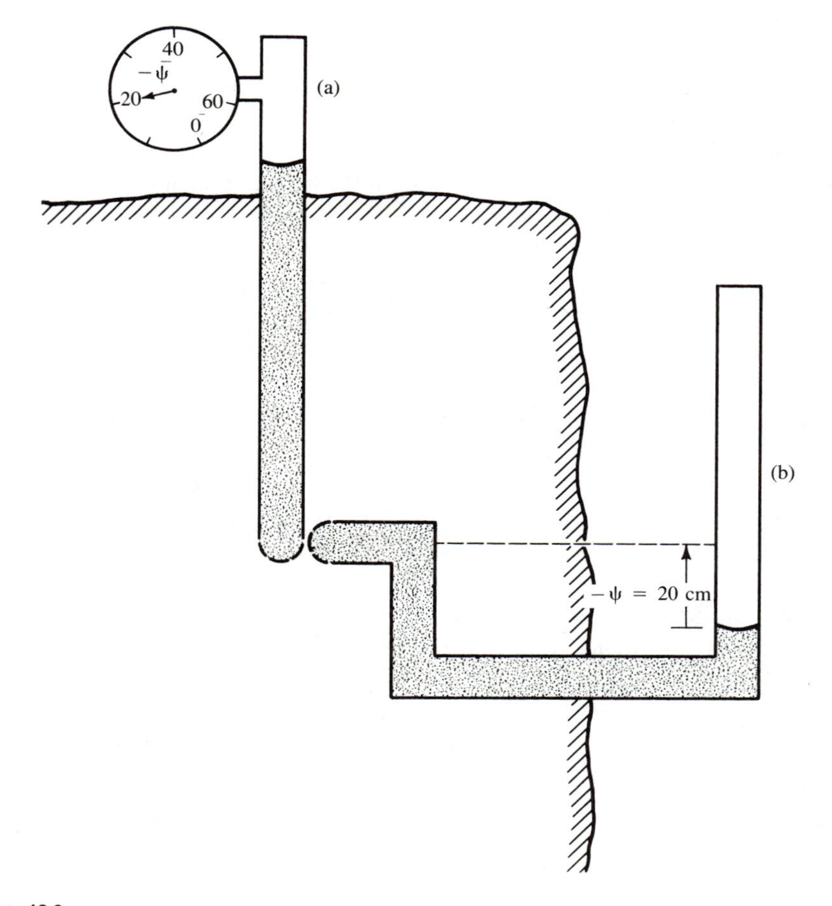

Figure 12.9
Schematic diagrams of tensiometers. (a) Typical version used in field application; vacuum gage registers a tension of 20 cm of water. (b) Modified version illustrating the physical interpretation of a negative pressure head. The water surface is drawn 20 cm below the point of measurement.

as the algebraic sum of the elevation, z, of the point above the datum and the gage reading ψ (which is of course treated as a negative number). The relationship between the strength of the vacuum and the tension head is also illustrated in Figure 12.9: If the tensiometer could be modified to the U-shaped design shown on the right, the tension head would be equal to the vertical distance to which the water surface in the tube is drawn below the level of the point of measurement.

THE EQUATION OF MOTION FOR GROUNDWATER FLOW

The equation of motion for groundwater flow is called *Darcy's law*. In most textbooks this law is presented as a phenomenological law developed from experiments (described more fully later) originally carried out by the French civil engineer Henri Darcy in 1856. It has been shown, however, that Darcy's law can be derived from the basic equations of fluid mechanics (see Hubbert, 1940; Daily and Harleman, 1966; Bear, 1972). Our goal here lies somewhere between these two approaches. We begin by applying the basic force-balance principle and our knowledge of the behavior of water as a viscous fluid in order to develop an equation of motion for flow within a small cylinder, which represents an idealized pore. This equation is then adapted to give an expression describing flow through a medium containing a large number of such pores. Finally, comparison of Darcy's experimental results with the equation developed for the idealized medium provides insight into the fundamental physical significance of Darcy's law.

Steady Laminar Flow in a Cylindrical Conduit Figure 12.10 defines the terms used in formulating an equation for flow through a single small cylindrical conduit. The radius of the conduit is a constant value R and its length is L; it slopes downward to the right at an angle ϕ. The vertical distance from the center of the conduit to the water surface at the left end is y_1 and the corresponding distance at the right end is y_2.

Motion in the situation illustrated can be induced by pressure-gradient and/or gravity forces, as discussed previously. The hydrostatic pressures at the left and right ends, P_1 and P_2, respectively, are

$$P_1 = \gamma y_1 \tag{12-18}$$

and

$$P_2 = \gamma y_2 . \tag{12-19}$$

We now focus attention on an arbitrary cylinder of radius r ($r < R$) centered within the conduit. The pressure forces at the ends of this cylinder, F_{p1} and F_{p2}, are then

$$F_{p1} = \gamma y_1 \pi r^2 \tag{12-20}$$

and

$$F_{p2} = \gamma y_2 \pi r^2. \tag{12-21}$$

The net pressure force F_p acting toward the right on the cylinder is then

$$F_p = \gamma \pi r^2 (y_1 - y_2). \tag{12-22}$$

The gravity force F_g acting toward the right on the cylinder is the component of the weight of the cylinder in that direction:

$$F_g = \gamma \pi r^2 L \sin \phi. \tag{12-23}$$

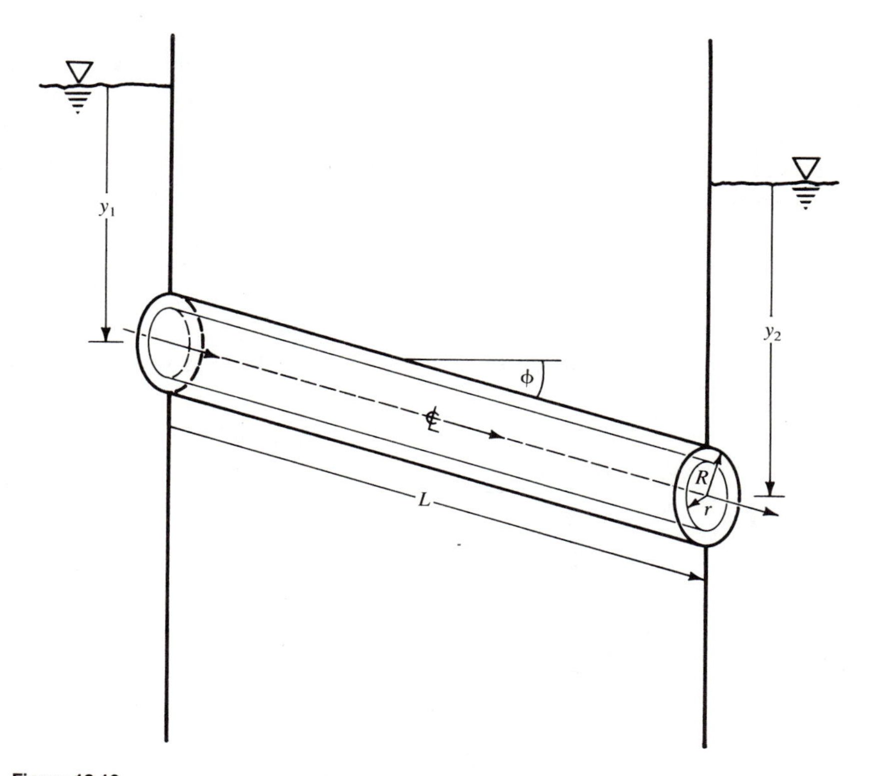

Figure 12.10
Definitions of terms for deriving the equation of motion for laminar flow through a cylindrical conduit (Equations 12-18–12-35).

Thus the total force F_r acting to the right is

$$F_r = F_p + F_g = \gamma \pi r^2[(y_1 - y_2) + L \sin \phi]. \tag{12-24}$$

In steady flow, the net force F_r causing the flow must be resisted by an equal force in the opposite direction. In general, viscous and turbulent forces provide this resistance, but as shown in Table 12.2, the viscous force is much greater than the turbulent force in the great majority of porous media flows. Thus to simplify matters we will neglect the turbulent force in the subsequent development.

In laminar open-channel flow (see Equations 6-8–6-10), viscous forces are transmitted by laminar sheets sliding over one another. The laminae are concentric cylinders in the present case, and Equation 4-5 is modified to give the shear force per unit area, τ, acting on the surface of these cylinders:

$$\tau = -\mu \frac{dv}{dr}, \tag{12-25}$$

where μ is the dynamic viscosity and dv/dr is the radial velocity gradient. The negative sign is required because r is measured from the center of the cylinder outward, so that v decreases as r increases. (Recall that the no-slip condition requires that the velocity be zero at the boundary of the conduit.) Multiplying τ by the area of the cylinder gives the total resisting force F_ℓ on that cylinder:

$$F_\ell = \tau \pi 2 r L = -\mu \frac{dv}{dr} 2\pi r L. \tag{12-26}$$

The balance of forces requires that $F_r = F_\ell$, so equating Equations 12-24 and 12-26 gives

$$\gamma \pi r^2[(y_1 - y_2) + L \sin \phi] = -\mu \frac{dv}{dr} 2\pi r L. \tag{12-27}$$

Equation 12-27 can be simplified and rearranged into a form that can be directly integrated:

$$\frac{\gamma}{2\mu}\left[\frac{(y_1 - y_2)}{L} + \sin \phi\right] \int r\, dr = -\int dv,$$

$$v = -\frac{\gamma}{4\mu}\left[\frac{(y_1 - y_2)}{L} + \sin \phi\right] r^2 + C, \tag{12-28}$$

where C is a constant of integration that is evaluated by invoking the boundary condition $v = 0$ when $r = R$. Thus

$$C = \frac{\gamma}{4\mu}\left[\frac{(y_1 - y_2)}{L} + \sin\phi\right]R^2, \tag{12-29}$$

and substituting Equation 12-29 into Equation 12-28 yields

$$v = \frac{\gamma}{4\mu}\left[\frac{(y_1 - y_2)}{L} + \sin\phi\right](R^2 - r^2) \tag{12-30}$$

as the equation for the velocity distribution of a steady laminar flow in a cylindrical conduit. This distribution has the shape of a paraboloid of revolution. Equation 12-30 was first derived by the French physiologist Jean-Louis-Marc Poiseuille (1799–1869) during his studies of the flow of blood in blood vessels, and flow of this type is called *Poiseuille flow*.

The next step is to use Equation 12-30 to develop an expression for the discharge Q_c from the conduit, where

$$Q_c = V_c A_c \tag{12-31}$$

and V_c is the mean velocity in the conduit and A_c is its cross-sectional area. We proceed by considering an annular element of the conduit with a very small thickness dr and located a distance r from the center (Figure 12.11). The velocity within such an element is essentially constant and is given by Equation 12-30. The cross-sectional area dA_c of the element is

$$\begin{aligned} dA_c &= \pi(r + \tfrac{1}{2}dr)^2 - \pi(r - \tfrac{1}{2}dr)^2 \\ &= 2\pi r\, dr. \end{aligned} \tag{12-32}$$

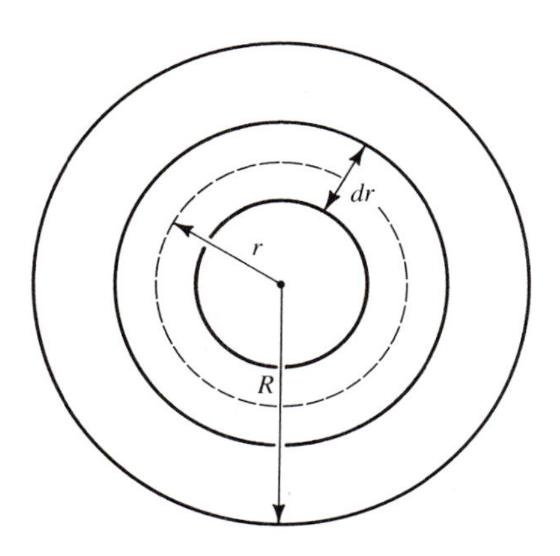

Figure 12.11
Annular element in a cylindrical conduit. Its area is dA_c (Equation 12-32).

Multiplying Equation 12-30 by Equation 12-32 gives the discharge dQ_c through the element:

$$dQ_c = \frac{\pi\gamma}{2\mu}\left[\frac{(y_1 - y_2)}{L} + \sin\phi\right](R^2 - r^2)r\,dr. \tag{12-33}$$

The discharge through the conduit is then found by integrating Equation 12-33 over R:

$$\begin{aligned} Q_c &= \frac{\pi\gamma}{2\mu}\left[\frac{(y_1 - y_2)}{L} + \sin\phi\right]\int_0^R (R^2 - r^2)r\,dr \\ &= \frac{\gamma\pi}{8\mu}\left[\frac{(y_1 - y_2)}{L} + \sin\phi\right]R^4. \end{aligned} \tag{12-34}$$

Finally, the mean velocity V_c for the conduit can be found by noting that $A_c = \pi R^2$ and substituting Equation 12-34 into Equation 12-31:

$$\begin{aligned} V_c \equiv \frac{Q_c}{A_c} &= \frac{\gamma\pi}{8\mu}\left[\frac{(y_1 - y_2)}{L} + \sin\phi\right]R^4\left(\frac{1}{\pi R^2}\right) \\ &= \frac{\gamma R^2}{8\mu}\left[\frac{(y_1 - y_2)}{L} + \sin\phi\right]. \end{aligned} \tag{12-35}$$

Steady Flow in a "Bundle" of Cylindrical Conduits Moving now to a more macroscopic level, we consider a simplified porous medium consisting of a large number of parallel conduits of radius R, each one subject to the same "upstream" and "downstream" water-surface elevations (Figure 12.12). The cross-sectional area of the medium is A, and if there are m conduits, the cross-sectional area A_p of all the pores is

$$A_p = mA_c = m\pi R^2. \tag{12-36}$$

The porosity n is

$$n = \frac{A_p}{A} = \frac{m\pi R^2}{A}, \tag{12-37}$$

and therefore

$$A = \frac{m\pi R^2}{n}. \tag{12-38}$$

The total discharge Q through the medium is equal to the sum of the discharges through the m pores, so

$$Q = mQ_c, \tag{12-39}$$

and with Equation 12-34, Equation 12-39 becomes

$$Q = \frac{m\gamma\pi R^4}{8\mu}\left[\frac{(y_1 - y_2)}{L} + \sin\phi\right]. \tag{12-40}$$

In order to compare this situation with Darcy's law, the final step is to divide

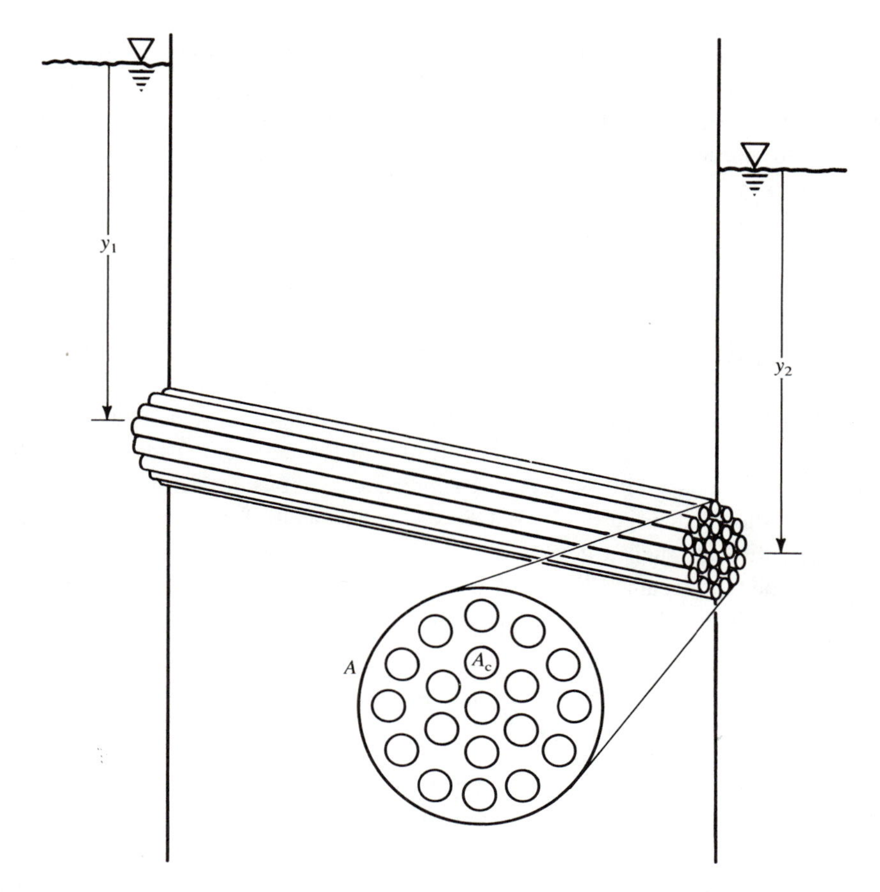

Figure 12.12
Definitions of terms for deriving the equation of flow in a ''bundle'' of cylindrical conduits of equal diameter (Equations 12-36–12-41).

Equation 12-40 by the cross-sectional area of the medium (Equation 12-38) in order to arrive at an expression for the discharge per unit area of the medium, or the *specific discharge* q_s:

$$q_s \equiv \frac{Q}{A} = \frac{\gamma n R^2}{8\mu}\left[\frac{(y_1 - y_2)}{L} + \sin\phi\right]. \tag{12-41}$$

Several aspects of Equation 12-41 should be particularly noted. First, it is of the same general form as the equations of motion developed for uniform open-channel flows (laminar, Equation 6-15; transitional, Equation 6-51; turbulent, Equations 6-48 and 6-49): The specific discharge is the product of the driving force ($\gamma[(y_1 - y_2)/L + \sin\phi]$) and the ''conductivity'' of the flow path ($nR^2/8\mu$). Note further that the flow rate is proportional to the first power of the gradient, as was true for laminar, but not transitional or turbulent, open-channel flows. However, the gradient for porous media flows apparently consists of two parts: a gravitational component ($\gamma \sin\phi$) as in open channels, plus a component due to a gradient of hydrostatic pressure $[\gamma(y_1 - y_2)/L]$. The conductivity term depends on both the fluid viscosity (μ) and the geometry of the flow path ($nR^2/8$), which is also true for uniform laminar open-channel flows (Equation 6-15).

Although Equation 12-41 was developed for a very simplified representation of a porous medium, we will see that it provides valuable insights into flow through real porous media.

Darcy's Law In the 1850s an engineer named Henri-Philibert-Gaspard Darcy (1803–1858) worked on the water supply system of the city of Dijon, France. This system included sand-bed filters for purification, and Darcy carried out a number of laboratory experiments concerning the rate of flow through sand. The essence of his conclusions, published in 1856, can be appreciated by considering the experimental apparatus shown in Figure 12.13. Darcy's experiments showed that the specific discharge through such a device was directly proportional to the difference ΔH between the ''upstream'' and ''downstream'' water levels and inversely proportional to the length L of the column. Thus

$$q_s = -K_D \frac{\Delta H}{L}. \tag{12-42}$$

He found further that the proportionality constant K_D depended on the grain size of the medium. The minus sign indicates that flow is considered to be in the positive direction when $\Delta H/L$ is negative.

Equation 12-42 is one form of Darcy's law. It has been found to describe flows in real porous media accurately, and is the starting point for all scientific and engineering treatments of such flows. Thus, if the development of Equation 12-41 has any value, we should be able to show its close correspondence to Darcy's law. Conversely, such a correspondence should also provide some insight into why Darcy's law *is* a law.

To explore the correspondence between Equations 12-41 and 12-42, we first modify the gradient terms in both equations to represent the gradients at a point rather than over finite distances. Thus

$$\frac{y_1 - y_2}{L} \longrightarrow -\frac{dy}{dl},$$

$$\sin\phi \longrightarrow -\frac{dz}{dl},$$

and

$$\frac{\Delta H}{L} \longrightarrow \frac{dH}{dl},$$

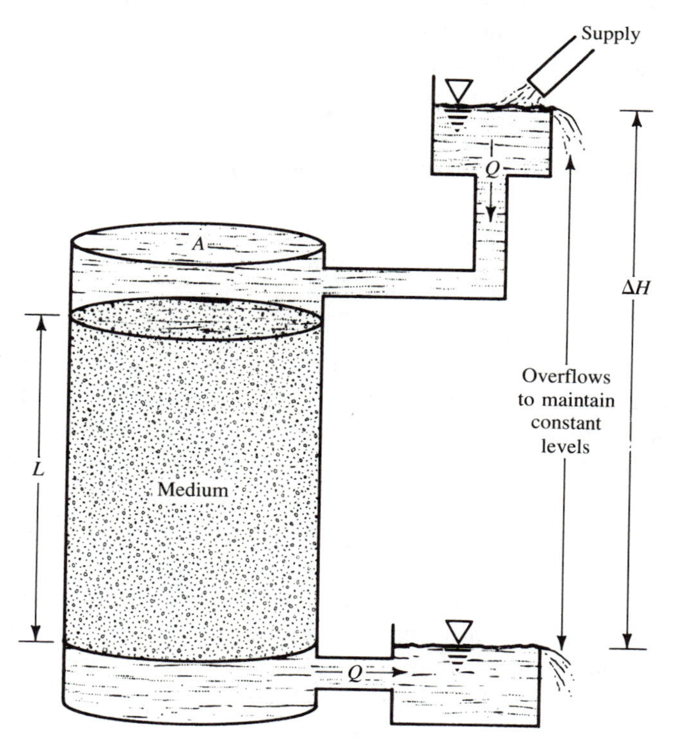

Figure 12.13
Schematic diagram of Darcy's experimental apparatus (see Equation 12-42).

where l is a distance measured along a streamline and z is the elevation above a horizontal datum. With these modifications, Equations 12-41 and 12-42 become

$$q_s = -\frac{\gamma n R^2}{8\mu}\left[\frac{dy}{dl} + \frac{dz}{dl}\right] \quad \text{(Poiseuille flow);} \tag{12-43}$$

$$q_s = -K_D \frac{dH}{dl} \quad \text{(Darcy's law).} \tag{12-44}$$

If the gradient terms in the two equations are identical, then

$$dH = dy + dz,$$

and therefore

$$H = y + z. \tag{12-45}$$

By examination of Equations 12-12 and 12-13a, we now see that H in Darcy's law is identical to the total potential head H_p, and that the two components of potential energy head, the elevation head and the pressure head, arise directly from consideration of the gravitational and pressure-gradient forces that give rise to flow in porous media.

Turning attention now to the terms that reflect the conductivity of the medium, we define for the Poiseuille equation a parameter analogous to K_D in Darcy's law:

$$K_p \equiv \frac{\gamma n R^2}{8\mu}. \tag{12-46}$$

Note initially that both K_D and K_p have the dimensions of a velocity, and are equal to the specific discharge when the potential energy gradient equals one. Beyond this, we have little information about K_D; based on his experiments, Darcy stated only that K_D depended "on the nature of the sand" (Todd, 1959, p. 45). Since there is reason to believe that Equations 12-43 and 12-44 are closely analogous, we can presumably get information about K_D by examining the components of K_p.

Clearly, K_p depends on properties of both the fluid (γ, μ) and the medium (n, R). In transferring from the simplified bundle of conduits to a real porous medium, γ, μ, and n should retain their meanings. However, R will not be a constant in a real medium. To account for this, we replace it with $\beta_1 R_m$, where R_m is the average pore radius and β_1 is a constant of proportionality that accounts for the effects of a range of pore sizes. We now write

$$K_D = \frac{\gamma n \beta_1^2 R_m^2}{8\mu}. \tag{12-47}$$

For a granular porous medium, however, this expression can be made more practical by relating the mean or median grain diameter d_m to R_m via another empirical coefficient, β_2:

$$R_m = \tfrac{1}{2}\beta_2 d_m, \tag{12-48}$$

so that now

$$K_D = \frac{\gamma n \beta_1^2 \beta_2^2 d_m^2}{32\mu}. \tag{12-49}$$

Next, we must recognize that other aspects of real porous media that we have not so far considered, such as grain shape and other details of actual pore geometry, will influence conductivity. We adjust for these effects by introducing a third coefficient, β_3, to obtain

$$K_D = \frac{\gamma n \beta_3 \beta_1^2 \beta_2^2 d_m^2}{32\mu}. \tag{12-50}$$

Finally, all the constants can be combined into a single term, β:

$$\beta \equiv \frac{\beta_1^2 \beta_2^2 \beta_3}{32},$$

so that

$$K_D = \frac{\beta \gamma n d_m^2}{\mu}. \tag{12-51}$$

Equation 12-51 shows how K_D, which is called the *hydraulic conductivity,* depends on the properties of the medium and of the fluid. The portion that is a function only of the medium is called the *intrinsic permeability* K_I:

$$K_I \equiv \beta n d_m^2; \tag{12-52}$$

so that

$$K_D = \frac{\gamma}{\mu} K_I. \tag{12-53}$$

As indicated in Chapter 4, the fluid properties are primarily a function of temperature. The ratio γ/μ changes from about 54,800 $cm^{-1}\,s^{-1}$ to 122,000 $cm^{-1}\,s^{-1}$, a factor of about 2, over the temperature range likely to be encountered in nature (0°C to 30°C). As shown in Table 12.1, n varies by a factor of about 10 for earth materials. Diameters of granular materials vary over a 10^6-fold range, from about 0.001 mm (clay) to about 1000 mm (boulders). Thus d_m^2 varies by a factor of approximately 10^{12}, and variations in grain size are responsible for most of the tremendous (10^{13}-fold) variation in hydraulic conductivity that is observed in nature (Figure 12.14).

Morris and Johnson (1967) have published laboratory measurements of hydraulic conductivity, porosity, and grain-size distribution that can be used to test the validity of Equation 12-51 and to explore the range of β values for natural media. The median grain sizes in their study ranged from 0.0035 mm to 8.5 mm, and the hydraulic conductivities from 4.7×10^{-9} mm s^{-1} to 19 mm s^{-1}. If Equation 12-51 is correct and if β were the same for all materials, a plot of log K_D versus $\log(\gamma n d_m^2/\mu)$ would define a straight line with a slope of one. Figure 12.15 is such a plot of Morris and Johnson's (1967) data, and indicates that the slope of the relationship is in fact near the theoretical value of one. Not surprisingly, however, β varies for different media as a result of variations in grain shapes and grain-size distributions, with values ranging between 10^{-6} and 10^{-3}.

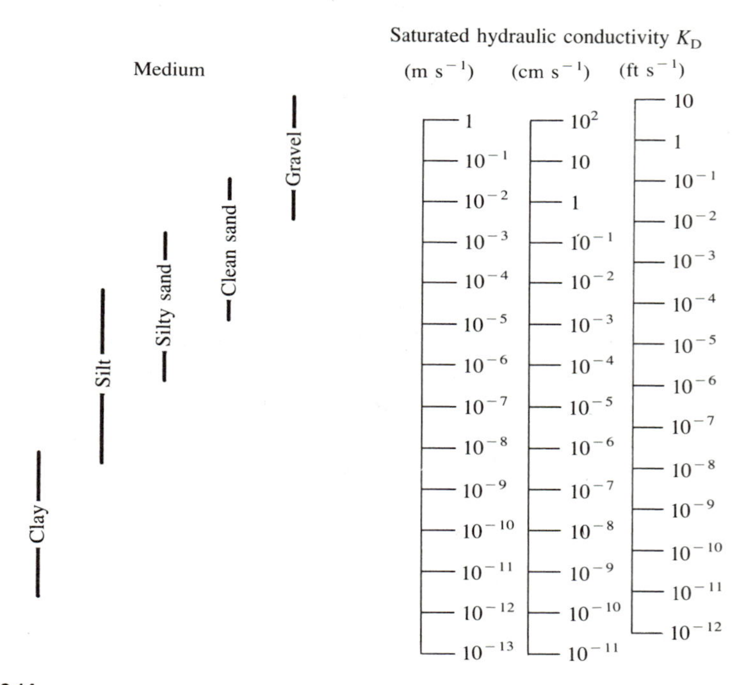

Figure 12.14
Range of values of the hydraulic conductivity K_D in natural porous media. After Freeze and Cherry (1979).

Range of Validity of Darcy's Law In our development of the equation for flow through a bundle of conduits, it was assumed that the only "upstream"-directed force resisting the "downstream"-directed driving forces of gravity and pressure was due to the viscosity of the fluid (Equation 12-26). This assumption is appropriate as long as the motion of the fluid "particles" is exactly parallel to the axis of the conduit, that is, as long as the flow is strictly laminar. As in open-channel flows, the

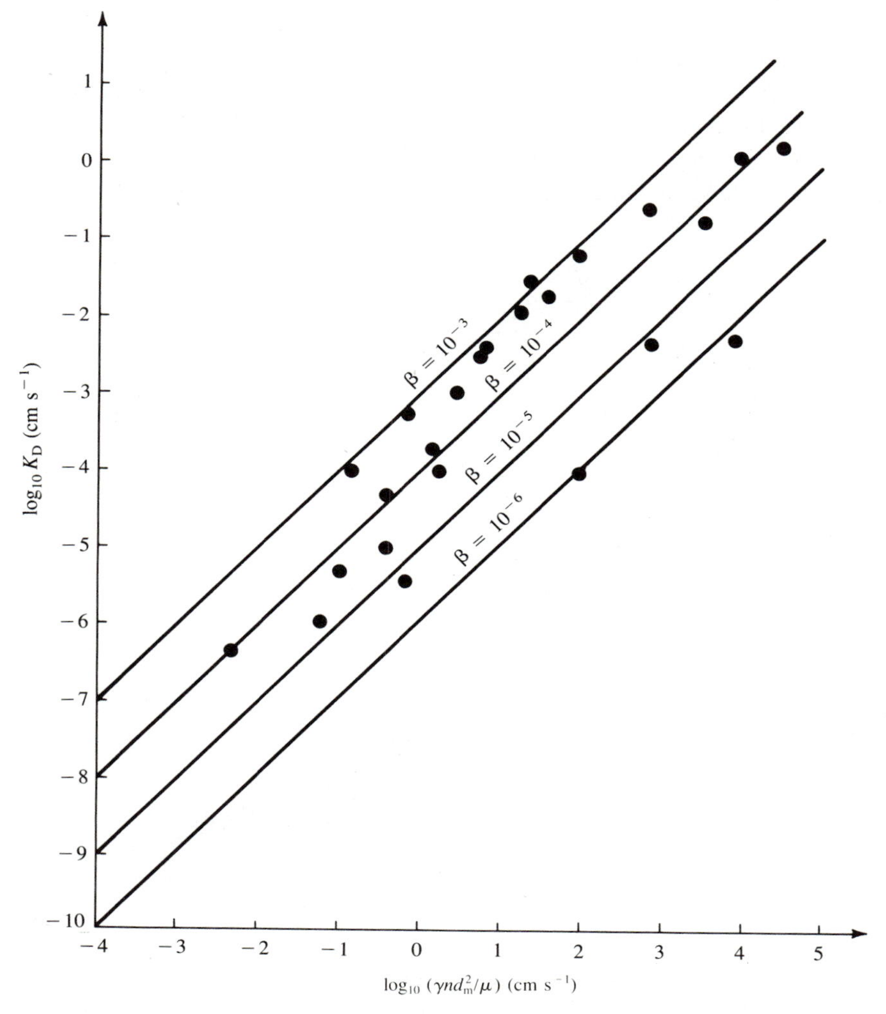

Figure 12.15
Plot of $\log_{10} K_D$ (cm s^{-1}) versus $\log_{10}(\gamma n d_m^2/\mu)$ (cm s^{-1}) for the data of Morris and Johnson (1967). Curves show relationships that would exist if β in Equation 12-51 had various constant values.

path of individual water particles becomes increasingly wavy as flow velocity increases (Figure 5.8), and movements of particles at right angles to the general flow direction cause a retarding effect on the flow. This resistance can be represented by a turbulent shear stress exerted in the upstream direction, which is added to the shear stress caused by viscous resistance. As in Equation 5-28 for open-channel flow, this force is proportional to the square of the velocity. When the velocity is very small, as it is for most natural porous media flows, these effects can be safely neglected. At higher velocities, however, the wavy motions and hence the turbulent forces become significant, and deviations from Darcy's law occur at velocities well below those required for turbulent flow.

The effects of turbulent forces can be detected in a plot of a friction factor f_p against the Reynolds number $\mathbb{R}_{pm}$ (Figure 12.16). These flow parameters have exactly the same physical significances for porous media flows as their analogs for open-channel flows, although their definitions are slightly different:

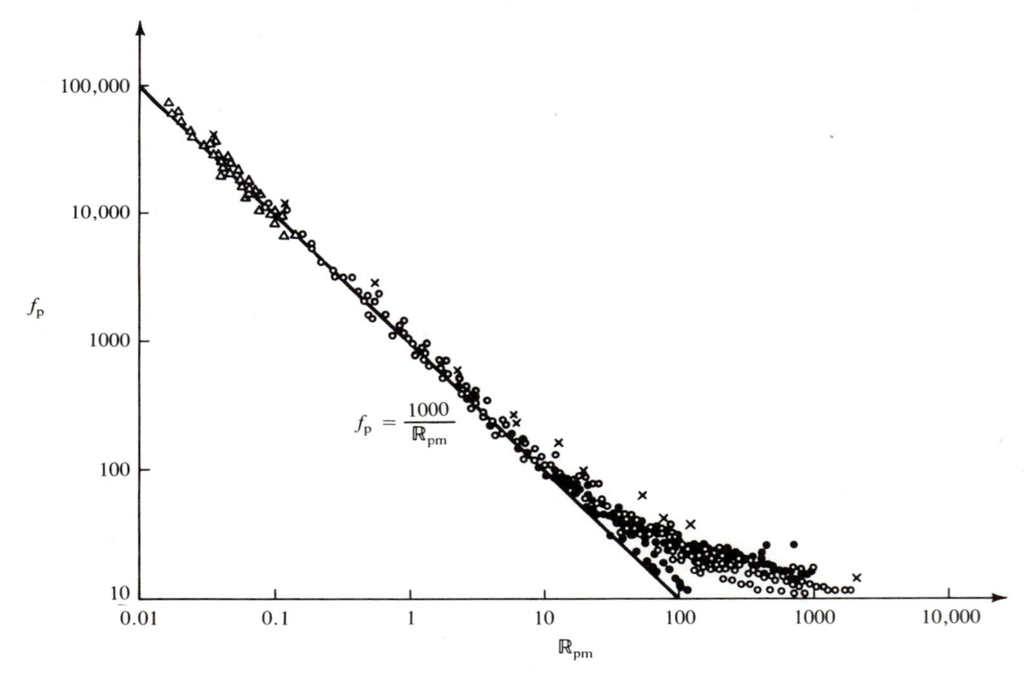

Figure 12.16
Plot of a friction factor f_p against the Reynolds number $\mathbb{R}_{pm}$ for a large number of groundwater flows. The factor f_p decreases with increasing $\mathbb{R}_{pm}$ in the laminar range ($\mathbb{R}_{pm} < 10$), and is independent of $\mathbb{R}_{pm}$ when flow is fully turbulent ($\mathbb{R}_{pm} > 1000$). Darcy's law begins to break down in the transitional regime ($10 < \mathbb{R}_{pm} < 1000$). From D. K. Todd, *Ground Water Hydrology,* p. 49. © 1959. Reprinted by permission of John Wiley and Sons, Inc., New York.

$$f_{\rm p} = \frac{gS_{\rm e}d_{\rm m}}{2q_{\rm s}^2}, \tag{12-54}$$

where S_e is the energy gradient ($\Delta H/L$) and the other terms are as previously defined (cf. Equation 8-7); and

$$\mathbb{R}_{\rm pm} \equiv \frac{q_{\rm s}d_{\rm m}}{\nu}, \tag{12-55}$$

where ν is kinematic viscosity.

Figure 12.16 shows that f_p and $\mathbb{R}_{pm}$ are related as

$$f_{\rm p} = \frac{1000}{\mathbb{R}_{\rm pm}} \tag{12-56}$$

at low Reynolds numbers, when flow is laminar. Above approximately $\mathbb{R}_{pm} = 10$, however, as the flow becomes transitional to the state of fully developed turbulence in which f_p is independent of $\mathbb{R}_{pm}$, Equation 12-56 no longer describes the relation. Thus a value of $\mathbb{R}_{pm} = 10$ is usually considered to be the upper limit of applicability of Darcy's law. Note that full turbulence does not become established until $\mathbb{R}_{pm} \simeq 1000$.

Under natural conditions $\mathbb{R}_{pm}$ only rarely exceeds 10. Exceptions may occur in flows through the relatively large openings that exist in some limestomes and volcanic rocks. When water is flowing radially to a well that is being pumped, however, there is considerable acceleration, and velocities may increase to the extent that $\mathbb{R}_{pm} > 10$ in the vicinity of the well.

THE EQUATION OF MOTION FOR SOIL-WATER FLOWS

Theory Combining Equations 12-12 and 12-44, we can express Darcy's law as

$$q_{\rm s} = -K_{\rm D}\frac{d(z + \psi)}{dl}. \tag{12-57}$$

From the previous discussion of energy in unsaturated media, we know that ψ is a function of the water content θ, with the exact functional relationship depending on the soil and to a lesser extent on its wetting history. This general dependence of ψ on θ is usually expressed by writing $\psi = \psi(\theta)$, so that Equation 12-57 becomes

$$q_{\rm s} = -K_{\rm D}\frac{d[z + \psi(\theta)]}{dl}. \tag{12-58}$$

The hydraulic conductivity K_D is also a function of θ, and we can gain some insight into the form of the relation between the two by considering again the flow through the bundle of conduits. After replacement of the gradient terms in Equation 12-34 by their derivative equivalents, the discharge Q_c through a single conduit of radius R is

$$Q_c = -\frac{\gamma \pi R^4}{8\mu} \frac{d[z + \psi(\theta)]}{dl} \tag{12-59}$$

if the conduit is full of water.

The cross-sectional area A_c of the conduit is

$$A_c = \pi R^2. \tag{12-60}$$

If the degree of saturation s is less than one, we can define an effective cross-sectional area A_c' as

$$A_c' = sA_c, \tag{12-61}$$

where A_c' is the area of the pore that contains water. With Equation 12-60, Equation 12-61 becomes

$$A_c' = s\pi R^2. \tag{12-62}$$

By rearrangement of Equation 12-60 we obtain

$$R = \left(\frac{A_c}{\pi}\right)^{1/2},$$

so it is reasonable to define an effective radius R' as

$$R' \equiv \left(\frac{A_c'}{\pi}\right)^{1/2}. \tag{12-63}$$

Substitution of Equation 12-62 into Equation 12-63 gives

$$R' = s^{1/2}R. \tag{12-64}$$

Since the discharge in Equation 12-59 is clearly controlled by the effective radius if the conduit is not full of water, that equation can be generalized by replacing R with R' $(= s^{1/2}R)$, so that it becomes

$$Q_c = -\frac{\gamma \pi s^2 R^4}{8\mu} \frac{d[z + \psi(\theta)]}{dl}. \tag{12-65}$$

Now Equations 12-36–12-39 can be utilized exactly as before to arrive at

$$q_s = -\frac{\gamma n s^2 R^2}{8\mu} \frac{d[z + \psi(\theta)]}{dl}. \tag{12-66}$$

Thus by analogy with our earlier discussion of saturated flow, the hydraulic conductivity K_{Du} for unsaturated flow in the bundle of conduits can be written as

$$K_{Du} = \frac{\beta_u \gamma n s^2 d_m^2}{\mu}, \tag{12-67}$$

where β_u is an empirical adjustment factor similar to β but taking into account additional factors relevant to unsaturated conditions.

From Equations 12-51 and 12-67,

$$\frac{K_{Du}}{K_D} = \left(\frac{\beta_u}{\beta}\right) s^2, \tag{12-68}$$

which indicates that the ratio of unsaturated hydraulic conductivity to saturated hydraulic conductivity for a given soil should be approximately proportional to the square of the degree of saturation. As shown in Figure 12.17, real porous media deviate from Equation 12-68, but this equation does capture the essential character of the relation to within a factor of 2 or so. In large part the differences occur because when saturation falls to the range of 0.2–0.4, the films of water become disconnected so that the hydraulic conductivity becomes essentially zero at considerably higher values than Equation 12-68 reflects.

Clearly θ is directly related to s for a given soil (Exercise 12-1), so that the hydraulic conductivity can be expressed as a function of water content, and $K_D = K_D(\theta)$. With this relation and Equation 12-58, the general equation of motion for flow through a porous medium, saturated or unsaturated, can be expressed as

$$q_s = -K_D(\theta) \frac{d[z + \psi(\theta)]}{dl}, \tag{12-69}$$

where the relations $K_D(\theta)$ and $\psi(\theta)$ must be determined by empirical measurements of the medium of interest. Hillel (1980) reviews several attempts at developing empirical equations relating hydraulic conductivity and water content, and describes in detail one theoretical approach to defining the relation.

Application A simple example of the use of Equation 12-69 and the principles of flow in unsaturated porous media can be given by reference to Figure 12.18. The water table is 160 cm above the datum and is horizontal, and the average height of the capillary fringe is 50 cm. Applying Equation 12-12, we find that the potential head at the water table is 160 + 0 = 160 cm. From Equation 12-14 the potential head at the top of the capillary fringe is 210 − 50 = 160 cm, so there is no gradient of potential energy in the saturated zone. At point A the tensiometer indicates a tension head of 320 cm, so the total head at that point is 440 − 320 = 120 cm. At point B the tension head is 480 cm, so the total head there is 520 − 480 = 40 cm. Thus there is a potential gradient from the top of the capillary fringe toward point A, and, assuming that the equipotential surfaces are approximately parallel, from point A upward.

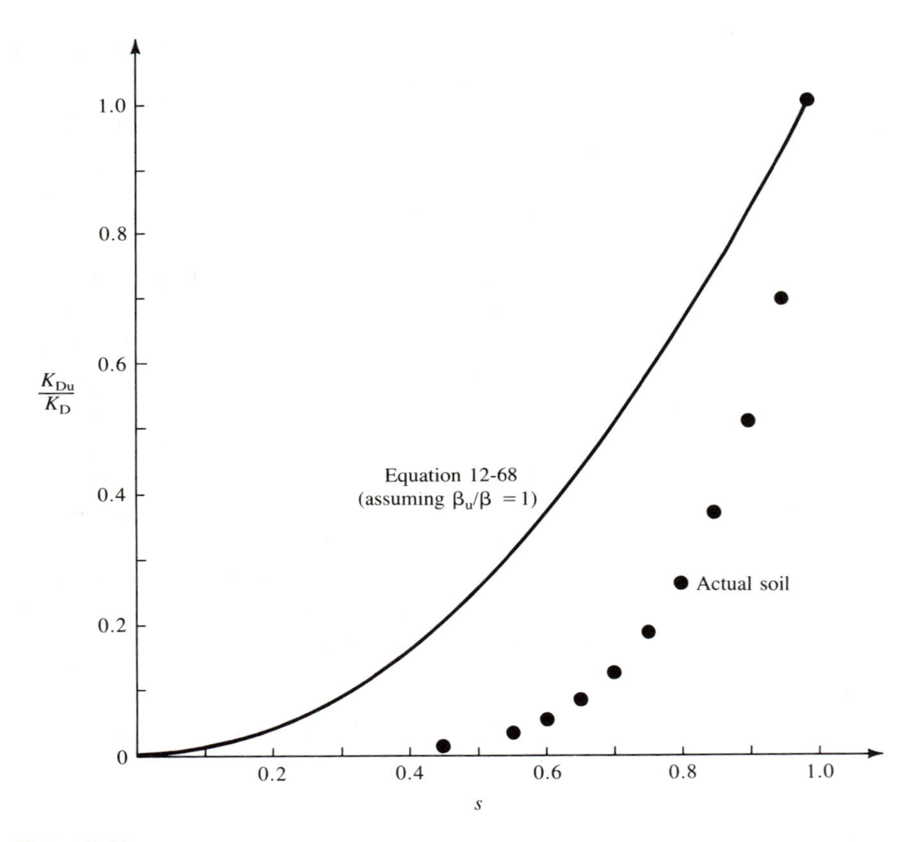

Figure 12.17
Comparison of K_{Du}/K_D versus s as computed from Equation 12-68 with the relation for an actual soil. Soil data from Hillel (1980).

If the soil in Figure 12.18 is homogeneous, the tensiometer readings indicate a general decrease in water content upward from the capillary fringe, such as might exist if transpiration were taking place above a shallow water table. Because of the relations among water content, tension, and hydraulic conductivity, the latter two must also decrease upward in this case. The application of Equation 12-69 will thus show a different flow rate at each elevation at any given time. Furthermore, if we consider a thin layer of the soil at any time, this flow-rate gradient means that the rate of flow into the layer from below must be different from the flow rate out through the top. Thus the water content in the layer must be changing with time,

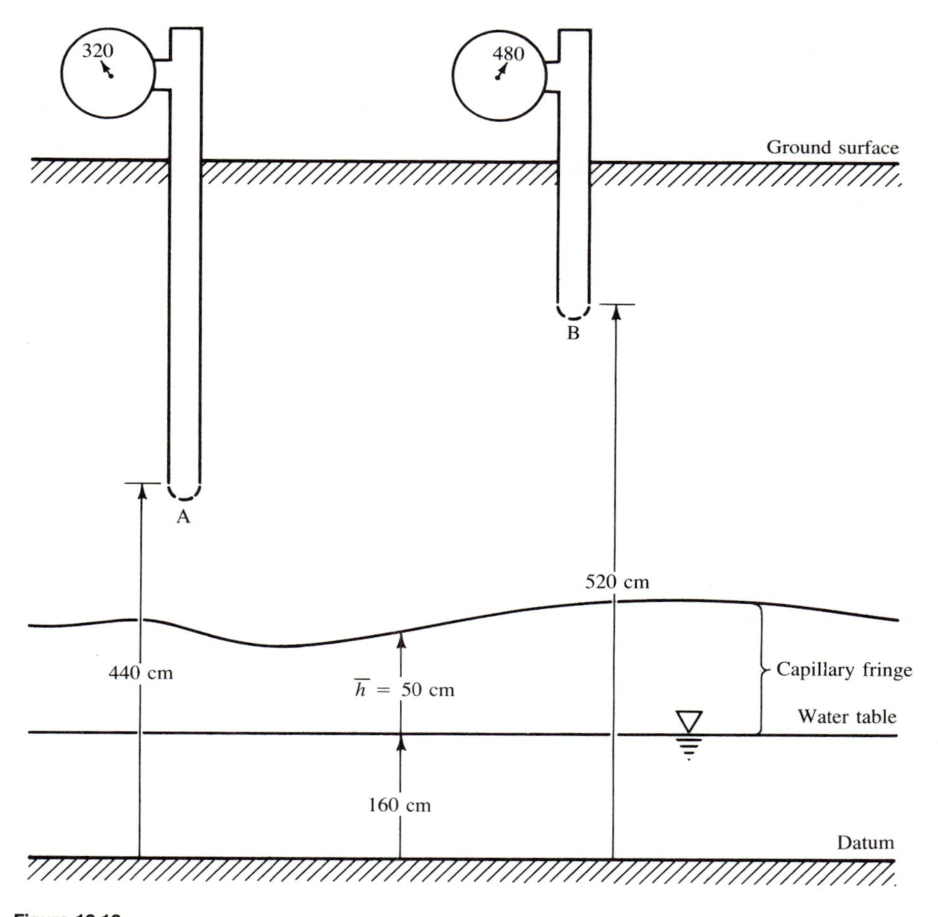

Figure 12.18
Example of energy relations in saturated and unsaturated zones in a typical field situation. Tensiometers measure tension at points A (320 cm) and B (480 cm); $\bar{h}$ is the average height of the capillary fringe. See computations in text.

with accompanying changes in the hydraulic conductivity and the tension gradient, as dictated by the nature of the appropriate $K_D(\theta)-\theta$ and $\psi(\theta)-\theta$ curves.

Thus it is clear that truly steady flow in unsaturated porous media is virtually impossible. The practical application of Equation 12-69 is therefore generally in the context of a model that can account at least approximately for the time and space changes that are inevitably associated with unsaturated flow, rather than in simple analytical computations. A brief introduction to the essentials of formulating such models is given at the end of this chapter.

COMPARISON OF GROUNDWATER AND SOIL-WATER FLOWS

Darcy's law, which has the form of a typical equation of motion as described in Chapter 3 (flow rate = potential gradient × conductivity), is the basic equation for both groundwater and soil-water flows. Its general form is given in Equation 12-69. With reference to that equation, and following Freeze and Cherry (1979), the fundamental distinguishing features of the two types of flow are summarized in Table 12.3.

Table 12.3

Groundwater flows	Soil-water flows
Soil pores are completely filled with water, so that the water content (θ) equals the porosity (n).	Soil pores are only partly filled with water, so that $\theta < n$.
Fluid pressures are greater than atmospheric, so the pressure head (ψ) is greater than zero.	Fluid pressures are less than atmospheric, so $\psi < 0$.
The hydraulic head ($H = z + \psi$, where z is the elevation above a datum) is measured with a piezometer.	The pressure head ψ is measured with a tensiometer, and H is computed by algebraically adding ψ and z.
The hydraulic conductivity (K_D) at any point remains constant.	The hydraulic conductivity K_D and pressure head ψ are both functions of θ.

MODELING FLOW THROUGH POROUS MEDIA

Recall from Chapter 11 that the complete dynamic equations for one-dimensional open-channel flow (the Saint-Venant equations) were derived from the continuity equation and the momentum equation. It turned out, however, that useful approximations of the complete equations could be written for flows in which accelerations and depth changes are negligible by using the continuity equation and replacing the complete momentum equation with a simple equation of motion of the Manning type. The combination of the continuity equation and the equation of motion resulted in a single equation describing the flow: Equation 11-87 (the kinematic-wave approximation).

The rates of change of depth and the convective and local accelerations in porous media flows are usually so small as to be negligible. Thus, as in the kinematic-wave approximation, the complete equation for flow can also be developed by combining the equation of motion (Darcy's law) with the continuity equation. The next part of this section develops the continuity equation for porous media flows, and since such flows can only rarely be described as one-dimensional, the development will consider all three spatial dimensions. The final section of this chapter describes how the equations of continuity and motion are combined in order to model flow through porous media.

The Continuity Equation To develop the continuity or conservation equation for porous media flows, the principles discussed in Chapter 3 are applied in the formulation of expressions for the amounts of water entering and leaving a representative volume element of the medium in a given time period and for the change in the amount of water stored in the volume during that period. Those expressions are then substituted into the general continuity equation:

$$\text{mass entering} - \text{mass leaving} = \text{change in mass stored.} \tag{12-70}$$

Figure 12.19 shows an isolated element of a porous medium in the shape of a rectangular parallelepiped. The sides of the element are oriented such that they are parallel to the x, y, and z axes of a three-dimensional Cartesian coordinate system, and have lengths dx, dy, and dz, respectively. These lengths are small, but the element is large enough to be representative of the medium as a whole (i.e., it is larger than several pores).

The flow of water through such an element can be in any direction, but the specific-discharge vector at any location can be resolved into vectors parallel to each of the axes, as shown in Figure 12.19. Thus the rate of flow into the element is the sum of the rates of flow into faces 1, 2, and 3, designated q_x, q_y, and q_z, respectively:

$$\text{rate of flow into element} = q_x + q_y + q_z. \tag{12-71}$$

Multiplying each of these inflow components by the area of the appropriate face, the time period being considered (designated dt), and the mass density of water (ρ) gives the mass of water entering the element during that period:

$$\begin{matrix}\text{mass of water} \\ \text{entering} \\ \text{volume element}\end{matrix} = \rho q_x \, dy \, dz \, dt + \rho q_y \, dz \, dx \, dt + \rho q_z \, dx \, dy \, dt. \tag{12-72}$$

In general, the specific discharge changes with distance, so if the rate of flow out of the element is resolved into components representing flow out of faces 4, 5, and 6, we have

$$\begin{matrix}\text{rate of flow} \\ \text{out of element}\end{matrix} = \left(q_x + \frac{\partial q_x}{\partial x}dx\right) + \left(q_y + \frac{\partial q_y}{\partial y}dy\right) + \left(q_z + \frac{\partial q_z}{\partial z}dz\right), \tag{12-73}$$

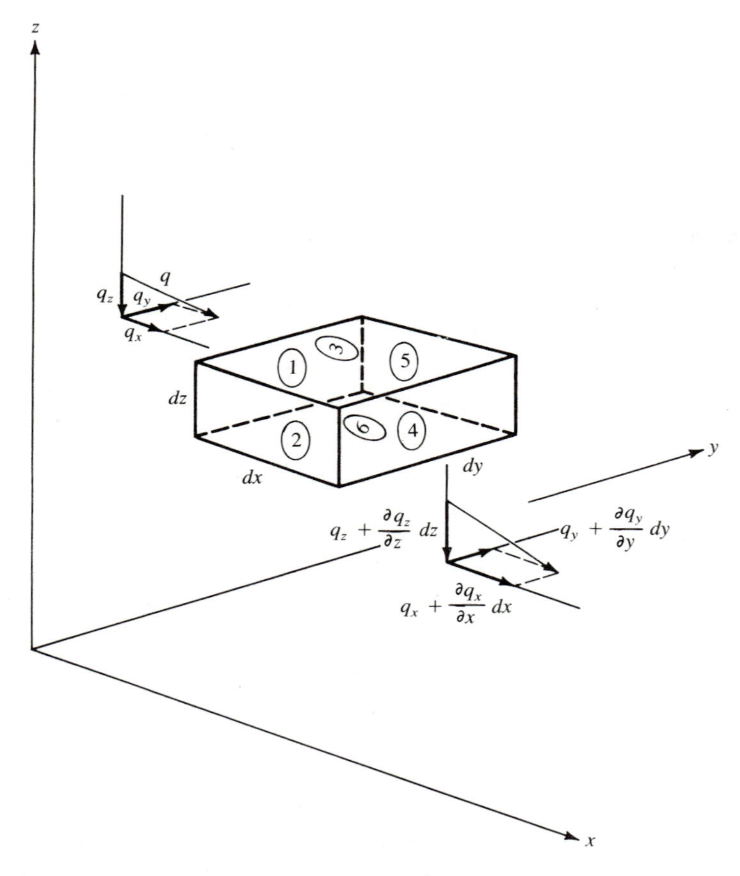

Figure 12.19
Definitions of terms for the derivation of the continuity equation for porous media flows (Equations 12-70–12-78).

and consequently

$$\begin{aligned}\text{mass of water leaving volume element} &= \rho\left(q_x + \frac{\partial q_x}{\partial x}dx\right) dy\, dz\, dt \\ &+ \rho\left(q_y + \frac{\partial q_y}{\partial y}dy\right) dz\, dx\, dt \\ &+ \rho\left(q_z + \frac{\partial q_z}{\partial z}dz\right) dx\, dy\, dt. \end{aligned} \tag{12-74}$$

The change in storage will be reflected by a change of mass of water in the volume. Multiplying the time rate of change of water content times the length of the time period, the volume of the element, and the mass density (assumed constant) gives this value:

$$\text{change in mass of water in volume element} = \rho\frac{\partial \theta}{\partial t}dt\, dx\, dy\, dz. \tag{12-75}$$

After substitution of Equations 12-72, 12-74, and 12-75 into Equation 12-70 and simplification, we arrive at

$$-\left(\frac{\partial q_x}{\partial x} + \frac{\partial q_y}{\partial y} + \frac{\partial q_z}{\partial z}\right) = \frac{\partial \theta}{\partial t} \tag{12-76}$$

as the three-dimensional continuity equation for a point in a porous medium. From the definitions of n, θ, and s (Equations 12-1, 12-5, and 12-6),

$$\theta = ns, \tag{12-77}$$

and application of the rules of derivatives gives

$$\frac{\partial \theta}{\partial t} = \frac{\partial (ns)}{\partial t} = n\frac{\partial s}{\partial t} + s\frac{\partial n}{\partial t} \tag{12-78}$$

as an alternative form of the right-hand side of Equation 12-76.

Equation 12-76 applies to steady and unsteady flows of soil water and confined and unconfined groundwater, but interpretations of the term on the right-hand side differ under the various conditions. Clearly, for steady-state conditions in all flows, $\partial\theta/\partial t = 0$. We have seen, however, that steady conditions do not occur in soil-water flows; in these flows it is assumed that porosity does not change with time, and the change in water content is interpreted straightforwardly as $\partial\theta/\partial t$ or as $n(\partial s/\partial t)$.

In groundwater flows the degree of saturation equals one, so $\partial s/\partial t = 0$ at all points except those through which the water table passes during the period dt. Thus for most points in unconfined flows and all points in confined flows,

$$\frac{\partial \theta}{\partial t} = \frac{\partial n}{\partial t}. \tag{12-79}$$

The only way that porosity can change with time is for the granular structure of the medium to change. Such changes do occur in response to changes in hydraulic head, which in turn cause the total weight supported by the medium to vary. If this weight increases, the grains may be forced together and the porosity thereby reduced. (The alarming ground subsidence in cities such as Houston, Texas, and Venice, Italy, is due to this effect, which was induced by high rates of groundwater extraction.) A decrease in weight carried by the medium may bring about some increase in porosity, but the changes in this direction are usually smaller, especially for finer-grained media.

Although changes in porosity are of small magnitude, they must be considered in the modeling of groundwater flows. The compressibility of some media, especially those of large grain size (such as gravels), is very small and is of approximately the same order of magnitude as the compressibility of water. In these cases both compressibilities must be included in formulating the continuity equation. More complete discussions of these aspects of porous media can be found in most groundwater texts; Freeze and Cherry (1979) give a particularly clear and comprehensive treatment.

Complete Equation of Porous Media Flow Darcy's law applies to flow in any direction, so

$$q_x = -K_{Dx}\frac{\partial H}{\partial x}, \tag{12-80a}$$

$$q_y = -K_{Dy}\frac{\partial H}{\partial y}, \tag{12-80b}$$

and

$$q_z = -K_{Dz}\frac{\partial H}{\partial z}, \tag{12-80c}$$

where K_{Dx}, K_{Dy}, and K_{Dz} are the hydraulic conductivities in the x, y, and z directions. Substituting Equations 12-80a–c into Equation 12-76 gives

$$\frac{\partial}{\partial x}\left(K_{Dx}\frac{\partial H}{\partial x}\right) + \frac{\partial}{\partial y}\left(K_{Dy}\frac{\partial H}{\partial y}\right) + \frac{\partial}{\partial z}\left(K_{Dz}\frac{\partial H}{\partial z}\right) = \frac{\partial \theta}{\partial t}. \quad \text{(12-81a)}$$

If the conductivities do not vary with location, Equation 12-81a can be simplified to

$$K_{Dx}\frac{\partial^2 H}{\partial x^2} + K_{Dy}\frac{\partial^2 H}{\partial y^2} + K_{Dz}\frac{\partial^2 H}{\partial z^2} = \frac{\partial \theta}{\partial t}. \quad \text{(12-81b)}$$

A further simplification is possible if the hydraulic conductivity K_D has the same value in all directions:

$$K_D\left(\frac{\partial^2 H}{\partial x^2} + \frac{\partial^2 H}{\partial y^2} + \frac{\partial^2 H}{\partial z^2}\right) = \frac{\partial \theta}{\partial t}. \quad \text{(12-81c)}$$

Equations 12-81 are three forms of the complete equation for flow in porous media. As written, these equations have two dependent variables (H and θ), four independent variables (x, y, z, and t), and up to three parameters (the K_D values). However, the number of dependent variables can be reduced to one by making use of relationships between H and θ that apply to groundwater and soil-water flows.

For groundwater flows, it can be shown that a linear relationship exists between porosity and a parameter of the medium called the *specific storage* S_s (see Freeze and Cherry, 1979), which is defined as the volume of water that is released from a porous medium when the hydraulic head is decreased by a unit amount, divided by the horizontal area of the medium. Because of this linear relationship and Equation 12-79,

$$\frac{\partial \theta}{\partial t} = \frac{\partial n}{\partial t} = S_s\frac{\partial H}{\partial t}. \quad \text{(12-82)}$$

If we use the simplest version of the complete equation (Equation 12-81c) as an example, the substitution of Equation 12-82 yields

$$K_D\left(\frac{\partial^2 H}{\partial x^2} + \frac{\partial^2 H}{\partial y^2} + \frac{\partial^2 H}{\partial z^2}\right) = S_s\frac{\partial H}{\partial t} \quad \text{(12-83)}$$

as the usual form of the complete equation for groundwater flows. As noted earlier, solutions to Equations 12-81 under steady conditions ($\partial\theta/\partial t = 0$) are the basis for graphical flow-net models of groundwater flows, as described by Cedergren (1967) and Freeze and Cherry (1979).

In soil-water flows it is usually necessary to use Equation 12-81a, since we have seen that such flows generally have gradients of water content, and hence of hydraulic conductivity, in the direction of flow. It is generally safe, however, to assume that the relation between hydraulic conductivity and water content is the same in all directions, so that

$$K_{Dx} = K_{Dy} = K_{Dz} = K_D(\theta). \tag{12-84}$$

If the z axis in Figure 12.19 is oriented vertically, we can write

$$H = z + \psi(\theta), \tag{12-85}$$

so that it is also true that

$$\frac{\partial H}{\partial x} = \frac{\partial z}{\partial x} + \frac{\partial \psi(\theta)}{\partial x} = 0 + \frac{\partial \psi(\theta)}{\partial x}, \tag{12-86a}$$

$$\frac{\partial H}{\partial y} = \frac{\partial z}{\partial y} + \frac{\partial \psi(\theta)}{\partial y} = 0 + \frac{\partial \psi(\theta)}{\partial y}, \tag{12-86b}$$

and

$$\frac{\partial H}{\partial z} = \frac{\partial z}{\partial z} + \frac{\partial \psi(\theta)}{\partial z} = 1 + \frac{\partial \psi(\theta)}{\partial z}. \tag{12-86c}$$

Substituting Equations 12-84 and 12-86 into Equation 12-81a then gives

$$\frac{\partial}{\partial x}\left[K_D(\theta)\frac{\partial \psi(\theta)}{\partial x}\right] + \frac{\partial}{\partial y}\left[K_D(\theta)\frac{\partial \psi(\theta)}{\partial y}\right] + \frac{\partial}{\partial z}\left[K_D(\theta) + K_D(\theta)\frac{\partial \psi(\theta)}{\partial z}\right] = \frac{\partial \theta}{\partial t} \tag{12-87}$$

as the operational form of the complete equation for soil-water flows. Equation 12-87 is commonly called the *Richards equation,* after the American soil physicist A. L. Richards who first derived it in 1931. Note that if we know the relations between $K_D(\theta)$ and θ and $\psi(\theta)$ and θ for the particular soil of interest, Equation 12-87 is effectively written in terms of a single dependent variable, θ.

It is sometimes helpful to transform Equation 12-87 into an alternate form that makes it analogous to equations of diffusion, such as were discussed in Chapters 3 (Equation 3-21) and 11 (Equation 11-75). This is done by defining a *hydraulic diffusivity* D_θ as the product of the hydraulic conductivity and the slope of the soil-water characteristic curve:

$$D_\theta(\theta) \equiv K_D(\theta)\frac{\partial\psi(\theta)}{\partial\theta}. \tag{12-88}$$

Note that $D_\theta(\theta)$ has the appropriate dimensions for a diffusivity, $[L^2T^{-1}]$. Using the symbol l to represent any of the three coordinate directions x, y, and z, we have

$$\frac{\partial\psi(\theta)}{\partial l} = \frac{\partial\psi(\theta)}{\partial\theta}\frac{\partial\theta}{\partial l}. \tag{12-89}$$

Thus Equations 12-88 and 12-89 can be substituted into Equation 12-87 to give

$$\frac{\partial}{\partial x}\left[D_\theta(\theta)\frac{\partial\theta}{\partial x}\right] + \frac{\partial}{\partial y}\left[D_\theta(\theta)\frac{\partial\theta}{\partial y}\right] + \frac{\partial}{\partial z}\left[D_\theta(\theta)\frac{\partial\theta}{\partial z}\right] + \frac{\partial K_D(\theta)}{\partial z} = \frac{\partial\theta}{\partial t}. \tag{12-90}$$

By further noting that

$$\frac{\partial K_D(\theta)}{\partial z} = \frac{\partial K_D(\theta)}{\partial\theta}\frac{\partial\theta}{\partial z}, \tag{12-91}$$

we see that Equation 12-90 can be written explicitly in terms of the single dependent variable θ:

$$\frac{\partial}{\partial x}\left[D_\theta(\theta)\frac{\partial\theta}{\partial x}\right] + \frac{\partial}{\partial y}\left[D_\theta(\theta)\frac{\partial\theta}{\partial y}\right] + \frac{\partial}{\partial z}\left[D_\theta(\theta)\frac{\partial\theta}{\partial z}\right] + \frac{\partial K_D(\theta)}{\partial\theta}\frac{\partial\theta}{\partial z} = \frac{\partial\theta}{\partial t}, \tag{12-92}$$

where $\partial K_D(\theta)/\partial\theta$ is the slope of the hydraulic-conductivity–water-content relation for the soil. Note that use of hydraulic diffusivity implies an alternate form of Darcy's law for soil-water flows:

$$q_l = -D_\theta(\theta)\frac{d\theta}{dl}, \tag{12-93}$$

which is precisely the form of the basic diffusion equation (Equation 3-21).

The advantage of using Equation 12-92 rather than Equation 12-87 is that the range of variation of $D_\theta(\theta)$ is less than the range of variation of $K_D(\theta)$, and the relation between $D_\theta(\theta)$ and θ can often be represented by a simple empirical equation (Hillel, 1980). However, the approach does not readily accommodate the hysteresis of the relation between $\psi(\theta)$ and θ that is particularly marked in fine-grained soils.

Solutions to the Complete Equations As with most of the equations that realistically describe physical processes, the various forms of the complete equation for flow through porous media cannot be solved analytically. In order to model a given flow situation, the equations must be rewritten in algebraic form as difference equations and solved by numerical methods on a digital computer. In general, these methods involve subdividing the flow region by means of a two- or three-dimensional grid, and iteratively solving the transformed equation for H or ψ at each grid intersection. If the problem involves groundwater flow, the appropriate values of K_D and S_s must be assigned to each grid point; if it involves soil-water flow, we must specify the relation between $\psi(\theta)$ and θ as well as that between $K_D(\theta)$ and θ [or $D_\theta(\theta)$ and θ] at each point. As with the Saint-Venant equations for open-channel flow (see Figure 11.8), we have to begin the computation by specifying the initial conditions (i.e., the values of H or ψ at each grid point when $t = 0$) and the boundary conditions (i.e., the head values at the grid points that represent the boundaries of the flow, and whether these boundaries are impermeable, or leaky, or surfaces at atmospheric pressure). Wang and Anderson (1982) provide an excellent introduction to the modeling of groundwater flows and the numerical solution of the equations of flow through porous media, and Hillel (1977) has compiled a number of approaches to numerical modeling of soil-water flows.

Exercises

12-1. Show that Equations 12-3 and 12-4 are correct.

12-2. By using the definitions of its terms, show that Equation 12-77 is correct.

12-3. Suppose you have a small cylindrical sample of a granular porous medium with a radius of 10 cm and a height of 20 cm. The mass density of the mineral grains is 2.65 g cm^{-3}. (This is the value for the mineral quartz, and is commonly used for earth materials.) Virtually all the moisture in the sample is driven out by heating it at 105°C for several hours. After this drying, the sample weighs 5528 g. What is its porosity?

12-4. Equation 12-9 provides a good estimate of the height of capillary rise in porous media, where R is replaced by $d_m/2$ and θ_c is taken as 0°. Using this information, compute the thickness of the capillary fringe for the following media. Assume a temperature of 10°C:

	Clay	Silt	Fine sand	Medium sand	Coarse sand	Gravel
d_m (cm)	0.0001	0.001	0.01	0.04	0.15	1.0

Note that the heights you computed are also estimates of the air-entry tension head for those media (Equation 12-10).

12-5. Suppose you were to "model" a soil pore as a cone with the dimensions shown in Figure 12.20a. Using the equation for the volume of a cone, you could compute the volume of water in the cone as the pore dries and the meniscus moves successively to the left and its radius decreases (Figure 12.20b). If the pore diameter d_m is 0.4 mm (typical of a medium sand), this relation gives these values:

	r (mm)											
	0	0.01	0.02	0.03	0.04	0.05	0.10	0.125	0.150	0.175	0.190	0.200
s	0	0.00012	0.0010	0.0034	0.0080	0.016	0.125	0.244	0.422	0.670	0.857	1.00

where r is the radius corresponding to given degrees of saturation s. Assuming a porosity of 0.3, compute the ψ versus θ relation for this soil by using Equation 12-13b. (Values of σ and γ are given in Table 4.2; assume a temperature of 10°C.) Plot $|\psi|$ versus θ using two-cycle semilogarithmic paper, with $|\psi|$ plotted on the logarithmic axis. Compare the form of your computed curve with soil-water characteristic curves measured for some actual soils and published in the soil physics literature.

12-6. See whether the ψ–θ relationship you derived in Exercise 12-5 can be well represented by the empirical Equation 12-17. To do this, assume that $\theta_r = 0$ (your plot for Exercise 12-5 should indicate that it is a very small number); ψ_{ae} is the value you computed in Exercise 12-5 for $s = 1.00$. Plot ψ/ψ_{ae} against n/θ on log–log paper (you will need four cycles to accommodate the n/θ values). If Equation 12-17 applies, the points should define a straight line; the value of λ is the inverse of the slope of this line.

12-7. Use the value $\beta = 10^{-4}$ to compute the approximate hydraulic conductivities of the following materials with Equation 12-51. Assume a temperature of 10°C.

	Clay	Silt	Fine sand	Medium sand	Coarse sand	Gravel
n	0.50	0.45	0.35	0.32	0.30	0.28
d_m (mm)	0.001	0.01	0.10	0.40	1.5	10

12-8. Some insight into the relative importance of contributions of pores of various sizes to the total flow through a porous medium can be obtained from Equation 12-34. Select any arbitrary value for the driving-force term in that equation and compute Q_c for pores of radii of 0.20, 0.40, 0.60, 0.80, and 1.00 mm. Suppose pores of those radii were present in equal proportions in a real medium. What percentage of the total flow through the medium would be occurring in pores of each of those sizes?

12-9. Use Equation 12-68 to construct a graph of the relation between K_{Du}/K_D and θ for the model in Exercise 12-5. Assume that $\beta_u/\beta = 1$.

12-10. The following table gives readings from two adjacent tensiometers inserted to the indicated depths in the soil represented by the ψ–θ curve you computed in Exercise 12-5.

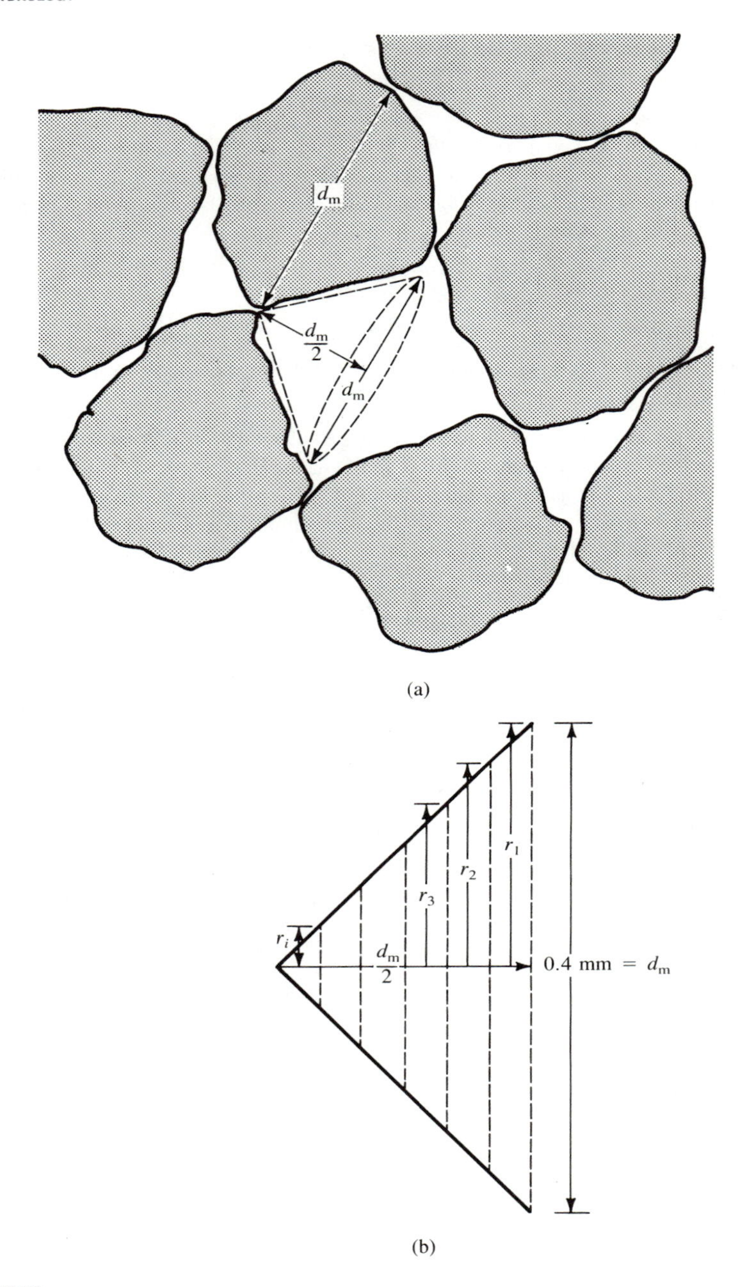

Figure 12.20
Model of water relations in a soil pore for Exercise 12-5. (a) Representation of a soil pore as a right circular cone; d_m is average pore diameter. (b) Successive values $r_1, r_2, \ldots, r_i$ of the radius of the meniscus as the degree of saturation s decreases.

Each set of readings was taken on a different day. For each case, determine whether the flow is upward or downward and give the water content at each tensiometer.

Tensiometer	Depth (cm)	Tensions (cm) Day 1	Day 2	Day 3	Day 4	Day 5
A	30	8.0	10.0	40.0	12.0	50.0
B	60	11.0	30.0	8.0	8.0	12.0

12-11. *Permeameters* are laboratory devices used in measuring hydraulic conductivities in samples of porous media. (These are described in detail in most groundwater texts.) The data in the accompanying table are actual permeameter measurements from a sample of the sphagnum moss that covers much of the surface of an Alaskan watershed. A volume of water V passes through the sample in time t under various hydraulic gradients. The sample is cylindrical, with a radius of 4.4 cm and a height (L) of 9.5 cm. The hydraulic head is measured at the two ends of the sample, and ΔH is the difference between them. The hydraulic gradient is thus $\Delta H/L$. (a) Use these data to compute K_D for this material at each value of $\Delta H/L$. (b) Plot K_D versus $\Delta H/L$ on log–log paper and determine the value of K_D that can be used in Darcy's law. (c) At what value of $\Delta H/L$ does the K_D versus $\Delta H/L$ relation change? The apparent decrease of K_D with $\Delta H/L$ above this value reflects the increasing influence of inertial forces and the inapplicability of Darcy's law at these higher flow rates.

V (cm^3)	t (s)	ΔH (cm)
100	21.9	7.95
100	25.2	6.1
100	29.6	5.0
100	38.4	3.1
100	39.1	3.0
100	56.9	1.7
100	97.4	0.8
70	100.0	0.4
42	100.0	0.2
33	100.0	0.2
57	120.0	0.3
16.8	100.0	0.1

APPENDIX A

Derivation of the Diffusion-Analogy Form of the Unsteady Gradually Varied Flow Equation (Equation 11-71)

To simplify the development, we consider the case when $q = 0$. Then the continuity equation (Equation 11-30) can be written as

$$w\frac{\partial Y}{\partial t} = -\frac{\partial Q}{\partial x}. \tag{A-1}$$

Also for simplicity, we can define

$$\mathcal{S} \equiv 1 - \frac{1}{S_0}\frac{\partial Y}{\partial x}. \tag{A-2}$$

With Equation A-2, Equation 11-69 becomes

$$Q = Q_n\mathcal{S}^{1/2}. \tag{A-3}$$

Both Q_n and $\mathcal{S}$ are functions of depth, so by the rules of derivatives of products,

$$\frac{\partial Q}{\partial Y} = Q_n\frac{\partial(\mathcal{S}^{1/2})}{\partial Y} + \mathcal{S}^{1/2}\frac{\partial Q_n}{\partial Y}. \tag{A-4}$$

Also from the rules of derivatives,

$$\frac{\partial(\mathcal{S}^{1/2})}{\partial Y} = \frac{\mathcal{S}^{-1/2}}{2}\frac{\partial \mathcal{S}}{\partial Y} \quad \text{(A-5)}$$

and

$$\frac{\partial \mathcal{S}}{\partial Y} = \frac{\partial \mathcal{S}}{\partial x}\frac{\partial x}{\partial Y}. \quad \text{(A-6)}$$

From the definition of $\mathcal{S}$ (Equation A-2),

$$\frac{\partial \mathcal{S}}{\partial x} = -\frac{1}{S_0}\frac{\partial^2 Y}{\partial x^2}. \quad \text{(A-7)}$$

Substituting Equation A-7 into Equation A-6 and the result into Equation A-5 yields

$$\frac{\partial(\mathcal{S}^{1/2})}{\partial Y} = \frac{\mathcal{S}^{-1/2}}{2}\left(\frac{1}{S_0}\right)\frac{\partial^2 Y}{\partial x^2}\frac{\partial x}{\partial Y}. \quad \text{(A-8)}$$

Since $\partial x/\partial Y = 1/(\partial Y/\partial x)$, Equation A-8 can equally well be written as

$$\frac{\partial(\mathcal{S}^{1/2})}{\partial Y} = \frac{\mathcal{S}^{-1/2}}{2}\left(\frac{1}{S_0}\right)\frac{\partial^2 Y}{\partial x^2}\left(\frac{1}{\partial Y/\partial x}\right). \quad \text{(A-9)}$$

Equation A-9 can now be substituted into Equation A-4:

$$\frac{\partial Q}{\partial Y} = -Q_n\left(\frac{1}{2\mathcal{S}^{1/2}S_0}\right)\frac{\partial^2 Y}{\partial x^2}\left(\frac{1}{\partial Y/\partial x}\right) + \mathcal{S}^{1/2}\frac{\partial Q_n}{\partial Y}. \quad \text{(A-10)}$$

Returning to Equation A-1, we call on the chain rule of derivatives in order to write

$$w\frac{\partial Y}{\partial t} = -\frac{\partial Q}{\partial Y}\frac{\partial Y}{\partial x}. \quad \text{(A-11)}$$

Substituting Equation A-10 into Equation A-11, we obtain

$$w\frac{\partial Y}{\partial t} = -\left[-Q_n\left(\frac{1}{2\mathcal{S}^{1/2}S_0}\right)\frac{\partial^2 Y}{\partial x^2}\left(\frac{1}{\partial Y/\partial x}\right) + \mathcal{S}^{1/2}\frac{\partial Q_n}{\partial x}\right]\frac{\partial Y}{\partial x}$$

$$= \left(\frac{Q_n}{2\mathcal{S}^{1/2}S_0}\right)\frac{\partial^2 Y}{\partial x^2} - \mathcal{S}^{1/2}\frac{\partial Q_n}{\partial Y}\frac{\partial Y}{\partial x}. \tag{A-12}$$

Further rearrangement and substitution of Equation A-2 yields

$$\frac{\partial Y}{\partial t} + \left(\frac{1}{w}\right)\left(1 - \frac{1}{S_0}\frac{\partial Y}{\partial x}\right)^{1/2}\frac{\partial Q_n}{\partial Y}\frac{\partial Y}{\partial x} - \left(\frac{1}{w}\right)$$

$$\frac{Q_n}{2[1 - (1/S_0)(\partial Y/\partial x)]^{1/2}S_0}\frac{\partial^2 Y}{\partial x^2} = 0. \tag{A-13}$$

Equation A-13 is the diffusion-analogy equation in its complete form. In order to transform it into its working form, we assume at this point that $\partial Y/\partial x$ (the change in depth with distance) is very small relative to the channel slope S_0. Thus from Equation A-2,

$$\mathcal{S} = 1 - \frac{1}{S_0}\frac{\partial Y}{\partial x} \simeq 1 \tag{A-14}$$

and Equation A-13 reduces to

$$\frac{\partial Y}{\partial t} + \frac{1}{w}\frac{\partial Q_n}{\partial Y}\frac{\partial Y}{\partial x} - \frac{Q_n}{2wS_0}\frac{\partial^2 Y}{\partial x^2} = 0 \tag{A-15}$$

which is identical to Equation 11-71.

APPENDIX B

Conditions under Which the Diffusion Term in Unsteady Gradually Varied Flow Can Be Neglected

Equation 11-75 gives the rate of decrease of the crest of a flood wave, $\partial Y/\partial t|_{\text{crest}}$, when there is no lateral inflow or outflow and acceleration terms can be neglected:

$$\left.\frac{\partial Y}{\partial t}\right|_{\text{crest}} = \frac{Q_{\text{n}}}{2wS_0}\frac{\partial^2 Y}{\partial x^2}. \tag{B-1}$$

Thus $\partial Y/\partial t$ is of negligible magnitude if the product of $Q_{\text{n}}/2wS_0$ and $\partial^2 Y/\partial x^2$ is negligible. As noted in Chapter 11, $Q_n/2wS_0$ is usually large in natural channels. We now want to gain some insight into the magnitude of $\partial^2 Y/\partial x^2$, to assist in judging when the diffusion term can be neglected and the kinematic-wave approximation (Equation 11-87) can be applied.

In order to simplify the development, we can generalize the shape of a flood wave as a sine wave, so that

$$Y = a \sin bx, \tag{B-2}$$

where a is the amplitude of the wave and b is related to the wavelength λ as

$$b = \frac{2\pi}{\lambda} \tag{B-3}$$

(see Figure B.1). Substituting Equation B-3 into Equation B-2 yields

$$Y = a \sin\left(\frac{2\pi}{\lambda} x\right). \tag{B-4}$$

Equation B-4 can now be differentiated twice:

$$\frac{dY}{dx} = \left(\frac{2\pi a}{\lambda}\right) \cos\left(\frac{2\pi}{\lambda} x\right);$$

$$\frac{d^2 Y}{dx^2} = \left(-\frac{4\pi^2 a}{\lambda^2}\right) \sin\left(\frac{2\pi}{\lambda} x\right). \tag{B-5}$$

From Figure B.1 we see that the crest of a sine wave occurs at $x = \lambda/4$. Substituting this value for x into Equation B-5 yields

$$\left.\frac{d^2 Y}{dx^2}\right|_{\text{crest}} = \left(-\frac{4\pi^2 a}{\lambda^2}\right) \sin\left(\frac{\pi}{2}\right). \tag{B-6}$$

From the characteristics of sine waves, $\sin(\pi/2) = 1$, so Equation B-6 becomes

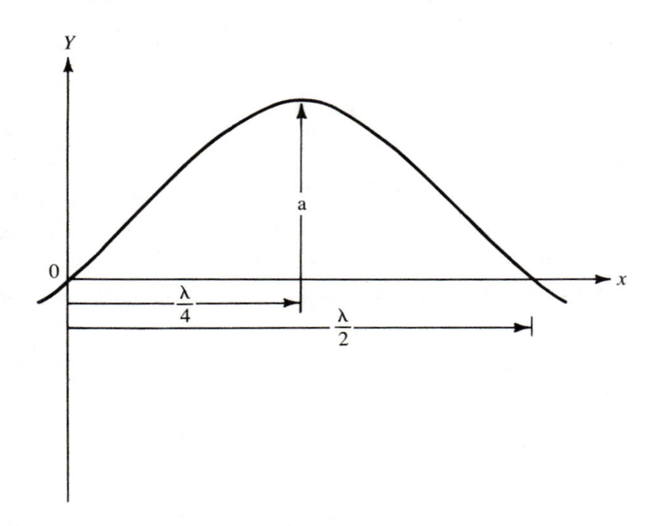

Figure B.1
Definitions of terms for modeling a flood wave as one-half a sine wave (Equations B-2–B-7).

$$\left.\frac{d^2Y}{dx^2}\right|_{\text{crest}} = -\frac{4\pi^2 a}{\lambda^2} = -\frac{39.48a}{\lambda^2} \simeq -\frac{40a}{\lambda^2}. \tag{B-7}$$

Equation B-7 can be used to estimate the magnitude of $\partial^2 Y/\partial x^2$ when the amplitude and wavelength of the flood wave can be estimated. (Note from Figure B.1 that the actual length of a flood wave is equal to $\lambda/2$.)

As an example, consider a river with a Manning's n of 0.033, a slope of 0.0023, and a constant width of 20 m. A flood wave of 3-m amplitude then would have a velocity V_k of

$$V_k = \left(\frac{5}{3}\right)\left(\frac{3^{2/3} \cdot 0.0023^{1/2}}{0.033}\right) = 5.1 \text{ m s}^{-1}.$$

If it takes 12 h for the wave to pass a given point, the value of λ is

$$\lambda = 5.1 \text{ m s}^{-1} \times 2 \times 12 \times 3600 \text{ s h}^{-1} = 440{,}000 \text{ m}.$$

The value of Q_n is computed from Manning's equation as

$$Q_n = \frac{3^{5/3} \cdot 0.0023^{1/2} \cdot 20}{0.033} = 182 \text{ m}^3 \text{ s}^{-1}.$$

Now we can compute $\partial Y/\partial t|_{\text{crest}}$ from Equation B-1, using Equation B-7 to approximate $\partial^2 Y/\partial x^2$:

$$\left.\frac{\partial Y}{\partial t}\right|_{\text{crest}} = -\frac{182}{2 \cdot 20 \cdot 0.0023}\frac{40 \cdot 3}{440000^2}$$
$$= -1.23 \cdot 10^{-6} \text{ m s}^{-1}.$$

Thus in one day's travel the peak would decrease by only about 0.1 m in this example, and the diffusion term could probably be neglected.

Glossary of Symbols

Symbol	Chapters	Definition and Dimensions
A	3	Any area $[L^2]$
	3, 5, 6, 9, 11	Wetted cross-sectional area of a channel $[L^2]$
	4	Area through which heat flow occurs $[L^2]$
	4, 6	Area parallel to flow direction, on which shear force acts $[L^2]$
	5	Horizontal area on which hydrostatic pressure acts $[L^2]$
	12	Cross-sectional area of a portion of a porous medium $[L^2]$
A_1	9	Wetted cross-sectional area of a channel at an upstream location $[L^2]$
A_2	9	Wetted cross-sectional area of a channel at a downstream location $[L^2]$

Symbol	Chapters	Definition and Dimensions
A_b	10	Cross-sectional area of a flow over a triangular weir $[L^2]$
A_c	12	Cross-sectional area of an idealized cylindrical pore $[L^2]$
A_p	12	Total area of pores in a cross section of a porous medium $[L^2]$
A_s	5	Area of an arbitrarily oriented face of an arbitrary volume of water $[L^2]$
A_x	5	Area of a vertical face of an arbitrary volume of water $[L^2]$
A_y	5	Area of a horizontal face of an arbitrary volume of water $[L^2]$
A_c'	12	Wetted cross-sectional area of a partially filled idealized cylindrical pore $[L^2]$
a	2, 3, 6, 7, 11	Acceleration $[LT^{-2}]$
	3	Empirical constant in a generalized regression equation [varies]
	11	Empirical constant relating width and discharge [varies]
	Appendix B	Amplitude of a flood wave $[L]$
B	3	Thermal quality of a snowpack [1]
	5, 7, 11	Arbitrary volume of water $[L^3]$
b	6	Proportionate distance above stream bed at which velocity equals mean velocity [1]
	11	Empirical exponent in relation between width and discharge [1]
	Appendix B	π times inverse of wavelength of a flood wave $[L^{-1}]$
b_1	3	Any regression coefficient [varies]
b_2	3	Any regression coefficient [varies]

Symbol	Chapters	Definition and Dimensions
b_n	3	Any regression coefficient [varies]
C	6, 8	Chézy roughness coefficient [1]
	6, 12	Constant of integration [varies]
C_1	8	Constant of integration [1]
C_2	8	Antilog of constant of integration $[ML^{-3}]$
C_b	10	Broad-crested weir coefficient [1]
C_D	8	Drag coefficient [1]
C_d	10	Discharge coefficient for a weir $[L^{1/2}T^{-1}]$
C_g	5, 11	Celerity of a gravity wave $[LT^{-1}]$
C_p	3, 4	Heat capacity $[L^2T^{-2}\Theta^{-1}]$
C_s	10	Sharp-crested weir coefficient [1]
C_t	8	Total suspended sediment concentration [1]
	10	Weir coefficient for a triangular sharp-crested weir [1]
c	2	Velocity of light $[LT^{-1}]$
	8	Concentration of suspended sediment at specified distance above the bed [1]
	11	Empirical constant relating mean depth and discharge [varies]
c_a	8	Concentration of suspended sediment at reference level [1]
c_{md}	8	Concentration of suspended sediment at mid-depth [1]
D	9	$1 - (Y_c/Y)^3$ [1]
	11	Open-channel hydraulic diffusivity $[L^2T^{-1}]$
D_A	3	Diffusivity of substance A in a medium $[L^2T^{-1}]$

Symbol	Chapters	Definition and Dimensions
D_H	3, 4	Diffusivity of heat energy $[L^2T^{-1}]$
D_m	8	Vertical diffusivity of momentum $[L^2T^{-1}]$
D_s	3	Generalized diffusivity $[L^2T^{-1}]$
	8	Vertical diffusivity of suspended sediment $[L^2T^{-1}]$
D_θ	12	Hydraulic diffusivity in a porous medium $[L^2T^{-1}]$
d	8	Diameter of sediment particles $[L]$
d_c	8	Diameter of largest sediment particle that can be eroded by a flow $[L]$
d_m	12	Average or median grain diameter of a granular porous medium $[L]$
E	2	Energy equivalent of a mass $[FL]$
	3	Voltage drop [volts]
E_ℓ	11	Energy expenditure in local acceleration $[FL]$
E_p	5	Total potential energy $[FL]$
E_{pC}	5	Total potential energy of parcel C $[FL]$
E_{pD}	5	Total potential energy of parcel D $[FL]$
E_{pg}	5, 7	Gravitational potential energy $[FL]$
E_{pp}	5	Pressure potential energy $[FL]$
E_v	7	Kinetic energy $[FL]$
E_{xC}	5	Nongravitational (pressure) potential energy of parcel C $[FL]$
E_{xD}	5	Nongravitational (pressure) potential energy of parcel D $[FL]$
e	6, 11	Base of natural logarithms, 2.7128 . . . [1]
	12	Void ratio [1]
F	2, 3, 7	Net force on a parcel of water $[F]$

Symbol	Chapters	Definition and Dimensions
F	4	Shear force $[F]$
F_D	8	Drag force on a sediment particle $[F]$
F_d	5, 6	Downslope force component $[F]$
F_{down}	3, 4	Total downward-acting force $[F]$
F_e	8	Total erosive force on a sediment particle $[F]$
F_g	8, 12	Gravity force $[F]$
F_L	8	Lift force on a sediment particle $[F]$
F_ℓ	11	Force expended in local acceleration $[F]$
	12	Total force acting toward the left $[F]$
F_n	7	Pressure force normal to the slope $[F]$
F_p	12	Pressure-gradient force $[F]$
F_{p1}	10, 12	Pressure force on an upstream cross section $[F]$
F_{p2}	10, 12	Pressure force on a downstream cross section $[F]$
F_r	12	Total force acting toward the right $[F]$
F_s	5	Pressure force on the s face of an arbitrary volume $[F]$
F_t	5, 6, 8, 10	Force due to turbulent frictional resistance (turbulent force) $[F]$
F_u	6	Upslope force component $[F]$
F_{up}	3, 4	Total upward-acting force $[F]$
F_v	5, 6, 8	Force due to viscous frictional resistance (viscous force) $[F]$
F_x	5	Pressure force on the x face of an arbitrary volume $[F]$
F_y	5	Pressure force on the y face of an arbitrary volume $[F]$

Symbol	Chapters	Definition and Dimensions
F^*	4	Surface-tension force on a slide-wire at equilibrium $[F]$
$\overleftarrow{F_f}$	5	Total frictional force $[F]$
$\overrightarrow{F_g}$	5	Gravity force $[F]$
$\overleftrightarrow{F_p}$	5	Pressure-gradient force $[F]$
$\overleftarrow{F_t}$	5	Force due to turbulent frictional resistance (turbulent force) $[F]$
$\overleftarrow{F_v}$	5	Force due to viscous frictional resistance (viscous force) $[F]$
$\dot{F}_C$	5	Coriolis force $[F]$
$\dot{F}_r$	5	Centrifugal force $[F]$
$\dot{F}_s$	5	Surface-tension force $[F]$
$\mathbb{F}$	throughout	Froude number $[1]$
$\mathbb{F}_1$	10	Froude number at an upstream cross section $[1]$
$\mathbb{F}^*$	6	Ratio of actual turbulent force to actual gravitational force in open-channel flow $[1]$
f	11	Empirical exponent in relationship between mean depth and discharge $[1]$
f_D	8	Darcy–Weisbach friction factor $[1]$
f_p	12	Friction factor for porous media flow $[1]$
g	throughout	Acceleration due to gravity $[LT^{-2}]$
H	3	Energy flux to a snowpack $[MT^{-1}]$
	3, 7, 9–11	Energy per weight of water (total head) at a cross section $[L]$
H_ℓ	11	Head loss due to local acceleration $[L]$
H_p	5, 9, 12	Potential energy per weight of water (potential or piezometric head) $[L]$
H_{p1}	9	Potential head at an upstream cross section $[L]$

Symbol	Chapters	Definition and Dimensions
H_{p2}	9	Potential head at a downstream cross section $[L]$
H_{pg}	7	Gravitational potential energy per weight of water (elevation head) $[L]$
H_{pp}	7	Pressure potential energy per weight of water (pressure head) $[L]$
H_s	7, 9–11	Specific head $[L]$
H_{s1}	10	Specific head at an upstream cross section $[L]$
H_{s2}	10	Specific head at a downstream cross section $[L]$
H_{sc}	10	Specific head at critical flow $[L]$
H_v	7, 9	Kinetic energy per weight of water (velocity head) $[L]$
H_{v1}	9	Velocity head at an upstream cross section $[L]$
H_{v2}	9	Velocity head at a downstream cross section $[L]$
H_w	10	Elevation of upstream water surface above weir crest (weir head) $[L]$
H_y	7	Total head at a point on a streamline $[L]$
h	2	Any distance $[L]$
	4, 12	Height of capillary rise $[L]$
	8	Average height of bed forms $[L]$
h_f	9	Head loss due to channel roughness and configuration $[L]$
h_j	10	Head loss through a hydraulic jump $[L]$
h_s	9	Head loss due to eddies (shock loss) $[L]$
h_w	11	Height (amplitude) of a gravity wave $[L]$
h_Δ	7, 9	Total head loss between two cross sections $[L]$

Symbol	Chapters	Definition and Dimensions
I	3	Quantity of mass $[M]$, energy $[ML^2T^{-2}]$, or momentum $[MLT^{-1}]$ entering a defined volume
	3	Electrical current [amperes]
J	8	Empirical factor relating dimensionless shear stress and sediment concentration [1]
K	6	Generalized resistance factor in uniform-flow equations [varies]
	8	Empirical exponent relating dimensionless shear stress and sediment concentration [1]
	9	Channel conveyance $[L^3T^{-1}]$
	11	$u_M S_0^{1/2}/n$ $[L^{1/3}T^{-1}]$
K_1	8	Proportionality constant relating gravitational force and particle diameter [1]
K_2	8	Proportionality constant relating critical erosion force and particle diameter [1]
K_c	3	Proportionality constant in a generalized equation of motion $[TL^{-1}]$ or $[ML^3T^{-3}\Theta^{-1}]$
K_D	3, 6, 12	Saturated hydraulic conductivity $[LT^{-1}]$
K_{Du}	12	Unsaturated hydraulic conductivity $[LT^{-1}]$
K_{Dx}	12	Hydraulic conductivity in the x direction $[LT^{-1}]$
K_{Dy}	12	Hydraulic conductivity in the y direction $[LT^{-1}]$
K_{Dz}	12	Hydraulic conductivity in the z direction $[LT^{-1}]$
K_I	3, 12	Intrinsic permeability $[L^2]$
K_p	12	Hydraulic conductivity in an idealized porous medium $[LT^{-1}]$

Symbol	Chapters	Definition and Dimensions
K_T	6	Constant relating turbulent shear stress to mean velocity [1]
k	6, 8	von Karman's constant [1]
	11	Empirical constant relating mean velocity and discharge [varies]
k_s	6, 8	Roughness height [L]
k_t	6, 8	Proportionality constant relating turbulent and gravitational forces in uniform turbulent flow [1]
k_v	6	Proportionality constant relating viscous and gravitational forces in uniform laminar flow [1]
k_θ	4	Molecular thermal conductivity [$MLT^{-3}\Theta^{-1}$]
L	7	Total length of a river [L]
	8	Sediment load [FT^{-1}]
	10	Length of a hydraulic jump [L]
	11	Length of a channel reach or overland-flow path [L]
	12	Length of an idealized cylindrical pore [L]
L_w	10	Downstream extent (length) of a weir crest [L]
l	5	Characteristic length of a flow [L]
	6	Mixing length [L]
	12	Distance along a streamline in a porous medium [L]
M	2–5	Mass [M]
	3	Snowmelt rate [LT^{-1}]
M_{in}	11	Mass of water entering a defined volume [M]

Symbol	Chapters	Definition and Dimensions
M_{out}	11	Mass of water leaving a defined volume [M]
m	7, 10	Mass of a parcel of water [M]
	8	Meander factor for estimating Manning's n [1]
	11	Empirical exponent relating mean velocity and discharge [1]
	12	Number of cylindrical conduits in a portion of idealized porous medium [1]
N	9	$1 - (Y_n/Y)^{10/3}$ [1]
n	3, 6–9, 11	Manning resistance (roughness) coefficient [1]
	12	Porosity [1]
n_0	8	Component of total resistance due to channel material [1]
n_1	8	Component of total resistance due to surface roughness [1]
	9	Manning resistance coefficient at an upstream cross section [1]
n_2	8	Component of total resistance due to channel variability [1]
	9	Manning resistance coefficient at a downstream cross section [1]
n_3	8	Component of total resistance due to flow obstructions [1]
n_4	8	Component of total resistance due to vegetation [1]
P	3	Average precipitation rate [LT^{-1}]
	3, 5, 6	Wetted perimeter [L]
	3, 5, 12	Hydrostatic pressure [FL^{-2}]
	4	Total pressure on plane tangent to meniscus [FL^{-2}]

Symbol	Chapters	Definition and Dimensions
P_1	10	Average pressure on upstream face of hydraulic jump $[FL^{-2}]$
	12	Pressure at upstream end of an idealized pore $[FL^{-2}]$
P_2	10	Average pressure on downstream face of hydraulic jump $[FL^{-2}]$
	12	Pressure at downstream end of an idealized pore $[FL^{-2}]$
P_a	4, 5, 12	Atmospheric pressure $[FL^{-2}]$
P_{abs}	12	Absolute pressure $[FL^{-2}]$
P_{ae}	12	Air-entry pressure $[FL^{-2}]$
P_{gage}	12	Gage pressure $[FL^{-2}]$
P_n	7	Pressure on a plane normal to streamline $[FL^{-2}]$
P_s	4	Pressure due to weight of suspended water on plane tangent to mensicus $[FL^{-2}]$
	5	Pressure on s face of an arbitrary volume $[FL^{-2}]$
P_u	12	Pressure in an unsaturated pore $[FL^{-2}]$
P_w	5	Component of absolute pressure due to overlying water $[FL^{-2}]$
P_x	5	Pressure on x face of an arbitrary volume $[FL^{-2}]$
P_y	5	Pressure on y face of an arbitrary volume $[FL^{-2}]$
p	3	Mechanical $[ML^2T^{-2}]$ or thermal $[\Theta]$ potential energy at a point
p_1	3	Mechanical $[ML^2T^{-2}]$ or thermal $[\Theta]$ potential energy at point 1
p_2	3	Mechanical $[ML^2T^{-2}]$ or thermal $[\Theta]$ potential energy at point 2

Symbol	Chapters	Definition and Dimensions
Q	3	Quantity of mass $[M]$, energy $[ML^2T^{-2}]$, or momentum $[MLT^{-1}]$ leaving a defined volume
	3, 6–11, Appendix A	Streamflow rate (discharge) $[L^3T^{-1}]$
	4	Heat-flow rate $[ML^2T^{-3}]$
	12	Discharge through a section of porous medium $[L^3T^{-1}]$
Q_1	9	Discharge at an upstream cross section $[L^3T^{-1}]$
Q_2	9	Discharge at a downstream cross section $[L^3T^{-1}]$
Q_A	3	Volume flow rate per unit area (flux density) of substance A $[ML^{-2}T^{-1}]$
Q_c	12	Discharge through an idealized cylindrical pore $[L^3T^{-1}]$
Q_d	8	Downward flux of suspended sediment $[ML^{-2}T^{-1}]$
Q_H	3	Flow rate per unit area (flux density) of heat energy $[MT^{-3}]$
Q_n	11, Appendixes A, B	Discharge of a uniform flow (normal discharge) $[L^3T^{-1}]$
Q_s	3	Flow rate per unit area (flux density) of mass $[ML^{-2}T^{-1}]$, momentum $[ML^{-1}T^{-2}]$, or energy $[MT^{-3}]$
	8	Bed-material sediment discharge $[FL^{-1}T^{-1}]$
Q_u	8	Upward flux of suspended sediment $[ML^{-2}T^{-1}]$
q	8, 10	Discharge per unit width $[L^2T^{-1}]$
	11	Lateral inflow discharge per unit of channel length $[L^2T^{-1}]$
q_b	8	Bed-load discharge per unit width $[FL^{-1}T^{-1}]$ or $[L^2T^{-1}]$

Symbol	Chapters	Definition and Dimensions
q_c	8	Threshold value of discharge per unit width to initiate bed-load movement $[L^2T^{-1}]$
q_l	12	Specific discharge in the l direction $[LT^{-1}]$
q_s	8	Suspended-load discharge per unit width $[L^2T^{-1}]$
q_s	12	Discharge per unit area of porous medium (specific discharge) $[LT^{-1}]$
q_x	3, 12	Specific discharge in the x direction $[LT^{-1}]$
q_y	12	Specific discharge in the y direction $[LT^{-1}]$
q_z	12	Specific discharge in the z direction $[LT^{-1}]$
R	3	Electrical resistivity [ohms]
	3, 5–9	Hydraulic radius $[L]$
	4, 12	Radius of a capillary tube $[L]$
	12	Radius of an idealized cylindrical pore $[L]$
R_1	9	Hydraulic radius at an upstream cross section $[L]$
R_2	9	Hydraulic radius at a downstream cross section $[L]$
R_m	12	Average or median pore radius $[L]$
R'	8	Effective hydraulic radius to balance skin friction $[L]$
	12	Effective radius of a partially filled idealized cylindrical pore $[L]$
R''	8	Effective hydraulic radius to balance form drag $[L]$
$\mathbb{R}$	5, 6	Reynolds number for open-channel flow $[1]$

Symbol	Chapters	Definition and Dimensions
$\mathbb{R}_e$	8	Erosive Reynolds number [1]
$\mathbb{R}_p$	8	Particle Reynolds number [1]
$\mathbb{R}_{pm}$	12	Reynolds number for porous media flow [1]
$\mathbb{R}^*$	6	Ratio of actual turbulent force to actual viscous force in a uniform open-channel flow [1]
r	3	Radius of a circle [L]
	5	Radius of curvature of a flow path [L]
	8	Radius of a spherical sediment particle [L]
	12	Radial distance from center of an idealized cylindrical pore [L]
r_1	12	Radius of curvature of a meniscus at location 1 [L]
r_2	12	Radius of curvature of a meniscus at location 2 [L]
r_3	12	Radius of curvature of a meniscus at location 3 [L]
r_m	12	Radius of curvature of a meniscus [L]
S	3	Quantity of mass [M], energy [ML^2T^{-2}], or momentum [MLT^{-1}], within a defined volume
S_0	throughout	Channel-slope tangent [1]
S_1	3	Quantity of mass [M], energy [ML^2T^{-2}], or momentum [MLT^{-1}] within a defined volume at the beginning of a time period
S_2	3	Quantity of mass [M], energy [ML^2T^{-2}], or momentum [MLT^{-1}] within a defined volume at the end of a time period
S_e	throughout	Tangent of slope of energy grade line (energy slope) [1]

Symbol	Chapters	Definition and Dimensions
S_{e1}	9	Energy-slope tangent at an upstream cross section [1]
S_{e2}	9	Energy-slope tangent at a downstream cross section [1]
S_s	9	Water-surface slope tangent [1]
	12	Specific storage [1]
$\overline{S}_e$	9	Average energy-slope tangent between two cross sections [1]
S_0'	8	Effective channel-slope tangent to balance skin friction [1]
S_0''	8	Effective channel-slope tangent to balance form drag [1]
$\mathcal{S}$	Appendix A	$1 - (1/S_0)(\partial Y/\partial x)$ [1]
s	3	Concentration of mass $[ML^{-3}]$, momentum $[ML^{-2}T^{-1}]$, or energy $[ML^{-1}T^{-2}]$
	12	Degree of saturation [1]
T	3, 4	Temperature $[\Theta]$
T_1	4	Temperature at plate 1 $[\Theta]$
T_2	4	Temperature at plate 2 $[\Theta]$
t	throughout	Time $[T]$
u	6	Generalized unit-adjustment factor [varies]
u_C	6, 8	Unit-adjustment factor in Chézy equation $[L^{1/2}T^{-1}]$
u_M	6–9, 11	Unit-adjustment factor in Manning equation $[L^{1/3}T^{-1}]$
V	throughout	Mean velocity in a vertical profile or at a cross section $[LT^{-1}]$
V_0	5	Free-stream velocity $[LT^{-1}]$

Symbol	Chapters	Definition and Dimensions
V_1	7, 10	Mean velocity at an upstream cross section $[LT^{-1}]$
	11	Mean velocity over left-bank floodplain $[LT^{-1}]$
V_2	7, 9, 10	Mean velocity at a downstream cross section $[LT^{-1}]$
	11	Mean velocity over central channel $[LT^{-1}]$
V_3	11	Mean velocity over right-bank floodplain $[LT^{-1}]$
V_{12}	3	Rate of movement of matter $[MT^{-1}]$ or energy $[ML^2T^{-3}]$ from point 1 to point 2
V_b	10	Mean velocity at the brink of a weir $[LT^{-1}]$
V_c	8	Threshold value of mean velocity for bed-load movement or sediment transport $[LT^{-1}]$
	10	Mean velocity at critical flow $[LT^{-1}]$
	12	Mean velocity in an idealized cylindrical pore $[LT^{-1}]$
V_g	11	Downstream velocity of a gravity wave $[LT^{-1}]$
V_k	11, Appendix B	Velocity of a kinematic wave $[LT^{-1}]$
V_s	12	Volume of solids in a portion of porous medium $[L^3]$
V_T	12	Total volume of a portion of porous medium $[L^3]$
V_v	12	Volume of void spaces in a portion of porous medium $[L^3]$
V_w	11	Apparent mean velocity at a cross section containing a solitary gravity wave $[LT^{-1}]$

Symbol	Chapters	Definition and Dimensions
V_w	12	Volume of water in a portion of porous medium $[L^3]$
$V*$	6, 8	Friction velocity $[LT^{-1}]$
$\overrightarrow{V}$	5	Mean velocity of a parcel of water $[LT^{-1}]$
v	throughout	Velocity at a point $[LT^{-1}]$
v_1	7	Velocity at point 1 $[LT^{-1}]$
v_2	7	Velocity at point 2 $[LT^{-1}]$
v_f	8	Fall velocity of a sediment particle $[LT^{-1}]$
$\overrightarrow{v}$	5	Velocity at a point $[LT^{-1}]$
W	7	Work done by flowing water $[FL]$
w	3, 5, 7, 9–11, Appendixes A, B	Water-surface width $[L]$
	8	Weight of a sediment particle $[F]$
w_1	9	Channel width at an upstream cross section $[L]$
	11	Width of flow on left-bank floodplain $[L]$
w_2	9	Channel width at a downstream cross section $[L]$
	11	Width of central channel $[L]$
w_3	11	Width of flow on right-bank floodplain $[L]$
w_b	10	Width of weir opening $[L]$
X	2	Arbitrary quantity [varies]
	5, 6	Down-channel dimension of a macroscopic volume of water $[L]$
X_1	5	Down-channel location of cross section 1 $[L]$
X_2	5	Down-channel location of cross section 2 $[L]$

Symbol	Chapters	Definition and Dimensions
X_{12}	3	Distance between points 1 and 2 [L]
x	2–4, 7, 12	Distance in the x direction [L]
	4	Distance over which surface-tension force acts [L]
	6, 7, 9, 11, Appendixes A, B	Down-channel distance [L]
x_1	3	Any independent variable [varies]
	7	Down-channel location of cross section 1 [L]
x_2	3	Any independent variable [varies]
	7	Down-channel location of cross section 2 [L]
x_n	3	Any independent variable [varies]
x^*	9	Channel distance required for convergence of estimated water-surface elevation to true elevation [L]
Y	2	Arbitrary quantity [varies]
	3, 5–11, Appendixes A, B	Total depth at a point or mean (hydraulic) flow depth [L]
Y_1	7, 10	Mean depth at an upstream cross section [L]
Y_2	7, 10	Mean depth at a downstream cross section [L]
Y_b	10	Mean depth at the brink of a weir [L]
Y_c	7, 8, 10	Mean or total depth at critical flow (critical depth) [L]
Y_{c1}	10	Critical depth at an upstream cross section [L]
Y_{c2}	10	Critical depth at a downstream cross section [L]
Y_n	9	Mean or total depth of uniform flow (normal depth) [L]

Symbol	Chapters	Definition and Dimensions
Y^*	5	Vertical distance from water surface to horizontal datum [L]
y	2	Distance traveled by a freely falling body [L]
	3	Any dependent variable [varies]
	3, 5–7	Vertical distance below water surface [L]
	4	Distance slide-wire moved to extend water surface [L]
	4, 5, 8	Normal distance above lower flow boundary [L]
	12	Vertical distance below piezometric surface [L]
	12	Distance in the y direction [L]
y_0	6, 8	Distance above channel bed at which logarithmic velocity profile reaches zero [L]
y_1	12	Vertical distance below water surface at upstream end of an idealized cylindrical pore [L]
y_2	12	Vertical distance below water surface at downstream end of an idealized cylindrical pore [L]
y_a	8	Reference level above bed for computation of suspended sediment concentration [L]
y_C	5	Distance below water surface of parcel C [L]
y_D	5	Distance below water surface of parcel D [L]
y_l	6	Thickness of laminar sublayer [L]
y_t	6	Distance of top of buffer zone above channel bottom [L]

Symbol	Chapters	Definition and Dimensions
Z	2	Arbitrary quantity [varies]
	3, 6, 7, 9, 11	Elevation of lower flow boundary above horizontal datum [L]
	8	v_f/kV^* [1]
Z_1	7, 10	Elevation of lower flow boundary above horizontal datum at upstream cross section [L]
Z_2	7, 10	Elevation of lower flow boundary above horizontal datum at downstream cross section [L]
Z_s	3	Elevation above sea level [L]
Z_w	10	Height of weir crest above channel bottom [L]
z	5, 7, 12	Elevation of a point above a horizontal datum [L]
	12	Distance in the z direction [L]
α	2	Arbitrary constant [1]
	6	Empirical exponent in a general uniform-flow equation [1]
	7, 9–11	Kinetic-energy coefficient [1]
α_1	7	Kinetic-energy coefficient at upstream cross section [1]
α_2	7, 9	Kinetic-energy coefficient at downstream cross section [1]
β	6	Empirical exponent in a general uniform-flow equation [1]
	12	Empirical constant relating intrinsic permeability to porosity and grain diameter [1]
β_1	12	Empirical constant to account for grain-size distribution [1]

Symbol	Chapters	Definition and Dimensions
β_2	12	Empirical constant to account for use of grain size to estimate pore size [1]
β_3	12	Empirical constant to account for pore shape [1]
β_b	10	Empirical constant relating critical depth and weir head for broad-crested weir [1]
β_s	10	Empirical constant relating brink depth and weir head for sharp-crested weir [1]
β_u	12	Empirical constant relating hydraulic conductivity to properties of fluid and medium for unsaturated porous medium [1]
γ	throughout	Weight density of water $[FL^{-3}]$
γ_s	8	Weight density of sediment particles $[FL^{-3}]$
ΔE	4	Change in thermal energy $[ML^2T^{-2}]$
ΔH	12	Difference in potential head $[L]$
ΔM	11	Change in mass within a defined volume $[M]$
ΔM_1	10	Differential increment of momentum at an upstream cross section $[MLT^{-1}]$
ΔM_2	10	Differential increment of momentum at a downstream cross section $[MLT^{-1}]$
Δm	10	Differential increment of mass $[M]$
ΔmV_1	10	Differential increment of momentum at an upstream cross section $[MLT^{-1}]$
ΔP	4	Pressure difference across a meniscus $[FL^{-2}]$
	5	Pressure difference across a macroscopic volume of water $[FL^{-2}]$
ΔT	4	Change in temperature $[\Theta]$

Symbol	Chapters	Definition and Dimensions
Δt	10	Differential increment of time $[T]$
Δx	4	Distance between two plates $[L]$
	9	Distance between upstream and downstream cross sections $[L]$
ΔY	5	Difference in depths at ends of a macroscopic volume of water $[L]$
	9	Difference in depths at adjacent cross sections $[L]$
δ	5	Boundary-layer thickness $[L]$
ε	5, 6, 8	Eddy viscosity $[ML^{-1}T^{-1}]$
θ	5	Angle between two faces of an arbitrary volume [1]
	5–7, 11	Channel slope angle [1]
	12	Volumetric water content [1]
θ_b	10	Vertex angle of a triangular weir [1]
θ_c	4, 12	Contact angle between meniscus and solid boundary [1]
θ_e	8	Dimensionless boundary shear stress [1]
θ_{ec}	8	Critical value of dimensionless boundary shear stress [1]
θ_r	12	Volumetric water content at which water films become discontinuous [1]
θ'	8	Dimensionless boundary shear stress due to skin friction [1]
θ''	8	Dimensionless boundary shear stress due to form drag [1]
λ	8	Wavelength of bedforms $[L]$
	11	Length of a gravity wave $[L]$
	12	Empirical exponent relating tension head and water content [1]

Symbol	Chapters	Definition and Dimensions
λ	Appendix B	Length of a flood wave $[L]$
λ_f	4	Latent heat of fusion $[L^2T^{-2}]$
λ_v	4	Latent heat of vaporization $[L^2T^{-2}]$
μ	throughout	Dynamic viscosity $[ML^{-1}T^{-1}]$
ν	throughout	Kinematic viscosity (diffusivity of momentum) $[L^2T^{-1}]$
π	throughout	3.14159 . . . [1]
ρ	throughout	Mass density of water $[ML^{-3}]$
ρ_A	3	Concentration of substance A $[ML^{-3}]$
ρ_s	8	Mass density of sediment particles $[ML^{-3}]$
ΣF	10	Net force on a hydraulic jump $[F]$
$\Sigma\overrightarrow{F}$	5	Net downstream-directed force $[F]$
ΣK	9	Total conveyance for a cross section $[L^3T^{-1}]$
σ	throughout	Surface tension of water $[FL^{-1}]$
τ	throughout	Shear stress (flux density of momentum) $[FL^{-2}]$
τ_0	6–8	Shear stress on flow boundary $[FL^{-2}]$
τ_{0c}	8	Critical boundary shear stress (critical tractive force) for erosion $[FL^{-2}]$
τ_u	6	Upslope component of shear stress $[FL^{-2}]$
τ_0'	8	Boundary shear stress due to skin friction $[FL^{-2}]$
τ_0''	8	Boundary shear stress due to form drag $[FL^{-2}]$
$\overleftarrow{\tau}$	5	Shear stress $[FL^{-2}]$
ϕ	5	Latitude [1]

Symbol	Chapters	Definition and Dimensions
ϕ	12	Slope angle of an idealized cylindrical pore [1]
χ	11	Any number [1]
ψ	12	Pressure (tension) head $[L]$
ψ_1	12	Pressure (tension) head at meniscus with radius r_1 $[L]$
ψ_2	12	Pressure (tension) head at meniscus with radius r_2 $[L]$
ψ_{ae}	12	Air-entry pressure (tension) head $[L]$
Ω	6	Generalized channel-resistance factor [1]
	7	Stream power per unit channel length $[FT^{-1}]$
ω	5	Angular velocity of earth's rotation $[T^{-1}]$
ω_A	7, 8	Stream power per unit bed area $[FT^{-1}L^{-1}]$
ω_{Ac}	8	Threshold value of stream power per unit bed area $[FT^{-1}L^{-1}]$
ω_w	7, 8	Stream power per unit weight of water $[LT^{-1}]$
ω_{wc}	8	Threshold stream power per unit weight of water $[LT^{-1}]$

References

Bagnold, R. A. 1956. Flow of cohesionless grains in fluids. *Royal Society of London Philosophical Transactions* **249** (964).

Bagnold, R. A. 1966. An approach to the sediment transport problem from general physics. U.S. Geological Survey Professional Paper 422-I.

Bagnold, R. A. 1977. Bed load transport in natural rivers. *Water Resources Research* **13,** 303–312.

Bailey, J. F. and H. A. Ray. 1966. Definition of stage–discharge relation in natural channels by step-backwater analysis. U.S. Geological Survey Water-Supply Paper 1869-A.

Barnes, H. H. 1967. Roughness characteristics of natural channels. U.S. Geological Survey Water-Supply Paper 1849.

Barton, J. R., and P. N. Lin. 1955. A study of sediment transport in alluvial channels. Colorado State University Hydrology Paper No. 55.

Bear, J. 1972. *Dynamics of Fluids in Porous Media.* New York: American Elsevier.

This is the most complete systematic treatment of porous media flow available in English. It employs advanced engineering mathematics throughout. The emphasis is on the theoretical foundations of solutions to practical problems, and the book is particularly useful to professionals and advanced students as a starting point for formulating models for solving groundwater flow problems. Each chapter concludes with a problem set testing both theoretical understanding and problem-solving ability.

Beven, K. 1979. On the generalized kinematic routing method. *Water Resources Research* **15,** 1238–1242.

Beven, K., K. Gilman, and M. Newson. 1979. Flow and flow routing in upland channel networks. *Hydrological Sciences Bulletin* **24,** 303–325.

Bishop, A. A., D. B. Simons, and E. V. Richardson. 1965. Total bed material transport. *American Society of Civil Engineers Proceedings, Journal of the Hydraulics Division* **97** (HY-2), 175–191.

Biswas, A. K. 1970. *The History of Hydrology.* New York: Elsevier.

Boyer, M. C. 1964. Streamflow measurement. In *Handbook of Applied Hydrology* (V. T. Chow, ed.). New York: McGraw-Hill.

Brakensiek, D. L. 1966. Storage flood routing without coefficients. U.S. Dept. of Agriculture, Agricultural Research Service Report ARS 41-122.

Brakensiek, D. L. 1967. A simulated watershed flow system for hydrograph prediction: A kinematic application. *Proceedings International Hydrology Symposium,* Colorado State University, Fort Collins.

Bray, D. I. 1979. Estimating average velocity in gravel-bed rivers. *American Society of Civil Engineers Proceedings, Journal of Hydraulics Division* **105** (HY-9), 1103–1122.

Brooks, R. H., and A. T. Corey. 1966. Properties of porous media affecting fluid flow. *American Society of Civil Engineers Proceedings, Journal of the Irrigation and Drainage Division* **92** (IR-2), 61–88.

Buchanan, T. J., and W. P. Somers. 1969. Discharge measurements at gaging stations. *Techniques of Water-Resources Investigations of the U.S. Geological Survey,* Book 3, Chapter A8.

Butler, S. S. 1957. *Engineering Hydrology.* Englewood Cliffs, New Jersey: Prentice-Hall.

Cedergren, H. R. 1967. *Seepage, Drainage, and Flow Nets.* New York: Wiley.

Chang, H. H. 1979. Minimum stream power and river channel patterns. *Journal of Hydrology* **41,** 303–327.

Chitale, S. V. 1973. Theory and relationship of river channel patterns. *Journal of Hydrology* **19,** 285–308.

Chow, V. T. 1959. *Open Channel Hydraulics.* New York: McGraw-Hill.

This basic text and reference on open-channel flow contains a clear and complete treatment of all aspects of the subject required for engineering design, supported by theoretical discussions. Major sections of the book are devoted to basic principles and to steady uniform, steady gradually varied and rapidly varied, and unsteady gradually and rapidly varied flows. The level of mathematics is that of seniors in earth and physical sciences. Numerous references to the European (and even Japanese) literature are provided. Sample problems are worked in the text, and each chapter concludes with a set of problems that are especially well designed for increasing the student's competence and understanding.

Colby, B. R. 1964a. Discharge of sands and mean velocity relationships in sand-bed streams. U.S. Geological Survey Professional Paper 462-A.

Colby, B. R. 1964b. Practical computation of bed-material discharge. *American Society of Civil Engineers Proceedings, Journal of the Hydraulics Division* **90** (HY-2), 217–246.

Colby, B. R., and C. H. Hembree. 1955. Computations of total sediment discharge, Niobrara River near Cody, Nebraska. U.S. Geological Survey Water-Supply Paper 1357.

Corbett, D. M. 1945. Stream-gaging procedure. U.S. Geological Survey Water-Supply Paper 888.

Cowan, W. L. 1956. Estimating hydraulic roughness coefficients. *Agricultural Engineering* **37,** 473–475.

Crawford, N. H., and R. K. Linsley. 1966. Digital simulation in hydrology: Stanford Watershed Model IV. Stanford University Dept. of Civil Engineering Technical Report No. 39.

Crow, E. L., F. A. Davis, and M. W. Maxfield. 1960. *Statistics Manual*. New York: Dover.

Daily, J. W., and D. R. F. Harleman. 1966. *Fluid Dynamics*. Reading, Massachusetts: Addison-Wesley.

This comprehensive engineering text is distinguished by a unified treatment of the basic physics and mathematics of fluid flow, beginning with the properties of fluids and hydrostatics. Aspects of flows of compressible and incompressible gases, as well as of liquids in channels, conduits, and porous media, are presented, with each subject developed from the fundamental flow equations. These equations are developed very clearly in the early chapters, and the developments are accessible to the upper-level student of hydrology and earth science. The book is well referenced and includes problems at the end of each chapter.

Davidian, J., P. H. Carrigan, Jr., and J. Shen. 1962. Flow through openings in width constrictions. U.S. Geological Survey Water-Supply Paper 1369-D.

Davis, K. S., and J. A. Day. 1961. *Water: The Mirror of Science*. New York: Doubleday–Anchor.

Designed for the intelligent layman, this book is completely devoted to an exploration of water as a substance. It uses the evolution of knowledge about water to convey an understanding of science as a process. The structure of the water molecule, the hydrogen bond, and the resulting properties of water are very clearly presented. The book is recommended for its philosophical aspects as well as for building a sound intuitive understanding of water as a substance.

Davis, S. N. 1969. Porosity and permeability of natural materials. In *Flow Through Porous Media* (R. J. M. DeWiest, ed.), Chapter 2. New York: Academic Press.

Day, J. A., and G. L. Sternes. 1970. *Climate and Weather*. Reading, Massachusetts: Addison-Wesley.

de Jong, F. J. 1967. *Dimensional Analysis for Economists*. Amsterdam: North Holland.

DeWiest, R. J. M. 1965. *Geohydrology*. New York: Wiley.

Dingman, S. L. 1971. Hydrology of the Glenn Creek watershed, Tanana River drainage, central Alaska. U.S. Army Cold Regions Research and Engineering Laboratory Research Report 297.

Dooge, J. C. I. 1983. On the study of water. *Hydrological Sciences Bulletin* **28,** 23–48.

This stimulating paper addresses a number of intriguing questions relative to the background of hydrology, including: What is the structure of bulk liquid water? Can water be treated as a continuum? Is water a Newtonian fluid? What is the scientific basis for the equations of water movement commonly used in hydrology? Can the various approaches to the study of water movement be reconciled with one another? What does the concept of water mean to the physical, biological, and social scientist?

Dorsey, N. E. 1940. *Properties of Ordinary Water Substance in All Its Phases: Water, Water-Vapor, and All the Ices*. New York: Reinhold.

Although now more than 40 years old, this remains the most comprehensive review and compilation of the properties of water in all its phases. In addition to the properties of liquid water examined in Chapter 4, there are extensive sections on strength and on acoustic, optical, and electrical properties. The various forms of ice and water vapor are similarly treated. For information on most properties, the student is advised to look first in a recent edition of the *Handbook of Chemistry and Physics* (Weast, 1971) for the latest information, but Dorsey is probably still the best source for many properties not included there. In addition, it is a model of how to write a critical review of the literature and makes fascinating reading; it even presents information on the viscosity of foggy air!

Draper, N. R., and H. Smith. 1981. *Applied Regression Analysis,* 2nd ed. New York: Wiley.

This text is completely devoted to regression analysis, and includes aspects often omitted in conventional statistics books. The essentials can be understood with a knowledge of algebra only. Parameter estimation, hypothesis testing, and computation of confidence intervals for equation parameters and for estimates made via regression equations are topics of particular interest. Problems are included.

du Boys, P. 1879. Études du régime du Rhône et l'action exercée par les eaux sur un lit á fond de graviers indéfiniment affouillable [Studies of the regime of the Rhone River and the force exerted by water on an erodible gravel bed]. *Annales des Ponts et Chausées* (Ser. 5), **18,** 141–195.

Dunne, T., and L. B. Leopold. 1978. *Water in Environmental Planning.* San Francisco: W. H. Freeman.

Eagleson, P. S. 1970. *Dynamic Hydrology.* New York: McGraw-Hill.

This text represents the most complete attempt to integrate basic mathematical physics and traditional hydrology. It includes chapters on atmospheric physics and physical oceanography as well. Although highly mathematical, it contains much that is of use to the practicing hydrologist, and is an excellent reference for that reason as well as for its unified treatment of theoretical aspects. Problems are given.

Einstein, A. 1926. Die Ursache der Maänderbildung der Flusslaufe und des sogenannten Beerschen Gesetzes [The cause of meander formation in rivers and of the so-called Beer's law]. *Naturwissenschaften* **14,** 223–224.

Einstein, H. A. 1950. The bedload function for sediment transportation in open channels. U.S. Dept. of Agriculture, Soil Conservation Service, Technical Bulletin No. 1026.

Einstein, H. A., and H. W. Shen. 1964. A study of meandering in straight alluvial channels. *Journal of Geophysical Research* **69,** 5239–5247.

Eisenberg, D., and W. Kauzmann. 1969. *The Structure and Properties of Water.* New York: Oxford University Press.

Engelund, F., and E. Hansen. 1967. *A Monograph on Sediment Transport in Alluvial Streams.* Copenhagen: Hydraulic Laboratory of Technical University of Denmark.

Freeze, R. A. 1974. Streamflow generation. *Reviews of Geophysics and Space Physics* **12,** 627–647.

This paper describes the integrated application of the basic equations of flow in a computer model of watershed behavior. It is an excellent review of fluvial hydrology, including unsaturated and saturated porous media flow, overland flow, and channel flow. The model is used to study the relative importance of sources of streamflow.

Freeze, R. A., and J. A. Cherry. 1979. *Groundwater*. Englewood Cliffs, New Jersey: Prentice-Hall.

This comprehensive text on groundwater contains extensive treatment of water quality as well as the more traditional subjects of groundwater physics, hydraulics, and geology. The opening discussions of the physics of flow in saturated and unsaturated porous media are lucid, and are recommended reading for building a solid understanding of the essential concepts. Problems are provided.

Garde, R. J., and M. L. Albertson. 1961. Bed load transport in alluvial channels. *La Houille Blanche* **3.**

Garde, R. J., and K. G. Ranga Raju. 1978. *Mechanics of Sediment Transportation and Alluvial Stream Problems*. New Delhi: Wiley Eastern.

Graf, W. H. 1971. *Hydraulics of Sediment Transport*. New York: McGraw-Hill.

Gray, D. M., ed. 1970. *Handbook on the Principles of Hydrology*. Port Washington, New York: Water Information Center.

This text covers the standard sequence of topics of an upper-level course in hydrology. The sections on river hydraulics and flood routing contain good discussions of the practical aspects of open-channel flow. There are also clear, concise treatments of flow in saturated and unsaturated porous media.

Gray, D. M., and J. M. Wigham. 1970. Peak flow–rainfall events. In *Handbook on the Principles of Hydrology* (D. M. Gray, ed.), Section VIII. Port Washington, New York: Water Information Center.

Gregory, K. J., and D. E. Walling. 1973. *Drainage Basin Form and Process*. New York: Halstead Press–Wiley.

Heggen, R. J. 1983. Thermal dependent physical properties of water. *Journal of Hydraulic Engineering* **109** (2), 298–302.

Hillel, D. 1977. *Computer Simulation of Soil-Water Dynamics*. Ottawa, Canada: International Development Research Centre.

Hillel, D. 1980. *Fundamentals of Soil Physics*. New York: Academic Press.

This is an authoritative and comprehensive treatment of soil physics, written at a level that is readily accessible to upper-level earth-science students. Principles of saturated and unsaturated porous media flow are clearly presented. The book is well written and is highly recommended as a reference and as a textbook. Problems are given at the end of each chapter.

Horton, R. E. 1945. Erosional development of streams and their drainage basins. *Geological Society of America Bulletin* **56,** 275–370.

Hubbert, M. K. 1940. The theory of groundwater motion. *Journal of Geology* **48,** 785–822.

The two seminal publications in the history of the study of groundwater flow are Darcy's reports of his experiments (1856) and this relatively recent theoretical analysis. In this elegantly reasoned classic paper, Hubbert derived the physical basis for Darcy's law and showed for the first time how hydraulic conductivity depends on the properties of the porous medium and of the fluid. He also showed that a number of then widely accepted hydrogeologic analyses were physically incorrect.

Hutchinson, G. E. 1957. *A Treatise on Limnology,* Vol. 1, Pt. 1: Geography and Physics of Lakes. New York: Wiley.

This is an extremely comprehensive scholarly survey of the geographic, physical, and chemical aspects of lakes. It includes detailed discussions of the properties of water and of the hydromechanics of lakes, and considerable additional material of interest to hydrologists. Although parts of it (particularly those dealing with lake chemistry) are now out of date, it remains a valuable reference. Over 1200 books and articles are cited, many from the European literature.

Ippen, A. T. 1950. Channel transitions and controls. In *Engineering Hydraulics* (H.Rouse, ed.), pp. 496–588. New York: Wiley.

Ipsen, D. C. 1960. *Units, Dimensions, and Dimensionless Numbers.* New York: McGraw-Hill.

This book contains theoretical—even philosophical—but concise treatments of the nature of physical variables, relationships, units, dimensions, and dimensionless numbers. Much of it is very practically oriented, however, with excellent discussions of units and dimensions in thermodynamics, heat transfer, and electricity, as well as mechanics. It also describes techniques for conversion of units and equations and for dimensional analysis. The mathematical level is well within the grasp of the serious earth-science student. Most chapters conclude with problems to test the reader's understanding.

Jellinek, H. H. G. 1972. The ice interface. In *Water and Aqueous Solutions* (R. A. Horne, ed.). New York: Wiley–Interscience.

Johnson, M. 1964. Channel roughness in steep mountain streams. Paper presented at the Annual Meeting of the American Geophysical Union.

Kalinske, A. A. 1947. Movement of sediment as bed load in rivers. *American Geophysical Union Transactions* **28** (4), 615–620.

Kikkawa, H., and T. Ishikawa. 1978. Total load of bed materials in open channels. *American Society of Civil Engineers Proceedings, Journal of the Hydraulics Division* **104** (HY-7), 1045–1059.

Kindsvater, C. E., and R. W. Carter. 1959. Discharge characteristics of rectangular thin-plate weirs. *American Society of Civil Engineers Transactions* **124,** 772–801.

Lane, L. J., and D. A. Woolhiser. 1977. Simplifications of watershed geometry affecting simulation of surface runoff. *Journal of Hydrology* **35,** 173–190.

Langbein, W. B., and L. B. Leopold. 1964. Quasi-equilibrium states in channel morphology. *American Journal of Science* **262,** 782–794.

Larkin, J., J. McDermott, D. P. Simon, and H. A. Simon. 1980. Expert and novice performance in solving physics problems. *Science* **208,** 1335–1342.

Lawler, E. A. 1964. Flood routing. In *Handbook of Applied Hydrology* (V. T. Chow, ed.), Section 25-II. New York: McGraw-Hill.

Leopold, L. B., and T. Maddock, Jr. 1953. Hydraulic geometry of stream channels and some physiographic implications. U.S. Geological Survey Professional Paper 252.

Leopold, L. B., and M. G. Wolman. 1957. River channel patterns: braided, meandering, and straight. U.S. Geological Survey Professional Paper 282-B.

Leopold, L. B., M. G. Wolman, and J. P. Miller. 1964. *Fluvial Processes in Geomorphology.* San Francisco: W. H. Freeman.

Lighthill, M. J., and G. B. Whitham. 1955. On kinematic waves: I. Flood movement in long rivers. *Royal Society of London Proceedings* **229** (Ser. A), 281–316.

This is the first comprehensive treatment of the important phenomenon of kinematic waves. After developing the basic physics of such waves, the authors show that the behavior of floods is to a large extent governed by the kinematic-wave equations. The discussion is largely theoretical, but the writing is clear and interesting and the mathematics does not go beyond the level of calculus.

Linsley, R. K., M. A. Kohler, and J. L. H. Paulhus, 1949. *Applied Hydrology*. New York: McGraw-Hill.

McDonald, J. E. 1952. The Coriolis effect. *Scientific American* **186** (5), 72–78.

Meyer-Peter, E., and R. Müller. 1948. Formulas for bed load transport. *International Association for Hydraulic Research Proceedings, 2nd Congress,* Stockholm, 39–65.

Michel, B. 1971. Winter regime of rivers and lakes. U.S. Army Cold Regions Research and Engineering Laboratory Cold Regions Science and Engineering Monograph III-Bla.

Miller, A., and J. C. Thompson. 1970. *Elements of Meteorology*. Columbus, Ohio: Charles E. Merrill.

Morgali, J. R. 1963. Hydraulic behavior of small drainage basins. Stanford University Dept. of Civil Engineering Technical Report No. 30.

Morris, D. A., and A. I. Johnson. 1967. Summary of hydrologic and physical properties of rock and soil materials, as analyzed by the Hydrologic Laboratory of the U.S. Geological Survey, 1948–1960. U.S. Geological Survey Water-Supply Paper 1839-D.

Morris, E. M., and D. A. Woolhiser. 1980. Unsteady one-dimensional flow over a plane: Partial equilibrium and recession hydrographs. *Water Resources Research* **16,** 335–360.

Pankhurst, R. C. 1964. *Dimensional Analysis and Scale Factors*. New York: Reinhold.

This is a complete but compact treatment of the theory and application of dimensional analysis in all branches of physics. It is designed as a monograph for undergraduate physics students, and since the mathematics is not advanced, it is a useful text for self-education of the student of hydrology and earth sciences.

Pauling, L. 1960. *The Nature of the Chemical Bond,* 3rd ed. Ithaca, New York: Cornell University Press.

Quade, W. 1967. The algebraic structure of dimensional analysis. In *Dimensional Analysis for Economists* (F. J. de Jong). Amsterdam: North Holland. (Originally published in German in 1961.)

Ragan, R. M. 1966. Laboratory evaluation of a numerical routing technique for channels subject to lateral inflows. *Water Resources Research* **2,** 111–122.

Raudkivi, A. J. 1976. *Loose Boundary Hydraulics,* 2nd ed. New York: Pergamon Press.

Remson, I., and J. R. Randolph. 1962. Review of some elements of soil-moisture theory. U.S. Geological Survey Professional Paper 411-D.

Ross, B. B., D. N. Contractor, and V. O. Shanholtz. 1979. A finite-element model of overland and channel flow for assessing the hydrologic impact of land-use change. *Journal of Hydrology* **41,** 11–30.

Rouse, H. 1936. Discharge characteristics of the free overfall. *Civil Engineering* **6** (7), 257–260.

Rouse, H. 1937. Modern conceptions of mechanics of fluid turbulence. *American Society of Civil Engineers Transactions* **102.** 463–505.

Rouse, H. 1946. *Elementary Mechanics of Fluids*. New York: Wiley.

This is the textbook version of Rouse's (1961) more comprehensive 1938 treatise. It is a pioneering attempt to establish a course in fluid mechanics as part of the basic engineering curriculum. The mathematics does not go beyond calculus. The book remains a valuable reference, and is especially useful for self-study because of the worked examples and the problems and questions for discussion at the end of each chapter.

Rouse, H. 1961. *Fluid Mechanics for Hydraulic Engineers*. New York: Dover. (Originally published in 1938.)

Through the first quarter of the twentieth century hydraulic engineering was based largely on empirical relations. This is the first textbook to provide a sound theoretical foundation for attacking engineering problems. It begins with a concise discussion of dimensional analysis, following which are major sections on the fundamentals of hydromechanics; the mechanics of fluid resistance, including chapters on open-channel flow and sediment transport; and the mechanics of dynamic waves. The book is authoritative and clearly written and remains a valuable text and reference. The mathematics is readily accessible to the upper level earth-science student.

Rouse, H., and S. Ince. 1963. *History of Hydraulics*. New York: Dover.

This is the only treatise on the history of the formulation of the underlying principles of fluid motion. It begins with a survey of the practical knowledge of hydraulics in antiquity and covers the succeeding periods into the twentieth century. The contributions of Bernoulli, Darcy, Chézy, Manning, St. Venant, and the many others whose names are linked to the subject are clearly described in their historical perspective. Many excerpts and figures from the original works are reproduced. The treatment is concise yet comprehensive, and provides a simultaneous education in fluid mechanics as well as in the history of this important branch of science and engineering.

Rovey, E. W., D. A. Woolhiser, and R. E. Smith. 1977. A distributed kinematic model of upland watersheds. Colorado State University Hydrology Paper No. 93.

Rubey, W. W. 1933. Settling velocities of gravel, sand, and silt particles. *American Journal of Science* (Ser. 5), **25,** 325–338.

Shapiro, A. H. 1961. *Shape and Flow.* Garden City, New York: Doubleday–Anchor.

This book is designed to provide the intelligent layperson with an explanation of fluid resistance. It is a companion to a film on the same subject, and is notable for its excellent illustrations (mostly photographs) and clear, straightforward explanations of many phenomena encountered in everyday life. It shows, for example, why streamlining reduces resistance under some conditions but increases it under others, and why golf balls with dimpled surfaces travel farther than smooth ones.

Shields, A. 1936. Anwendung der Ähnlichkeitsmechanik und der Turbulenzforschung auf die Geschiebebewegung [Application of similitude and turbulence research to bed-load movement]. *Mitteilungen der Prüssischen Versuchanstalt für Wasserbau und Schiffbau* (Berlin), Heft 26.

Shulits, S. 1935. The Schoklitsch bed load formula. *Engineering* **139,** 644–646, 687.

Sibulkin, M. 1983. A note on the bathtub vortex and the earth's rotation. *American Scientist* **71,** 352–353.

Simons, D. B., and E. V. Richardson. 1966. Resistance to flow in alluvial channels. U.S. Geological Survey Professional Paper 422-J.

This is the most complete empirical study of resistance, bed forms, and sediment transport in open-channel flows. The experiments were carried out on a large flume into which sediment could be introduced. They showed that the occurrence of various bed forms in alluvial channels can be predicted on the basis of the stream power and sediment size. Useful empirical approaches were developed to the difficult and somewhat circular problem of predicting fluid resistance when sediment transport is occurring and bed forms are changing.

Simons, D. B., and F. Senturk. 1977. *Sediment Transport Technology*. Fort Collins, Colorado: Water Resources Publications.

Smith, A. A. 1980. A generalized approach to kinematic flood routing. *Journal of Hydrology* **45,** 71–89.

Stillinger, F. H. 1980. Water revisited. *Science* **209,** 451–455.

This is an authoritative and understandable review of the current state of knowledge of the atomic and molecular structures of water. In particular, it discusses critical recent research on the intermolecular structure of water in the liquid state, a subject that is still being clarified. This article is required reading for any hydrologist seeking a fundamental understanding of water as a substance.

Strelkoff, T. 1969. One-dimensional equations of open-channel flow. *American Society of Civil Engineers Proceedings, Journal of the Hydraulics Division* **95** (HY-3), 861–876.

This is the most complete and careful derivation of the Saint-Venant equations available in English. The assumptions involved in their derivation are clearly identified.

Strelkoff, T. 1970. Numerical solution of Saint-Venant equations. *American Society of Civil Engineers Proceedings, Journal of the Hydraulics Division* **96** (HY-1), 223–252.

Taft, M. I., and A. Reisman. 1965. The conservation equations: A systematic look. *Journal of Hydrology* **3,** 161–179.

This clearly written paper gives a systematic view of the equations of conservation of mass, energy, and momentum. The assumptions underlying the formulations of those equations are stated, and each is written in its most general form. A diagram shows how simplifications of the general equations can be systematically related. Selected simplifications are examined in detail, and fields of application of each are indicated. Fundamental similarities of many seemingly unrelated physical problems are revealed, and the paper is thus a great aid in identifying the patterns, connections, and analogies that are essential to physical analysis.

Task Committee for Preparation of Sediment Manual. 1971. Sediment transportation mechanics: H. Sediment discharge formulas. *American Society of Civil Engineers Proceedings, Journal of the Hydraulics Division* **97** (HY-4), 523–567.

Task Force on Bed Forms in Alluvial Channels. 1966. Nomenclature of bed forms in alluvial channels. *American Society of Civil Engineers Proceedings, Journal of the Hydraulics Division* **92** (HY-3), 51–64.

Taylor, E. S. 1974. *Dimensional Analysis for Engineers*. New York and London: Oxford (Clarendon Press).

Todd, D. K. 1959. *Ground Water Hydrology*. New York: Wiley.

Toffaleti, F. B. 1969. Definitive computations of sand discharge in rivers. *American Society of Civil Engineers Proceedings, Journal of the Hydraulics Division* **95** (HY-1), 225–248.

Tracy, H. J. 1957. Discharge characteristics of broad-crested weirs. U.S. Geological Survey Circular 397.

Trincher, K. 1981. The mathematic–thermodynamic analysis of the anomalies of water and the temperature range of life. *Water Research* **15,** 433–448.

This is a detailed examination of the several anomalies in the properties of water, including the density maximum at 3.98°C. These anomalies are explained on the basis of varying proportions of three structural components that constitute a ''quasi-colloidal'' system.

U.S. Army Corps of Engineers. 1969. *Water Surface Profiles,* HEC Training Document, U.S. Army Corps of Engineers Hydrologic Engineering Center.

van Hylckama, T. E. A. 1979. Water, something peculiar. *Hydrological Sciences Bulletin* **24,** 499–507.

This is an entertaining and factual review of many of the unusual properties of water, showing how they are related to the structure of the water molecule.

Vanoni, V. A., ed. 1975. *Sedimentation Engineering* (American Society of Civil Engineers Manual and Report of Engineering Practice No. 54). New York: American Society of Civil Engineers.

Verruijt, A. 1970. *Theory of Ground Water Flow.* New York: Gordon and Breach.

Viessman, W., Jr., J. W. Knapp, G. L. Lewis, and T. E. Harbaugh. 1977. *Introduction to Hydrology.* New York: IEP–Dunn-Donnelly.

This is a practically oriented hydrology text, especially valuable for its detailed presentations of the most widely used hydrologic models. Theoretical aspects of open-channel and ground-water flow are clearly and concisely developed as foundations for the models. Extensive problem sets are provided at the end of each chapter.

Vittal, N., K. G. Ranga Raju, and R. J. Garde. 1973. Sediment transport relations using concept of effective shear stress. *International Association for Hydraulic Research, International Symposium on River Mechanics Proceedings,* Bangkok, Thailand.

Wallis, J. R. 1965. Multivariate statistical methods in hydrology—comparison using data of known functional relationship. *Water Resources Research* **1,** 447–461.

Wang, H. F., and M. P. Anderson. 1982. *Introduction to Groundwater Modeling.* San Francisco: W. H. Freeman.

Watson, R. L. 1969. Modified Rubey's law accurately predicts sediment settling velocities. *Water Resources Research* **5,** 1147–1150.

Weast, R. C., ed. 1971. *Handbook of Chemistry and Physics,* 52nd ed. Cleveland, Ohio: Chemical Rubber Co.

This ''ready reference book of chemical and physical data'' is the standard compilation of properties of all substances for professional scientists. It is published every two years.

Weinmann, P. E., and E. M. Laurenson. 1979. Approximate flood routing methods: A review. *American Society of Civil Engineers Proceedings, Journal of the Hydraulics Division* **105** (HY-12), 1521–1536.

This clearly written article gives a very useful overview and classification of flood routing methods. Beginning with the Saint-Venant equations, it shows the assumptions that are involved in a wide range of simplified routing models, and hence provides a guide for selecting methods appropriate to particular situations.

Winkler, R. L., and W. L. Hays. 1975. *Statistics—Probability, Inference, and Decision*. New York: Holt, Rinehart, and Winston.

This basic text treats regression analysis in the context of a comprehensive development of probability and statistics. The writing is lucid and directed at the nonmathematician user of statistics. It is an excellent book for developing a sound intuitive grasp of the subject. Each chapter concludes with an extensive set of exercises.

Wooding, R. A. 1965. A hydraulic model for the catchment–stream problem. *Journal of Hydrology* **3,** 254–267.

Woolhiser, D. A., and J. A. Liggett. 1967. Unsteady one-dimensional flow over a plane—The rising hydrograph. *Water Resources Research* **3,** 753–771.

Yalin, M. S. 1972. *Mechanics of Sediment Transport*. New York: Pergamon Press.

Yalin, M. S. and E. Karahan. 1979. Inception of sediment transport. *American Society of Civil Engineers Proceedings, Journal of the Hydraulics Division* **105** (HY-11), 1433–1443.

Yang, C. T. 1971. On river meanders. *Journal of Hydrology* **13,** 231–253.

Yang, C. T., C. C. S. Song, and M. J. Woldenburg. 1981. Hydraulic geometry and minimum rate of energy dissipation. *Water Resources Research* **17,** 1014–1018.

Yang, C. T., and J. B. Stall. 1976. Applicability of unit stream power equation. *American Society of Civil Engineers Proceedings, Journal of the Hydraulics Division* **102** (HY-5), 559–568.

Yevjevich, V. 1972. *Probability and Statistics in Hydrology*. Fort Collins, Colorado: Water Resources Publications.

This book is an important reference for hydrologists because it covers many aspects of statistics that are not treated in conventional texts. A chapter is devoted to regression analysis. Problems and points for discussion are included at the end of each chapter.

Answers to Selected Exercises

Chapter 2

2-1 293,000 gal min^{-1}; 18.5 m^3 s^{-1}; 473,000 ac · ft yr^{-1}.

2-2 120 cm s^{-1}; 4.0 ft s^{-1}; 1.2 m s^{-1}.

2-3 0.0081119 cal cm^{-2} s^{-1}; 339.56 W m^{-2}; 2933.8 J cm^{-2} day^{-1}.

2-4 448 ha; 1110 ac; 4,480,000 m^2.

2-5 430 ft^3 s^{-1}; 12.0 m^3 s^{-1}; 12,000 liters s^{-1}.

2-6 143,000 m^3; 5,050,000 ft^3; 37,800,000 gal.

2-7 0.00121 cm s^{-1}; 1110 ft^3 s^{-1} mi^{-2}; 1.72 in. h^{-1}.

2-8 7800 Pa; 1.13 lb in.$^{-2}$; 78,000 dyn cm^{-2}; 2.30 in. Hg; 78.0 mb.

2-9 1.00 Btu lb^{-1} °F^{-1}.

2-10 2.4×10^{-5} cal cm^{-2} s^{-1} °C^{-1}; 1.02 W m^{-2} K^{-1}.

2-11 9,790.

2-12 11,600.

2-13 553,000.

Chapter 3

3-1 $U_c = n(2.78 \times 10^{-4}a + 1.16 \times 10^{-5}b)$.

3-2 $E = 0.038(1 + 0.2W)(e_w - e_a)$.

3-3 $R = (0.085 + 0.0259q_i)P$.

3-4 $Q = \left(\frac{481}{0.386A + 170} + 0.219\right)A$.

3-5 No change, even though equation does not appear homogeneous as written.

3-6 Equation is homogeneous, so no change.

3-7 $M = K(0.477u_b)(0.396T_a + 1.40T_d + 32.0) + 1.32T_a + 23.6$.

3-8 Equation is homogeneous, so no change.

3-9 $C = 1150Q^{0.7}$.

Chapter 4

4-3 At $T = 0°C$, velocity at moving plate ($y = 1.00$ cm) is 2.80 cm s^{-1}; at $T = 30°C$, velocity at moving plate ($y = 1.00$ cm) is 6.27 cm s^{-1}.

4-5

$2R$ (cm)	10^{-4}	10^{-3}	10^{-2}	10^{-1}
at $T = 0°C$, h (cm)	3080	308	30.8	3.08
at $T = 30°C$, h (cm)	2900	290	29.0	2.90

4-6 For paraffin, $\theta_c = 107.5°$, so $\cos\theta_c = -0.301$. If $T = 10°C$, $\sigma = 74.2$ dyn cm^{-1}, thus $h = -0.455$ cm and $\Delta P = -446$ dyn cm^{-2}.

4-7 $h = (2\sigma\cos\theta_c)/\gamma d$.

4-8 $\Delta P = (2\sigma\cos\theta_c)/d$.

4-9 Equivalent $R = 5.05 \times 10^{-5}$ cm; tension is 2,940,000 dyn cm^{-2} (2.90 atm).

4-10 12.0°C.

Chapter 5

5-1 157 atm.

5-2 31.5 atm.

5-3 Pressure is 43.3 lb ft^{-2}, adequate for hydrants.

5-4

Flow	$\mathbb{R}$	$\mathbb{F}$
Trickle	1100	5.1
Parking lot	150	0.71
Model stream	3800	0.14
Small stream	190000	0.23
Medium stream	1500000	0.23

Large river	3800000	0.14
Gulf Stream	1500000	0.0071

5-6 $R = wY/(w + 2Y)$;

w/Y	5	10	20	40
Y/R	1.4	1.2	1.1	1.05

(w is width; Y is mean depth; R is hydraulic radius.)

5-7 $A = Y_{max}^2 \tan(\theta/2)$; $Y = Y_{max}/2$; $R = (Y_{max}/2)\sin(\theta/2)$; $w = 2Y_{max}\tan(\theta/2)$; ($A$ is the area, Y_{max} the maximum depth, θ the vertex angle, R the hydraulic radius, and w the water-surface width).

5-9 Flow is critical at approximately $Y_{max} = 2.5$ m.

Chapter 6

6-1 The general formula is $V = V_{max} - 2.5V^*$ where V_{max} is the maximum (surface) velocity and the other symbols have their usual meanings. Thus we have to know the mean depth (hydraulic radius) and slope in order to compute V^* and use the formula.

6-2 Show that $V^* = 2.5V^* \ln(1.49y_0/y_0)$.

6-3 Show that

$$\frac{2.5V^* \ln(0.8Y/y_0) + 2.5V^* \ln(0.2Y/y_0)}{2} = 2.5V^* \ln(0.4Y/y_0).$$

6-4

	Flow				
	1	2	3	4	5
y_l (m)	1.21×10^{-4}	6.10×10^{-5}	1.71×10^{-4}	1.71×10^{-5}	1.56×10^{-5}

6-5

	Flow				
	1	2	3	4	5
y_l (m)	1.91×10^{-5}	1.91×10^{-5}	3.13×10^{-5}	1.71×10^{-4}	9.90×10^{-6}

All are less than k_s, so all are rough.

6-6

	Flow				
	1	2	3	4	5
Mean Velocity (m s^{-1})	4.90	4.26	1.79	0.368	6.37

6-7

	Flow				
	1	2	3	4	5
V (m s^{-1})	1.28	4.02	1.12	0.0169	7.59
$\mathbb{F}$	0.408	0.908	0.651	0.0171	1.08

6-8

Z (m)	0	0.5	1.0	1.5	2.0	2.5	3.0	3.5	4.1
Q (m^3 s^{-1})	0	0.900	5.71	15.7	29.9	45.8	64.3	97.4	139

6-9 Equation 6-49: $Q = (u_M/n)Y^{5/3}S_0^{1/2}w$ (assuming that $Y = R$); Equation 6-17: $Q/w = q = \gamma Y^3 S_0/3\mu$; Equation 6-51: $Q/w = q = (u_M/n)Y^2 S_0^{1/2}$.

Chapter 7

7-1 $Y_c = \alpha V^2/(g \cos \theta)$; $\mathbb{F} = \alpha^{1/2}V/(gY \cos \theta)^{1/2}$.

7-2 Definition of H_s: $H_s \equiv Y + V^2/2g$. When $\mathbb{F} = 1$, $V^2 = gY$. Thus $H_s = Y + gY/2g = 3Y/2$.

7-3 Rearranging Equation 7-22 yields

$$Q^2 = 2gw^2H_sY^2 - 2gw^2Y^3.$$

Since $Q \geq 0$, the maximum of Q^2 = the maximum of Q. Thus to find the maximum, take the derivative with respect to Y and set the results equal to zero:

$$\frac{d(Q^2)}{dY} = 4gw^2H_sY - 6gw^2Y^2 = 0;$$

$$4gw^2H_sY = 6gw^2Y^2;$$

$$H_s = \tfrac{3}{2}Y,$$

which is true when $\mathbb{F} = 1$ (Exercise 7-2). If this is a maximum rather than a minimum,

$$\left.\frac{d^2(Q^2)}{dY^2}\right|_{H_s=\frac{3}{2}Y} < 0.$$

To evaluate: $d^2(Q^2)/dY^2 = 4gw^2H_s - 12gw^2Y$, and when $H_s = \frac{3}{2}Y$, $d^2(Q^2)/dY^2 = 6gw^2Y - 12gw^2Y$. This is always less than zero because g, w, and Y are always greater than zero.

7-4

Section	Heads (m)				
	Elevation	Pressure	Velocity	Specific	Total
0	6.39	3.00	0.033	3.033	9.423
1	6.19	2.20	0.061	2.261	8.451
2	4.19	1.98	0.075	2.055	6.245
3	3.60	0.80	0.459	1.259	4.859
4	2.02	0.75	0.522	1.272	3.292
5	1.60	1.05	0.267	1.317	2.917
6	0	2.30	0.056	2.356	2.356

7-5 $Y_c = 0.838$ m.

7-6 (b)

Q (m^3 s^{-1})	2	4	6	8	10	12
Y_c (m)	0.357	0.566	0.742	0.899	1.04	1.18

7-7 (c) Both flows are subcritical. (d) Alternate depths are 1.99 m for the first flow and 1.18 m for the second flow.

Chapter 8

8-3 From Figure 8.4, $v_f = 0.01$ m s^{-1} when $d = 0.1$ mm at 10°C. For this temperature, $\nu = 1.31 \times 10^{-6}$ m^2 s^{-1} (Table 4.2). Putting these values into Equation 8-21 gives $\mathbb{R}_p = 0.76$.

8-4

Criterion	Flow a	b	c	d	e	f	g	h
Tractive force	E	E	0	0	E	E	E	0
Velocity	E	0	0	0	E	E	E	0

E means erosion, 0 no erosion.

8-5

	Flow a	b	e	f	g
$\tau_0 V$ (N m^{-1} s^{-1})	5.21	4.40	1.57	3.34	1.77
Bed form	Dunes or transition	†	Ripples	Dunes	Dunes

†Not in Figure 8.13.

8-8

	Flow a	b	c	d	e	f	g	h
Z	0.80	12.8	16.2	15.3	0.91	2.91	0.095	18.1

8-10

	Flow a	b	e	f	g
Sediment concn. (ppm)	880	338	447	264	16,000

8-11

	Q (m^3 s^{-1}) 5	10	50	100
Sediment concn. (ppm)	171	306	758	1030
Sediment load (metric tons day^{-1})	0.0208	0.0750	0.923	2.51

Approximate equations: $C_t = 80Q^{0.56}$; $L = 0.0024Q^{1.52}$, where C_t is sediment concentration (ppm), L is sediment load (metric tons day^{-1}), and Q is discharge (m^3 s^{-1}).

Chapter 9

9-1 (a)

	Q (m^3 s^{-1}) 0.1	1	10	100	1000
Y_n (m)	0.0257	0.102	0.407	1.62	6.45
Y_c (m)	0.0165	0.077	0.356	1.66	7.69

(c) Critical flow occurs at $Q = 74.7$ m^3 s^{-1}.

9-2

	Distance from initial point (ft)					
	0	100	200	400	600	1000
Water-surface elevation†	853.35	853.35	853.40	853.45	853.52	853.76

†In feet above sea level.

9-3

	Distance from initial point (m)					
	0	5	10	20	50	100
Water-surface elevation†	22.61	22.93	23.00	23.14	23.29	23.70

†In meters above sea level.

9-4

	Distance from initial point (m)				
	0	100	200	500	1000
Water-surface elevation†	159.92	159.92	159.92	159.97	161.12

†In meters above sea level.

9-5 (a) Overbank section K = 666 (metric units); channel section K = 4460 (metric units). (b) S_e = 0.00180.

9-6 (a) Left overbank section K = 20,400 (English units); channel section K = 213,000 (English units); right overbank section K = 36,300 (English units). (b) S_e = 0.00200.

9-7

	Flow									
	a	b	c	d	e	f	g	h	i	j
Type	M2	M2	M1	M1	M1	M1	M2	M2	M2	M1
x^* (m)	911	135	957	16.1	73,900	668	628	1240	679	11.6

Chapter 10

10-1

	Flow								
	a	b	c	d	e	f	g	h	i
$\mathbb{F}_1$	0.392	0.319	1.62	1.52	0.381	0.271	0.0745	1.21	0.379
Y_2 (m)	2.45	0.291	2.16	1.01	0.572	2.22	6.20	0.382	1.47
V_2 (m s^{-1})	3.38	0.901	5.67	4.01	1.63	2.67	0.371	7.11	0.413
$\mathbb{F}_2$	0.691	0.534	1.23	1.28	0.691	0.573	0.0475	3.67	0.109

	j	k	l
$\mathbb{F}_1$	0.424	1.52	0.224
Y_2 (m)	0.926	0.274	1.41
V_2 (m s^{-1})	0.363	5.25	0.496
$\mathbb{F}_2$	0.120	3.21	0.133

10-2

	Flow							
	a	b	c	d	e	f	g	h
$\mathbb{F}_1$	0.614	0.799	4.43	1.20	1.60	0.673	4.40	0.753
Y_2 (m)	0.344	0.179	0.392	0.425	0.0594	0.0553	0.248	0.147
V_2 (m s^{-1})	1.51	0.782	6.72	4.87	3.13	0.613	6.93	0.634
$\mathbb{F}_2$	0.822	0.591	3.43	2.39	4.11	0.832	4.44	0.528

10-3

	Flow									
	a	b	c	d	e	f	g	h	i	j
Y (m)	1.63	0.37	1.01	0.21	6.78	0.73	1.16	1.94	0.81	0.16
Y_c (m)	0.43	0.19	0.38	0.11	1.24	0.25	0.55	0.91	0.37	0.07
$\mathbb{F} = V/(gY)^{1/2}$	0.14	0.36	0.23	0.37	0.08	0.20	0.33	0.32	0.31	0.31

10-5

	Flow					
	a	b	c	d	e	f
L (m)	11.1	7.90	8.21	9.16	3.95	6.11
Y_2 (m)	2.56	2.06	1.99	2.01	0.873	1.72
V_2 (m s^{-1})	2.74	3.10	2.62	2.26	1.49	3.20
h_j (m)	0.293	0.0568	0.150	0.322	0.139	0.0146
Head loss via Manning's equation (m)	0.148	0.0459	0.116	0.301	0.685	0.0273

10-7 (a)

	Flow					
	a	b	c	d	e	f
Q (m^3 s^{-1})	65.4	0.884	17.8	0.0494	322	34.2

(b)

	H_w (m)							
	1.0	2.4	3.0	4.0	5.0	6.0	7.0	8.0
Q (m^3 s^{-1})	17.5	65.4	92.7	148	215	293	382	483

10-8

	H_w (m)									
	0.1	0.2	0.3	0.4	0.5	0.6	0.7	0.8	0.9	1.0
Q (m^3 s^{-1})	0.12	0.33	0.61	0.96	1.36	1.81	2.32	2.87	3.47	4.12

10-9

	H_w (m)									
	0.1	0.2	0.3	0.4	0.5	0.6	0.7	0.8	0.9	1.0
Q (m^3 s^{-1})	0.06	0.16	0.29	0.45	0.62	0.82	1.04	1.27	1.52	1.79

10-10

	H_w (m)									
	0.1	0.2	0.3	0.4	0.5	0.6	0.7	0.8	0.9	1.0
Q (m^3 s^{-1})	0.004	0.02	0.07	0.14	0.24	0.38	0.55	0.77	1.03	1.35

10-11 Triangular weirs are more sensitive, that is, a given change in discharge is reflected in a much greater change in weir head than is the case for rectangular weirs. However, rectangular weirs have a higher discharge capacity for a given head.

Chapter 11

11-1 Approximate equations are $w = 3.50Q^{0.15}$, $Y = 0.70Q^{0.34}$, and $V = 0.41Q^{0.51}$, where w and Y are in feet, V is in feet per second, and Q is in cubic feet per second.

11-2

	Y_{max} (m)					
	0.2	0.4	0.6	0.8	1.0	1.2
w (m)	1.90	2.10	2.30	2.50	2.70	2.90
Y (m)	0.21	0.44	0.68	0.93	1.19	1.45
V (m s^{-1})	0.32	0.48	0.61	0.72	0.81	0.90
Q (m^3 s^{-1})	0.13	0.44	0.95	1.66	2.60	3.79

Approximate equations are $w = 2.32Q^{0.15}$, $Y = 0.70Q^{0.57}$, and $V = 0.62Q^{0.29}$.

11-3 The width-to-depth ratio tends to increase, because $b > f$ for downstream hydraulic geometry relations.

11-5 (a) Equation 11-14 is

$$C_g = \left[\frac{2g(Y + h_w)^2}{2Y + h_w}\right]^{1/2}$$

so

$$C_g^2 = \frac{2g(Y + h_w)^2}{2Y + h_w}.$$

Dividing both sides by gY yields

$$\frac{C_g^2}{gY} = \frac{2(y + h_w)^2}{Y(2y + h_w)} = \frac{2(Y^2 + 2h_wY + h_w^2)}{2Y^2 + h_wY}.$$

Now divide both numerator and denominator by Y^2:

$$\frac{C_g^2}{gY} = \frac{2[1 + 2(h_w/Y) + (h_w/Y)^2]}{2 + h_w/Y};$$

and note that $1 + 2(h_w/Y) + (h_w/Y)^2 = [1 + (h_w/Y)]^2$. Therefore

$$\frac{C_g^2}{gY} = \frac{2[1 + (h_w/Y)]^2}{2 + h_w/Y}.$$

11-6 (b) $V_k = \frac{3}{2}V$.

11-7 (c) $V_k = 3V$.

11-8

	% flood-plain filling										
	0	10	20	30	40	50	60	70	80	90	100
V_k (m s^{-1})	0.87	0.94	1.02	1.13	1.25	1.40	1.59	1.85	2.21	2.74	3.60

11-10 (a) Q_n is computed via the Manning equation as $Q_n = u_M w S_0^{1/2} Y^{5/3}/n$, so $Q_n/2wS_0 = u_M Y^{5/3}/2nS_0^{1/2}$.

(b)

	Flow								
	a	b	c	d	e	f	g	h	i
D (m^2 s^{-1})	136	64.5	21.6	822	88.2	5.27	29800	1.53	8.05

	Flow			
	j	k	l	m
D (m^2 s^{-1})	1.30×10^{-3}	5.28×10^{-4}	1.93×10^{-3}	1.13×10^{-3}

11-11

Flow	$\partial Y/\partial t$ (m s^{-1})	$\partial Y/\partial x$ (m m^{-1})	L (m)
a	3.67×10^{-7}	3.02×10^{-7}	388
b	3.42×10^{-7}	3.94×10^{-7}	251
c	3.02×10^{-6}	3.70×10^{-6}	107
d	3.12×10^{-6}	1.87×10^{-6}	1230
e	4.50×10^{-5}	1.04×10^{-4}	86.2
f	1.26×10^{-5}	3.30×10^{-5}	5.40
g	3.28×10^{-5}	2.94×10^{-5}	6030
h	1.45×10^{-7}	1.74×10^{-6}	0.67
i	5.31×10^{-8}	1.68×10^{-7}	9.70
j	2.99×10^{-8}	1.63×10^{-6}	6.86×10^{-4}
k	5.28×10^{-8}	7.04×10^{-6}	1.72×10^{-4}
l	1.93×10^{-7}	2.07×10^{-5}	6.40×10^{-4}
m	4.52×10^{-6}	7.53×10^{-5}	1.57×10^{-4}

Chapter 12

12-3 $n = 0.332$.

12-4

Estimated height (cm)	Clay	Silt	Fine sand	Medium sand	Coarse sand	Gravel
			Medium			
Capillary fringe	3030	303	30.3	7.6	2.0	0.30

12-5

r (mm)	0	0.010	0.020	0.030	0.040	0.050
θ	0	3.7×10^{-5}	3.0×10^{-4}	0.0010	0.0024	0.0048
$\lvert\psi\rvert$ (cm)	∞	151	75.5	50.3	37.8	30.3

r (mm)	0.100	0.125	0.150	0.175	0.190	0.200
θ	0.0375	0.0732	0.121	0.201	0.257	0.300
$\lvert\psi\rvert$ (cm)	15.1	12.1	10.1	8.65	8.00	7.57

12-6 $\lambda \simeq 0.302$.

12-7

	Clay	Silt	Fine sand
		Medium	
K_D (cm s^{-1})	1.87×10^{-8}	1.87×10^{-6}	1.87×10^{-4}

	Medium sand	Coarse sand	Gravel
K_D (cm s^{-1})	3.00×10^{-3}	4.22×10^{-2}	1.87

12-8

R (mm)	0.20	0.40	0.60	0.80	1.00
Total discharge %	0.10	1.6	8.3	26.1	63.8

12-9

R (mm)	0	0.010	0.020	0.030	0.040	0.050
θ	0	3.7×10^{-5}	3.0×10^{-4}	1.0×10^{-3}	2.4×10^{-3}	4.8×10^{-3}
K_{Du}/K_D	0	1.4×10^{-8}	1.0×10^{-6}	1.2×10^{-5}	6.4×10^{-5}	2.6×10^{-4}

R (mm)	0.10	0.125	0.15	0.175	0.19	0.20
θ	0.038	0.073	0.12	0.20	0.26	0.30
K_{Du}/K_D	0.016	0.060	0.18	0.45	0.73	1.0

12-10

	Day (Flow)				
	1 (A → B)	2 (A → B)	3 (B → A)	4 (A → B)	5 (B → A)
θ at A	0.256	0.133	0.002	0.071	0.001
θ at B	0.100	0.050	0.256	0.256	0.071

12-11 (b) $K_D \simeq 0.26$ cm s^{-1}. (c) The break in slope of the line fitted to the plot of points indicates that the K_D versus $\Delta H/L$ relation changes at about $0.04 \leq \Delta H/L \leq 0.05$.

Index